Agroecosystems: An Ecological Perspective

Agroecosystems: An Ecological Perspective

Editor: Bella Harper

Callisto Reference
www.callistoreference.com

Callisto Reference,
118-35 Queens Blvd., Suite 400,
Forest Hills, NY 11375, USA

Visit us on the World Wide Web at:
www.callistoreference.com

ISBN: 978-1-64116-291-3 (Hardback)

The publisher's policy is to use permanent paper from mills that operate a sustainable forestry policy. Furthermore, the publisher ensures that the text paper and cover boards used have met acceptable environmental accreditation standards.

Printed in China.

Cataloging-in-Publication Data

Agroecosystems : an ecological perspective / edited by Bella Harper.
p. cm.
Includes bibliographical references and index.
ISBN 978-1-64116-291-3
1. Agricultural ecology. 2. Agriculture--Environmental aspects. 3. Ecology. I. Harper, Bella.
S589.7 .A37 2020
577.55--dc23

Preface

This book has been a concerted effort by a group of academicians, researchers and scientists, who have contributed their research works for the realization of the book. This book has materialized in the wake of emerging advancements and innovations in this field. Therefore, the need of the hour was to compile all the required researches and disseminate the knowledge to a broad spectrum of people comprising of students, researchers and specialists of the field.

The basic unit of study in agroecology is known as an agroecosystem. It is defined as a functionally and spatially coherent unit of agricultural activity. It comprises of both the living as well as the nonliving components which are involved in that unit, along with their interactions. An agroecosystem can be considered as a subset of a conventional ecosystem, with the human activity of agriculture at its core. The region around the agricultural activity, which is impacted by this activity, is also included in the agroecosystem. The farming methods which are employed currently have resulted in over-stretched water resources, high levels of erosion and reduced soil fertility. Farming within agroecosystems has been suggested as the way forward for mainstream agriculture. This book aims to shed light on some of the unexplored aspects of agroecosystems by examining it through the perspective of ecology. It also includes contributions of experts and scientists which will provide innovative insights into this field. Those in search of information to further their knowledge will be greatly assisted by this book.

At the end of the preface, I would like to thank the authors for their brilliant chapters and the publisher for guiding us all-through the making of the book till its final stage. Also, I would like to thank my family for providing the support and encouragement throughout my academic career and research projects.

Editor

Chapter 11 **Adoption of Agroforestry Systems by Smallholders in Brazilian Amazon** 93
Gisele do Socorro Santos Pompeu, Leonilde Santos Rosa, Mônica Mota Santos, Rozi Silva Modesto and Thiago Almeida Vieira

Chapter 12 **Effects of Crude Oil on the Growth of *Brachiaria Mutica* and *Leucaena Leucocephala* and on Soil and Plant Macronutrients** 101
María del Carmen Rivera-Cruz, Eduardo Maldonado-Chávez and Antonio Trujillo-Narcía

Chapter 13 **Botanical and Chemical Composition of Common Grass Species Around Dip–Tank Areas in Semi–Arid Communal Rangelands** 111
Solomon T. Beyene and Victor Mlambo

Chapter 14 **Nutritional Composition and Antinutritional Factors of Little-Known Species of *Vigna*** 121
Chinnamadasamy Kalidass and Veerabahu Ramasamy Mohan

Chapter 15 **Chemical Composition, *In Vitro* Rumen Fermentation, and Digestibility of Selected Browse Species by Buffaloes (*Bubalus Bubalis* L.) at a Semi-Deciduous Forest** 135
Álvaro Ojeda, Carlos Canelones, Dorian Muñoz and José L. Gil

Chapter 16 **Energy use Analysis for Rice Production in Nasarawa State, Nigeria** 146
H. Y. Ibrahim and H. I. Ibrahim

Chapter 17 **Technology Transfer Networks on Papaya Production with Transitional Growers** 153
Octavio Cano-Reyes, Juan A. Villanueva-Jiménez, Juan L. Reta-Mendiola and Arturo Huerta-De-la-Peña

Chapter 18 **Parthenium Hysterophorus L: Distribution, Impact, and Possible Mitigation Measures in Ethiopia** 162
Jemal Abdulkerim-Ute and Behailu Legesse

Chapter 19 **Potential of Carbon Storage of Rubber (*Hevea Brasiliensis* Müll. Arg.) Plantations in Monoculture and Agroforestry Systems** 174
José Alfredo Orjuela-Chaves, Hernán J. Andrade C. and Yeraldine Vargas-Valenzuela

Chapter 20 **Diversty and Abundance of Arbuscular Mycorrhizal Fungi Across Different Land use Types in a Humid Low Land Area** 184
Zerihun Belay, Mauritz Vestberg and Fassil Assefa

Permissions

List of Contributors

Index

Table of Contents

Preface VII

Chapter 1 **Determinants of Agricultural Productivity Growth in Sub-Sahara Africa** 1
Ajao Olajide A.

Chapter 2 **Morphological Diversity of Native Maize in the Humid Tropics** 9
Fernando López-Morales, Oswaldo Rey Taboada-Gaytán, Abel Gil-Muñoz, Pedro Antonio López and Delfino Reyes-López

Chapter 3 **Activity of Leaf-Cutting Ant Atta *Sexdens Piriventris* Submited to High Dilution Homeopathic Preparations** 22
Alexandre Giesel, Mari. I. C. Boff, Paulo. A. S. Gonçalves and Pedro Boff

Chapter 4 **Allelic and Genotypic Frequency of Kappa Casein Gene in Double Purpose Cattle** 31
N. G. Cortés-López, S. del Moral, José A. Rueda, C. Luna-Palomera, C. A. Meza-Herrera and J. Abad-Zavaleta

Chapter 5 **Fertility Status of Eutric Nitisol and Fertilizer Recommendation using Numass** 39
Abebe Nigussie, Gebremedihin Ambaw and Endalkachew Kissi

Chapter 6 **Effect of Soybean Plant Populations on Yield and Productivity of Cassava and Soybean Grown in a Cassava-Based Intercropping System** 48
E. U. Mbah and E. Ogidi

Chapter 7 **Effect Of Mycorrhizal Inoculants in the Development of Mexican Landrace Avocado Rootstocks** 56
Edgar Castro Alvarado, Ana Tztzqui Chávez Bárcenas, Pedro Antonio García Saucedo, Leovigilda Reyes Ramírez and Ana Elizabeth Bárcenas Ortega

Chapter 8 **An Effective *In Vitro* Slow Growth Protocol for Conservation of the Orchid *Epidendrum Chlorocorymbos* Schltr** 63
Guadalupe Lopez-Puc

Chapter 9 **Soil Aggregate Fractions and Organic Carbon Pools as Influenced by Tree Diversity in Forest Reserve** 71
H. M Lawal

Chapter 10 **Bush Encroachment Control Demonstrations and Management Implications on Herbaceous Species in Savannas** 80
A. Angassa, G. Oba and A. Tolera

Determinants of Agricultural Productivity Growth in Sub-Sahara Africa

Ajao Olajide A.

Agricultural economics and extension dept Ladoke akintola university of tech Ogbomoso-nigeria. E-mail: dayo_67@hotmail.com
**Corresponding Author*

SUMMARY

This study examined changes in agricultural productivity in Sub-Sahara Africa countries in the context of diverse institutional arrangements using Data Envelopment Analysis (DEA). From a time series which consists of information on agricultural production and means of production were obtained from FAO AGROSTAT and rainfall data from Steve O'Connell data base. The information was for a 43-year period (1961-2003); DEA method was used to measure Malmquist index of total factor productivity. A decomposition of TFP measures revealed that the observed increase in the TFP in the sub–Sahara Africa agriculture is due to technological change rather than efficiency change which is the main constrained of achieving higher level of TFP during the reference period. The study further examined the effect of land quality, malaria, education and selected governance indicators such as, control of corruption and government effectiveness on productivity growth. All the variables included in the model are significant with the exception of government effectiveness.

Keywords: Data Envelopment Analysis, Efficiency, Productivity, Sub-Saharan Africa

INTRODUCTION

Like many developing economies, agriculture is the dominant sector of the Sub-Saharan African (SSA) countries for growth, poverty alleviation, contribution to GDP, employment and incomes. The growth and development of this crucial sector is essential for the overall process of socioeconomic development in the region. For agricultural sector to achieve these objectives, various governments and institutions have sought strategies that would lead to higher levels of production and a key factor for a sustained increase of agricultural production is improvement of productivity, which is carried out through technological change and efficiency change. Hence, increasing agricultural productivity in Sub-Sahara Africa countries has received a wide spread attention in the literature e.g Fulginiti et al. (2004) on economic development and poverty alleviation. Since agricultural growth is linked to farm profit, there had been considerable research that examined the performance of this crucial sector in the region (e.g. Moock, 1973 and Lipton, 1988).

As observed by Nkamleu (2003), many African farmers are still using low yielding agricultural technologies, which lead to low productivity. Also, it is always argued that, relevant question for agricultural policy makers, is whether the agricultural sector can be made more efficient, by achieving more output with the current input level, or achieving the current output with less input usage than is currently observed. An important step in answering this question is to identify the behavior of productivity and its components.

The concept of efficiency is at the core of economic theory. The theory of production economics is concerned with optimization and this implies efficiency. The crucial role of efficiency in increasing agricultural output has been widely recognized by researchers and policy makers alike. It is no surprise; therefore, that considerable effort has been devoted to the analysis of the farm level efficiency in developing countries. An underlying premise behind much of this work is that if farmers are not making efficient use of the existing technology, their efforts designed to improve efficiency would be more cost effective than introducing new technologies as a means of increasing agricultural outputs (Bravo-ureta and Everson, 1994). The issue of determining the pattern and the efficiency of resource use in traditional farming arises in the context of formulating development strategies designed not only to raise the productivity of resources already committed to the farming but also to ensure that the newly created resources in the agricultural development efforts are allocated to areas and for enterprises in which their productivities are higher (Awoyemi, *et al.*, 2003).

In order to collectively raise productivity, global and regional productivity growth in agriculture has been the focus of intense research in the past few decades. Economists (e.g. Block, 1995) have examined the sources of productivity growth over time and the productivity differences among countries and regions over this period. Productivity growth in the agricultural sector is considered important in the Sub-Saharan region of Africa if agricultural sector output is to improve at a rate equal to or greater than the population growth rate to meet the demand for food and raw materials. Also, productivity performance in the agricultural sector is critical to improvement in the economic well being of these countries. Unlike previous studies that have measured and compared intercountry agricultural productivity which have been motivated by a variety of issues including identifying the primary sources of productivity growth and comparing the structural and productivity differences among the countries. This study however goes further to examine the influence of selected governance indicator on productivity growth in SSA.

The main policy emphases were to upgrade production systems and increasing agricultural commercialization. The model was heavily dependent on transfers of high levels of development assistance and highly expensive, it is therefore imperative to examine the growth witnessed in the sector after various reforms. Since output from agriculture can be broadly classified into two main groups: Crop and Livestock, the non-parametric DEA approach provides an attractive option. Therefore, this study used Data Envelopment Analysis (DEA) technique to calculate Malmquist TFP index numbers, hence, aims to provide understanding of agricultural productivity growth in sub-Sahara Africa.

This study addressed several questions such as: What is the status of agricultural productivity in SSA? Was the green revolution accompanied by declining productivity growth? Has SSA agricultural productivity declined sharply as perceived? Are there major differences in sub-regional productivity growth? And finally, what are the factors that determine the productivity growth? The broad objective of the study is to examine the performances of each country in Sub-Saharan African countries agricultural productivity.

MATERIALS AND METHODS

DEA is linear-programming methodology, which uses data on input and output quantities of a Decision Making Units (DMU) such as individual firms of a specific sectors to construct a piece-wise linear surface over data points. In this study, the countries were used as the DMU. The DEA method is closely related to Farrell's original approach (1957) and it is widely being regarded in the literature as an extension of that approach. This approach was initiated by Charnes et al.; (1978) and related work by Fare, Grosskopf and Lovell 1985). The frontier surface is constructed by the solution of a sequence of linear programming problems. The degree of technical inefficiency of each country, which represents the distance between the observed data point and the frontier, is produced as a by-product of the frontier construction method. The DEA method is a non-parametric approach in which the envelopment of decision-making units (DMU) can be estimated through linear programming methods to identify the "best practice" for each DMU. The efficient units are located on the frontier and the inefficient ones are enveloped by it.

DEA can either be input or output oriented depending on the objectives. The input-oriented method, defines the frontier by seeking the maximum possible proportional reduction in input usage while the output is held constant for each country. The output-oriented method seeks the maximum proportional increase in output production with input level held fixed. These two methods, that is, input-output oriented methods provide the same technical efficiency score when a constant return to scale (CRS) technology applies but are unequal when variable returns to scale (VRS) is assumed (Coelli and Rao, 2001). In this study, the output-oriented method will be used by assuming that in agriculture, it is common to assume output maximization from a given sets of inputs. The interpretation of CRS assumption has attracted a lot of critical discussion e.g. Ray and Desli, 1997, Lovell, 2001, but also monotonicity and convexity are debatable e.g. Cherchye, *et al.*, 2000.

Fare *et al.*, (1994) used Data Envelopment Analysis (DEA) methods to estimate and decompose the Malmquist productivity index. Four linear programs (LPs) must be solved for each DMU in this study (Country) to obtain the distances defined in equation (iii) and they are:

$$\text{(i)}\quad \left[d_o^t(x_t, y_t)\right]^{-1} = \mathrm{Max}_{\phi,\lambda}\phi,\quad s.t\quad -\phi y_{it} + Y_t\lambda \geq 0,\quad x_{i,t} - X_t\lambda \geq 0,\quad \lambda \geq 0$$

$$\text{(ii)}\quad \left[d_o^{t+1}(x_{t+1}, y_{t+1})\right]^{-1} = Max_{\phi,\lambda}\phi,\quad st\quad -\phi y_{i,t+1} + Y_{t+1}\lambda \geq 0,\quad x_{i,t+1} - X_{t+1}\lambda \geq 0,\quad \lambda \geq 0$$

$$\text{(iii)}\quad \left[d_o^t(x_{t+1}, y_{t+1})\right]^{-1} = \mathrm{Max}_{\phi,\lambda}\phi,\quad st\quad -\phi y_{i,t+1} + Y_t\lambda \geq 0,\quad x_{i,t+1} - X_t\lambda \geq 0,\quad \lambda \geq 0$$

$$\text{(iv)}\quad \left[d_o^{t+1}(x_t, y_t)\right]^{-1} = Max_{\phi,\lambda}\phi,\quad st\quad -\phi y_{i,t} + Y_{t+1}\lambda \geq 0,\quad x_{i,t} - X_{t+1}\lambda \geq 0,\quad \lambda \geq 0$$

Where λ is a N X 1 vector of a constant and ϕ is a scalar with $\phi \geq 1$

Over time best practice are natural and to include frontier shifts, that is, technical change, the Malmquist productivity index is a well established measure.

Malmquist productivity index

The Malmquist productivity index, as proposed by Caves, Christensen and Diewert (1982), allows one to describe multi-input, multi-output production without involving explicit price data and behavioral assumptions. The Malmquist Productivity Index identifies Total Factor Productivity (TFP) growth with respect to two time periods through a quantitative ratio of distance functions (Malmquist, 1953). Distance functions can be classified into input distance functions and output distance functions. Input distance functions look for a minimal proportional contraction of an input vector, given an output vector, while output distance functions look for maximal proportional expansion of an output vector, given an input vector. By using distance functions, the Malmquist Productivity Index can measure TFP growth without cost data, only with quantity data from multi-input and multi-output representations of technology. In this study, we use output distance functions. According to Hjalmarson and Veiderpass (1992), The Malmquist (quantity) index was originally introduced in a consumer theory context as a ratio between two deflation or proportional scaling factor deflating two quantity vectors onto the boundary of a utility possibility set. This deflation or distance function approach was later applied to the measurement of productivity in Caves, Christensen and Diewert (1992) in a general production function framework and in a non-parametric setting by Fare, Grosskopf, Lindgren and Roos (1992). The productivity change, that is TFP change (TFPCH) using technology of period *t* as reference is as follows:

$$M_o^t(x_t, y_t, x_{t+1}, y_{t+1}) = \left[\frac{d_o^t(x_{t+1}, y_{t+1})}{d_o^t(x_t, y_t)}\right] \quad \text{......... (v)}$$

Similarly, we can measure Malmquist productivity index with period t+1 as references as follows:

$$M_o^{t+1}(x_t, y_t, x_{t+1}, y_{t+1}) = \left[\frac{d_o^{t+1}(x_{t+1}, y_{t+1})}{d_o^{t+1}(x_t, y_t)}\right] \quad \text{...... (vi)}$$

in order to avoid choosing arbitrary period as reference, Fare *et al.*, (1994) specifies the Malmquist productivity index as the geometric mean of the above two indices

$$M_o(x_t, y_t, x_{t+1}, y_{t+1}) = \left[\frac{d_o^t(x_{t+1}, y_{t+1})}{d_o^t(x_t, y_t)} \frac{d_o^{t+1}(x_{t+1}, y_{t+1})}{d_o^{t+1}(x_t, y_t)}\right]^{1/2} \quad \text{(vii)}$$

equation (vi) can be decomposed into the following two components namely efficiency change index (EFFCH) which measures the catching up components measuring efficiency change in relation to the frontier at different time. The second component is the geometric average of both components and measures technical change (TECHCH) which measure the technology shift between period t and *t+1*. The first component in TECHCH measures the position of unit *t+1* with respect to the technologies in both periods. The second component also estimates this for unit *t*. If the TECHCH is greater (or less) than one, then technological progress (or regress) exists.

$$EFFCH = \frac{d_o^{t+1}(x_{t+1}, y_{t+1})}{d_o^t(x_t, y_t)} \quad \text{........................ (viii)}$$

and

$$TECHCH = \left[\frac{d_0^t(x_{t+1}, y_{t+1})}{d_o^{t+1}(x_{t+1}, y_{t+1})} X \frac{d_o^t(x_t, y_t)}{d_o^{t+1}(x_t, y_t)} \right]^{1/2} \quad \ldots(ix)$$

Data and Data Source

The study was based on the data that were drawn from the FAO web site (AGROSTAT) and it covers a period of 43 years (1961-2003). The following are some of the main features of the data series that were used for the study. The data consists of information on agricultural production (Crop and Livestock index) and means of production such as total rural population and total agricultural area for each of the selected countries were drawn from FAO statistic database, while rainfall was obtained from Steve O'Connell's website. To examined the determinants of TFP and impact of political economy on agricultural productivity, the study considered the following variables: Conflict (International Peace Research Institute, Oslo); corruption and government effectiveness (Governance Matter II); Land Quality (Peterson, 1987); Malaria and Education which was used as a proxy for quality of labor (Center of International Development, Harvard University).

Measurement of Variable

Output: Crop and Livestock index
Input: (a) Total agricultural area (1000ha)
(b) Total rural population (1000)
(c) Rainfall (weighted)
(d) Irrigation (1000ha)

The institutional quality index, that is, corruption and government effectiveness reflects the statistical compilation perception of the quality of governance. They are measured in units ranging from -2.5 to 2.5 with higher values indicate better governance outcomes (see Kaufman *et al*, 2001). For the purpose of this study, a threshold of positive value was used to allow for the use of a dummy variable (1, 0) for better governance or otherwise.

Since a significant amount of variation in TFP level may be explained by variations in the quality of land and since comparison of agricultural productivity among nations would not be meaningful unless differences in land quality is taken into account. Also for econometric purposes, the use of land variable unadjusted for quality could cause bias of estimates of coefficients. Land quality index of Peterson (1987) which is determined by the predicted value of agricultural land per acre in a country was divided by the average value of land per acre across all countries (see Peterson, 1987). Land quality index of Peterson (1987) is determined by the predicted value of agricultural land per acre in a country divided by the average value of land per acre across all countries, which are 126 countries in Peterson's study. The log of the value of agricultural land per acre is a weighted lines function of (a) non-irrigated crop land as a percent of all agricultural land (b) irrigated land as a percent of arable crop land, and (c) the log of the long-run average annual precipitation .

The effect of conflict was also examined and quantified as minor, intermediate and war depending on the number of deaths per year using a dummy variable. Minor armed conflict: at least 25 battle-related deaths per year and fewer than 1000 battle-related death during the course of conflict. Intermediate armed conflict: at least 25 battle-related deaths per year and an accumulated total of at least 1000 deaths. War: at least 1000 battle-related deaths per year.

The effect of malaria was also examined and calculated in equal-area cylindrical projection. (See Pampana, and Russell. 1955. *Malaria: A World Problem*)

Other variable included in the determinant model is life expectancy at birth and openness variable. Life expectancy at birth is measured in years. This is a broad indicator of the health of the population, which has been shown in earlier study to be a significant predictor of future economic growth. The openness variable measures the proportion of years within the interval 1965-1990 in which an economy is open to international trade by the criteria in Sachs and Warner (1995).

Model specification for determinant of agricultural productivity growth

The first two objectives were achieved, that is the technical efficiency and the TFP by solving equation (iv)-(vi) and for the last objective, that is, the determinants of productivity growth an Ordinary Least Square (OLS) estimation techniques was used to examined the effect of the above selected variables on agricultural productivity growth.

$Y = f(X_i, e)$................................(x)

Where Y is the TFPCH index, that is, Malmquist Productivity Index and; i = 1, 2.........8
X_1 represents conflict
X_2 represents corruption
X_3 represents land quality
X_4 represents malaria
X_5 represents education
X_6 represents government effectiveness
X_7 represents life expectancy at birth
X_8 represents openness

RESULTS AND DISCUSSION

Fare *et al.* (1984) made known that the output distance function is the equivalent of the inverse of Farell's measure of output efficiency. This study used malmquist index to measure the productivity growth of agricultural sector for thirty–six countries in sub–Saharan African countries between 1961-2003. The method used constructed the best – practice frontier in agricultural production for the sampled countries and later compared according to regions which sub-Sahara African countries was broken into, that is, (West Africa; Southern Africa; East Africa; Horn of Africa; Central Africa; and Indian Ocean). Malmquist productivity indexes as well as efficiency change and technological change components for each country in the sample were calculated. Since this index is based on discrete time, each country has an index for every pair of years.

Table 1a presents mean Malmquist Indices by country for the period of study while Table 1b shows annual means. Table 1b reveals that Burkina Faso, Cote d'Ivoire, Kenya and Djibouti as the four countries with the highest TFP growth. Burkina Faso has about 4.3% average growth in TFP; this is due to 8% growth in technical efficiency and 3.6% growth in technological change. Though, Kenya, Djibouti and Cote d'Ivoire equally have TFP growth of 3.9%, for Kenya and Cote d'Ivoire, technological change accounted for 3.2 and 3.4% while efficiency marginally changed by 0.07 and 0.05% respectively. For Djibouti, technical progress solely accounted for the TFP growth. The findings further revealed that Lesotho, Sierra – Leone and Swaziland had negative TFP growth of 0.16, 0.13 and 0.03% respectively. A decline of 1.8; 3.9 and 1.8% in the technical efficiency were responsible for the decline TFP.

Recall that the value greater than one implies increasing productivity and less than one productivity decrease from period t to period $t + l$. The mean values of TFP change ranged from 0.937 to 1.05; 0.985 to 1.098; 0.948 to 1.105; 0.937 to 1.089 and 0.937 to 1.075 for the whole period and sub–periods 1 to 4 respectively. The average TFP growth over the whole period was 1.8% per annum according to mean TFP change (TFPCH). The mean values for the 1st and 2nd sub–periods are 1.027 and 1.015 respectively. This implies that overall TFP growth is improving in the two periods and the highest TFP growth was observed in 1980 when TFP change is 10.5%. This is most likely due to various reforms, programs and policies adopted by countries in sub–Sahara.

Table 1a. Malmquist index summary of country means

Country	Effch	Techch	Tfpch
Angola	0.999	1.019	1.018
Botswana	0.999	1.014	1.013
Burkina Faso	1.008	1.036	1.043
Burundi	0.984	1.031	1.015
Cameroon	0.990	1.030	1.020
Cape Verde	1.000	1.024	1.024
Chad	0.999	1.020	1.019
Côte d'Ivoire	1.005	1.034	1.039
Djibouti	1.000	1.039	1.039
Gabon	0.995	1.040	1.035
Gambia	1.000	1.002	1.001
Ghana	0.992	1.029	1.021
Guinea	0.988	1.035	1.022
Guinea Bissau	0.988	1.027	1.015
Kenya	1.007	1.032	1.039
Lesotho	0.982	1.002	0.984
Madagasca	0.983	1.034	1.016
Malawi	0.994	1.027	1.020
Mali	1.003	1.023	1.027
Mauritania	1.000	1.013	1.013
Mauritius	0.997	1.008	1.005
Mozambique	0.989	1.025	1.014
Namibia	0.996	1.016	1.011
Niger	1.006	1.011	1.017
Nigeria	1.004	1.027	1.031
Rwanda	0.998	1.022	1.020
Sao Tome	1.000	1.007	1.007
Senegal	0.985	1.028	1.012
Sierra Leone	0.961	1.027	0.987
Sudan	1.004	1.025	1.029
Swaziland	0.982	1.016	0.997
Tanzania	1.003	1.026	1.029
Togo	0.985	1.015	1.000
Uganda	0.981	1.022	1.003
Zambia	0.998	1.024	1.023
Zimbabwe	0.990	1.034	1.024
Mean	0.994	1.023	1.017

It is important to examine the main cause of improved productivity. The level of TFP of the agricultural sector can be improved either by change in technical efficiency or a shift in production frontier (technological change). Table 2 shows the comparison between technical efficiency change and technological change for the regions considered in the study.

Table 1b. Malmquist index summary of annual means

Year	Effch	Techch	Tfpch
1962	0.996	1.025	1.021
1963	1.062	0.956	1.016
1964	0.942	1.156	1.089
1965	0.944	1.043	0.985
1966	1.087	0.907	0.986
1967	1.062	0.935	0.993
1968	0.872	1.223	1.067
1969	1.060	0.941	0.997
1970	1.066	1.030	1.098
1971	1.007	1.006	1.013
1972	0.952	1.066	1.015
1973	1.116	0.911	1.017
1974	0.995	0.952	0.948
1975	1.077	0.966	1.041
1976	0.935	1.104	1.033
1977	0.996	1.041	1.037
1978	1.050	0.909	0.954
1979	0.845	1.190	1.005
1980	0.695	1.591	1.105
1981	1.217	0.821	0.999
1982	1.057	0.988	1.044
1983	1.187	0.905	1.074
1984	0.622	1.518	0.944
1985	1.332	0.746	0.994
1986	0.988	1.059	1.046
1987	0.865	1.209	1.046
1988	0.890	1.053	0.937
1989	1.114	0.896	0.998
1990	1.026	1.061	1.089
1991	1.091	0.917	1.001
1992	1.207	0.891	1.075
1993	0.936	1.002	0.937
1994	1.010	0.983	0.992
1995	1.078	0.924	0.996
1996	0.829	1.268	1.051
1997	1.053	0.956	1.006
1998	0.878	1.172	1.029
1999	0.882	1.099	0.968
2000	1.161	0.904	1.050
2001	0.997	1.027	1.024
2002	0.969	1.095	1.061
2003	0.994	0.997	0.992
Mean	0.994	1.023	1.017

Note that Effch (efficiency change); Techch (technological change); Tfpch (total factor productivity change)

Table 2. Comparison between technical efficiency change and technological change

Regions	EFFCH > TECHCH	TECHCH > EFFFCH
Central Africa		**
Horn of Africa		**
Southern Africa	**	
East Africa		**
Indian Ocean Island		**
West Africa		**

** Indicates Yes

Since efficiency change and technological change are the components measure of TFP, these two mean (efficiency change and technological change) values were then compared to know the source of TFP growth in the sample period. On the average the efficiency change decreases by 0.06% while the technological change increased on the average by 2.3%. This suggests that the observed increase in the TFP in the sub–Sahara Africa agriculture is due to technological change rather than efficiency change which is the main constrained of achieving higher level of TFP during the reference period. It was further observed from the table that the technological change was responsible for the TFP growth in the entire region. Horn of Africa which comprises Djibouti and Sudan is the only region with both efficiency and technological change growth. This region experienced a TFP growth of 3.4%, followed by East Africa (2.3%); Central Africa (1.94%) and West Africa (1.8%). These other regions though showed technological progress without any significant improvement in the relative deviation from their corresponding frontier over the reference period. This implies that for West Africa, the region deteriorates by 0.06% in catching up with the frontier. The regional discrepancies in technical progress are obviously higher than the efficiency improvement; hence, regional TFP growth is largely determined by the technical progress.

Determinants of productivity growth

Understanding factors that influence growth within and across the countries can be used to formulate policies that will enhance productivity. Several factors have been identified in the literature as the most important sources of productivity change in agriculture: research and development, extension services, education, infrastructure and government programmes but ignores institutional quality. This study however examined the effect of land quality, malaria, education and selected governance indicators such as, control of corruption and government effectiveness on productivity growth.

Table 3: OLS showing effects selected variables on TFP

Variable	Coefficient	t-ratio
Constant	1.054	97.35
Conflict	-3.3E-03**	-1.66
Corruption	-8.4E-03*	-2.50
Land quality	-1.74E-04*	-2.22
Malaria	-1.49E-02*	-2.30
Education	-4.91E-03*	-2.29
Government effectiveness	-9.74E-04	-0.29
Life expectancy at birth	3.85E-04	1.97
openness	6.71E-05	.44

R-square = 0.49
F = 2.28
* and ** indicates significance at 5 and 10 % probability level

Statistical results provided by table 3 show that statistically, the model performed fairly well, with an R-squared of 0.49 suggesting we are capable of explaining close to half the variation in TFP and F of 2.28 suggests the overall significance of all the explanatory variables. All the variables included in the model are highly significant at 5 % probability level with the exception of government effectiveness, and trade openness, conflicts is however, significant at the 10 % level. They equally performed well in terms of expected relationship with TFP except education and land quality index which unexpectedly had an inverse relationship with TFP. Improper operation and agricultural practice coupled with the fact that there is an extensive grazing, low rainfall, small share of land in crops and small share of cropland that is irrigated might probably responsible for land quality index not transformed into productivity growth. The inverse relationship of education with the TFP suggests that agricultural sector is still unimprovedand that a large majority of rural dwellers in SSA are yet to have access to basic education though exposed to informal agricultural training. To summarize the findings from the analysis, corruption, land quality, malaria and education have the highest level of significance (5 %) and malaria have the expected signs and are highly significant. These factors are highly prevalent in SSA especially in the 1980s and 1990s. Malaria prevalence is still very high and is responsible for over 50 percent of the child mortality which invariably have effect on the availability of family labor. Corruption is still very rampant as many SSA countries ranked high as compared to other continents. Likewise, the continent has been ridden with various forms of conflict. Life expectancy at birth is positively related to the productivity growth. This is expected because as the SSA country life expectancy improves through improvement in public health delivery and eradication of diseases which will in turn have effect on productivity.

CONCLUSIONS

This study presents some important findings on level and trends in SSA agricultural productivity and further examined the political economics of agricultural productivity in SSA between 1961 and 2003. The findings revealed that the TFP growth was observed for all the countries except Madagascar, Cote d'Ivoire and Sierra Leone and the sources of the growth was found to be technological progress rather than efficiency change. In terms of sub-region, the horn of Africa which comprises Djibouti and Sudan is the only region with both efficiency and technological change growth and has a positive annual productivity change of about 0.08% given the technology.

The effect of land quality, malaria, education and selected governance indicators such as, control of corruption and government effectiveness on productivity growth was examined. It was observed that all variables included in the model have significant impact on the TFP except government effectiveness and openness. If corruption, conflict and malaria incidence are properly addressed, the negative impact they have on growth would probably have been eliminated.

Given the fact that agriculture is a very important sector in SSA, the findings from this study revealed that activities of rural development has not really transformed into effective action, hence, policy implication of these findings are significant in that SSA and foreign aid agencies should channel their resources in such a way that an average rural dwellers will have access to unfettered and quality education to improve the existing man power and capacity building. Corruption should be tackled effectively especially as it regards money laundering by top government officials and genuine efforts should be taken in reducing or eradicating conflicts in the region.

ACKNOWLEDGEMENTS

I wish to express my deep appreciation to AERC for the financial support to carry out this research and also grateful to the resource persons and Prof. Ann Veiderpass for various comments and suggestions that have helped the evolution of this study from its very beginning to this stage.

REFERENCES

Awoyemi, T.T, Amao, J.O. and Ehirim, N.C. 2003. Technical Efficiency in Aquaculture in Oyo State, Nigeria Indian Journal of Agricultural Economics 58:812-813.

Block, S.A. 1995. The recovery of Agricultural Productivity in sub-Saharan Africa. Journal of Food Policy 20: 385-405.

Bravo-ureta, B.E and Evenson, R.E. 1994. Efficiency in Agricultural Production: The Case of Peasant Farmers in Eastern Paraguay. The Journal of International Association of Agricultural Economics 10: 27-28.

Caves, D. W., Christensen, L.R. and Diewert, W.E. 1992. The Economic Theory of Index Numbers and the Measurement of Input, Output, and Productivity. Econometrica 50: 1393–1414.

Cherchye, L., Kuosmanen, T. and Post, G.T. 2000. What is the Economic Meaning of FDH, A Reply to Thrall. Journal of Productivity Analysis 13: 259–263.

Coelli, T.J. and Rao, D.S. 2001. Implicit Value shares in Malmquist TFP Index Numbers CEPA Working Papers, No. 4/2001, School of Economics, University of New England, Armidale, pp. 27.

Fare, R., Grosskopf, S., and Lovell, C.K. 1994. Production Frontier. Cambridge University Press, Cambridge, UK.

Fare, R., Grosskopf, S., Lindgren, B., and Roos, P. 1992. Productivity Changes in Swedish Pharmacies 1980-1989: A Non-parametric Approach. Journal of Productivity Analysis, 3: 85-101.

Farrell M.J. 1957. The measurement of Productive Efficiency, Journal of the Royal Statistical Society. 120: 252-290.

Kaufmann, D., Kraay, A. and Zoido-Lobaton, P. 2002. Governance Matters II: Updated Indicators for 2000/01. World Bank Policy Research Department Working Paper.

Lipton, M. 1988. The Place of Agricultural Research in the Development of Sub- Saharan Africa. World Development. 16: 1231-57.

Lovell, C.A.K. 2001. The Decomposition of Malmquist Productivity Indexes. Journal of Productivity Analysis 20: 437–458.

Malmquist, S. 1953. Index Numbers and Indifference Surfaces. Trabajos de Estadistica 4: 209–242.

Moock, P.R. 1973. Managerial Ability in Small Farm Production: An Analysis of Maize Yields in the Vihiga Division of Kenya. Ph.D. dissertation, Columbia University 920–936.

Nkamleu G.B., Gokowski, J. and Kazianga, H. 2003. Explaining The Failure of Agricultural Production in Sub-Saharan Africa Contributed paper selected for presentation at the 25th International Conference of Agricultural Economists, August 16-22, 2003, Durban, South Africa.

Pampana, E.J., and Russell, P.F. 1955. Malaria: A World Problem. WHO, Geneva.

Peterson, W. 1987. International Land Indexes. Staff Paper P87-10 Department of Agricultural And Applied Economics, University of Minnesotta.

Ray, S.C. and Desli, E. 1997. Productivity Growth, Technical Progress and Efficiency Change in Industrialized Countries: Comment. American Economic Review 87: 1033–1039.

Sachs, J.D. and Warner A.M. 1995. Sources of Slow Growth in African Economies. Journal of African Economies 6: 335-376.

2

Morphological Diversity of Native Maize in the Humid Tropics of Puebla, Mexico

Fernando López-Morales[1], Oswaldo Rey Taboada-Gaytán[2*], Abel Gil-Muñoz[2], Pedro Antonio López[2], Delfino Reyes-López[3]

[1]*Programa de Maestría en Estrategias para el Desarrollo Agrícola Regional. Colegio de Postgraduados Campus Puebla, México.*

[2]*Colegio de Postgraduados, Campus Puebla. Km. 125.5 Carretera Federal México-Puebla, Santiago Momoxpan, San Pedro Cholula, Puebla. CP. 72760. Email: toswaldo@colpos.mx*

[3]*Facultad de Ingeniería Agrohidráulica, Benemérita Universidad Autónoma de Puebla. CP. 73695. Teziutlán, Puebla.*

** Corresponding author*

SUMMARY

The most recent studies on maize in the humid tropics of Puebla were done in 1997, and consisted of agronomic evaluations. Thus, the objectives of this study were to assess the level of genetic diversity present among the native populations of maize, cultivated in this region, through their morphological characterization, and to determine the degree of association between these maize populations and the Tuxpeño race. Therefore, in the 2009-2010 autumn-winter season, 93 native maize populations, six improved varieties, and a racial control (Tuxpeño type) were evaluated in two localities, under a lattice experimental design 10×10 with two replications. The results of the analysis of variance suggested that genetic diversity was present in the populations evaluated, represented by a large morphological diversity. With cluster and principal component analyses, such variation was grouped in six groups, based on vegetative, phenological, tassel, ear and grain traits, as well as on several indexes. The maize landraces cultivated in the humid tropics of Puebla can be regarded as part of the Tuxpeño race, since most of them kept similarity with such race.

Key words: Agronomic traits; morphological characterization; racial similarity; native populations; Tuxpeño race.

INTRODUCTION

In Mexico, the area annually planted with maize (*Zea mays* L.) reaches 8 million ha, 86 % of which are located in rainfed areas in the sub-humid tropical, temperate-humid, and sub-humid regions of central and southwestern Mexico (Bellon *et al.*, 2009; Kato *et al.*, 2009). In these areas, maize is grown for self-sustenance and tradition; thus explaining why farmers protect, maintain and use the genetic diversity that they have shaped (Ángeles *et al*., 2010). As a result of the continuous interaction between the environmental diversity, the genetic variation present in maize and farmer's management (which includes an empirical breeding process), in several regions of the country a large genetic diversity has been generated in this crop. This diversity is perceptible through the variation of both agronomical and morphological characteristics, as well as by the large number of uses given to the products of this plant (Kato *et al*., 2009).

Maize diversity in Mexico was initially classified by Wellhausen *et al*. (1951) in 25 races and several sub-races. Since then, several multivariate techniques have been used to have a better classification of the races and sub-races of maize (López *et al*., 2005). Nowadays, the world's maize diversity has been grouped in more than 300 races (Brown and Goodman, 1988; Serratos, 2009), 59 of which have been described in Mexico (Sánchez *et al*., 2000; Kato *et al*., 2009) thus representing a 20 % of the total racial diversity. This is one of the reasons why Mexico has been considered a center of origin, domestication and diversity for this crop (Kato *et al*., 2009). One of the most important races from an agronomic perspective is the Tuxpeño race; this is because of its high grain yield, drought tolerance and disease and pest resistances. These attributes have turned this race into one of the most worldwide used materials in breeding programs (Brown and Goodman, 1988; Bellon *et al*., 2005; Wen *et al*., 2012). The Tuxpeño race is distributed along the coast of the Gulf of Mexico, from Yucatan to northeastern Mexico, at altitudes ranging from 0 to 500 m; however, this race has demonstrated a wide adaptation to other regions, including the humid tropics of the state of Puebla (Wellhausen *et al*., 1951; Muñoz, 2005).

According to Herrera *et al*. (2000), most of the genetic diversity of Mexico's native maize is represented by the local populations that are planted in agricultural fields throughout the country. The studies on morphological diversity of maize in the humid tropics of Puebla started in 1997; they proved the existence of a large diversity of types and colors among the native maize populations from the ecological niches of Ayotoxco, Hueytamalco and Acateno (Muñoz, 2005). For the first niche, Gil *et al*., (2004) found differences in grain color: 166 white, 20 yellow, 8 pinto and 4 blue. It was also found that with respect to days to silking, all materials were classified as 'ultra-early', with 60 to 77 days to silking. Regarding grain yield, it was found that the best native maize materials surpassed by 35 % the improved varieties used as controls. Strictly speaking, both works were just agronomic evaluations, therefore it was deemed convenient to conduct a study to quantify the level of genetic diversity present in the native populations of maize grown in the region (via their morphological characterization) and to determine the degree of association between such populations of maize and the Tuxpeño race.

According to the numbers of the Sistema de Información Agroalimentaria y Pesquera (SIAP), in 2010 in the state of Puebla, 606,534 ha of maize for grain were sown, mainly under rainfed conditions (92 %).The region where the largest area is sown and where the highest yields are achieved are the high valleys; however, there are other regions, such as the humid tropics of Puebla, where maize is also important. The humid tropics of Puebla encompass an area of 1,574.95 km^2 which represents approximately 4 % of the state area (Yanes, 2011). The agricultural area of those municipalities with at least 35 % of their territory within the humid tropics is of 1,192.59 km^2 (INEGI, 2009), while the area devoted to maize production for grain is of 24,927 ha. In the municipalities of San José Acateno, Hueytamalco, Ayotoxco and Tenampulco (Figure 1), located in the northeastern portion of the state, and where this research took place, maize is produced under rainfed conditions, on an area of 4,300 ha, with a production volume of four thousand tons and an average yield of 1.3 t ha^{-1} (SIAP, 2010).

Several studies on native maize diversity in the state of Puebla have been conducted, but none of them has been carried out in the humid tropics of Puebla, addressing both the morphological diversity of native maize and their level of association with the racial types reported for that region. Therefore, the objectives of this study were: 1) to assess the genetic diversity present in the humid tropics of Puebla through a morphological characterization of the native maize populations there cultivated; and 2) to determine whether or not a correspondence exists between those maize populations and the Tuxpeño racial type reported for that region.

MATERIALS AND METHODS

Plant Material

Between May and September 2008, a collection of native maize was made in the municipalities of Tenampulco, Ayotoxco de Guerrero, Hueytamalco and San José Acateno. A total of 250 accessions was obtained. The explored region is located in the humid tropics of Puebla State, Mexico, between 20° 08' and 20° 12' N and 97° 19' and 97° 23' W. The altitude of the four municipalities ranges from 10 to 1,900 m. For evaluation purposes, a set of 93 native maize populations collected from 93 to 372 masl, was selected (Table 1). These accessions were evaluated along with an accession representative of the Tuxpeño race (Vera – 39) provided by the International Maize and Wheat Improvement Center (CIMMYT) and six improved varieties: AF09A and AF07A (CIMMYT's experimental hybrids, recommended for the region); CP-560, CP-562 and CPVM-301, from Colegio de Postgraduados, and the hybrid PROGRANO 988, produced by 'Productores de Grano del Norte', included for its aptitude for corn husk production.

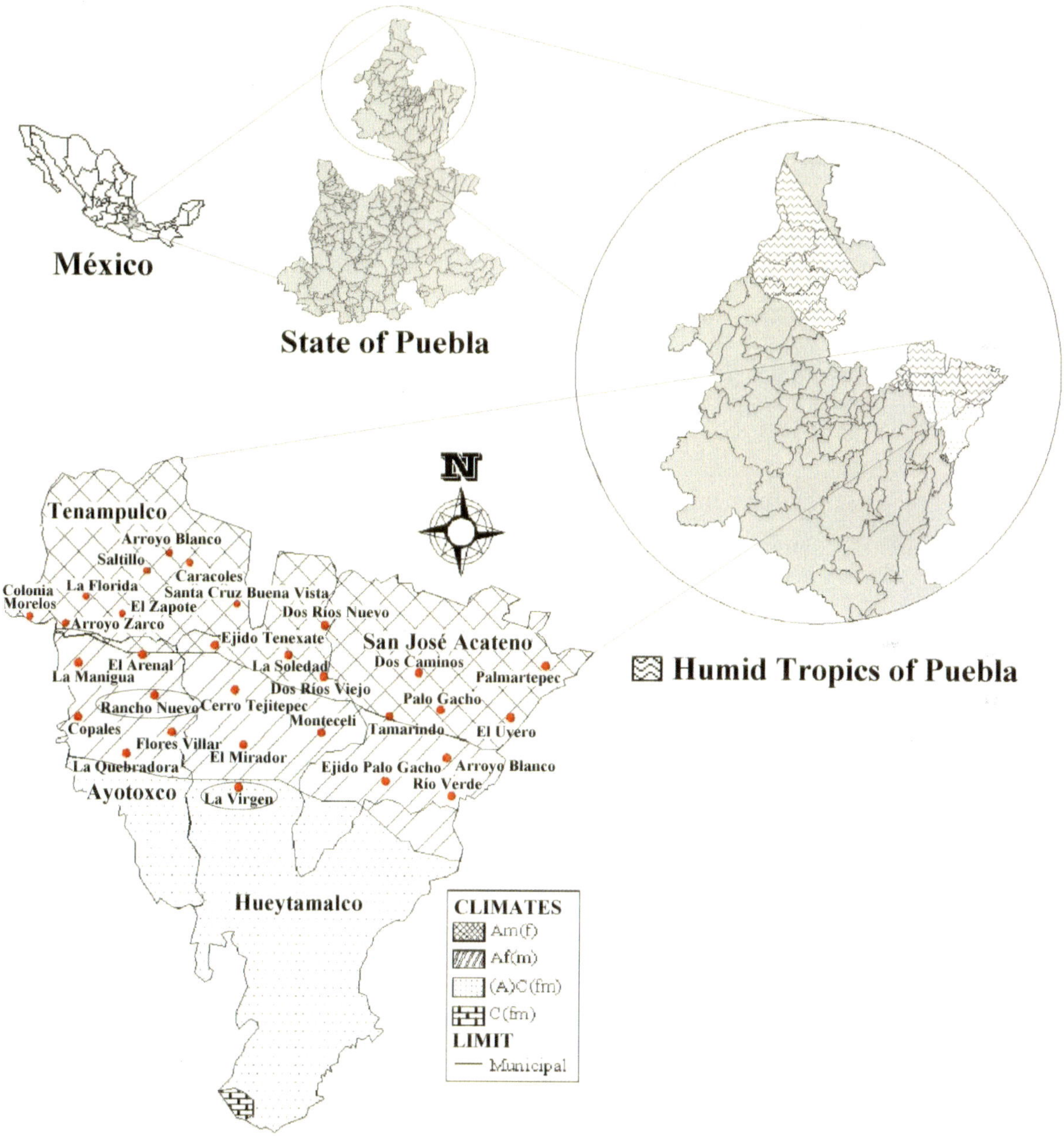

Figure 1. Climate types distribution and location of collection sites in four municipalities of the humid tropics of the state of Puebla.

Table 1. Number of accessions, maize grain color, and maximum and minimum altitudes of the collection sites in four municipalities from the humid tropics of Puebla.

Municipality	Grain color†				Total of accessions	Site altitude (masl[1])	
	WH	YE	BL	RE		Minimum	Maximum
Ayotoxco	14	5	5	5	29	181	352
Hueytamalco	17	2	2	0	21	143	372
Acateno	17	1	3	0	21	151	319
Tenampulco	17	1	3	1	22	93	253
Total	65	9	13	6	93		

†WH=white, YE=yellow, BL=blue, RE=red; [1]= meters above sea level

Localization of experiments

In Puebla, the humid tropics can be found in two parts of the state: one in the south (11 % of the area), and the other in the north, north-east (with 89 % of the area). This study was conducted within the latter, which includes 15 municipalities. Two climate types are found: Af(m), tropical rain forest, with rains all year, and Am(f), tropical monsoon with abundant summer rains.

Two experiments were planted, one in Rancho Nuevo, municipality of Ayotoxco de Guerrero, located at 20° 06' N, 97° 25' W, and 250 masl, and another in La Virgen, municipalty of Hueytamalco, located at 19° 56' N, 97° 17' W and 420 masl (INEGI, 2010). In these communities, the average annual temperature is of 23 and 21 °C, and the average annual precipitation of 3000 and 2550 mm, respectively (INEGI, 2009). Climates are tropical rain forest, Af(m), in the first town, and semi warm with rains all year, (A)C(fm), in the second (García, 1973).

Planting dates were: December 9th and 10th, 2009 (Rancho Nuevo and La Virgen, respectively) in "year after year" fields, with a slope lower than 10 %. Sowing was made according to local practices: the stover from the previous crop was incorporated and after a moderate rain, sowing took place under the following procedure: with a 'coa' (a large stick with a sharp, pointy end in one extreme), in the place corresponding to each hill, every 50 cm, a small hole was opened in the soil (10 cm deep) and three seeds were placed. After four weeks, thinning was made in order to have 34 plants per experimental plot. The experimental plot consisted of two rows, five meters long and 0.8 m apart. Experiments were conducted exclusively under rainfed conditions. Treatment distribution was made under a lattice design 10×10 with two replications (Martínez, 1988).

Crop management

Crop management was similar in both experiments and according to the traditional practices in the region, except for fertilization, which was applied 70 days after sowing using a 128-92-00 (N-P-K) dose. Pest control for armyworm (*Spodoptera* spp.) and cutworm (*Agrotis* spp.) was performed 20 days after sowing, using a cypermethrin insecticide, at a dose of 250 $ml \cdot ha^{-1}$. Throughout all the cycle weeds were controlled manually to avoid competition with maize.

Evaluated variables

Based on the studies of Caballero and Cervantes (1990), Herrera *et al.* (2000) and Hortelano *et al.* (2008), at each experimental plot, several vegetative, reproductive, tassel, totomoxtle, grain and ear characteristics were recorded. The indexes reported by Sánchez and Goodman (1992) were also calculated. The vegetative characteristics measured in every experimental plot, on a sample of five representative plants, with complete competence were: plant height, height of insertion of top ear (cm), and number of leaves below and above the top ear. The phenological traits evaluated were: days to 50 % anthesis and to 50 % silking; considered as the number of days elapsed from sowing date up to the moment in which 50 % of the plants from the experimental plot were shedding pollen and had exposed stigmas in the upper ear, respectively; floral asynchrony was the result of subtracting days to silking from days to anthesis. Characteristics measured on the tassel were: tassel length, length of the branched segment, length of the central branch, peduncle length (all in cm), and number of primary branches. On the bract that covers the ear (corn husk), ear coverage was evaluated, considered as the distance (in centimeters) from the tip of the ear cob up to the tip of the bract; other traits measured were the length of the bract that covers completely the ear, and the number and total weight of the bracts. The variables recorded in the ear were: length, diameter, number of rows and number of grains per row in two opposite rows. All tassel and ear data were taken in the same five plants where the vegetative characteristics were measured. The grain characteristics evaluated were thickness, length and width of 10 grains taken from the central part of the ear, and weight of 100 grains. Finally, several indexes were calculated: plant height divided by upper ear height, leaves below divided by leaves above the ear, and ear length divided by its diameter. The product of multiplying kernel length by kernel width by kernel thickness provided the average kernel volume (in cm^3).

Statistical analysis

A combined analysis of variance was performed for all measured variables using the software Statistical Analysis System (SAS Institute, 2002) version 9.0. Afterwards, a simple correlation analysis was conducted, which served to identify those highly correlated ($|\geq 0.7|$) variables; from each pair, only the variable considered more informative was selected. Thus, out of the 40 variables measured, only 21 were considered for the multivariate analysis. With the average values per population across environments and the 21variables previously selected, a data matrix was generated for the principal component analysis (Jarvis *et al.*, 2006). Based on the data generated, the populations were located in a three dimensional plane whose axes were the first three main components. Additionally, a cluster analysis was conducted (Franco and Hidalgo, 2003) using the matrix of Euclidean distances; the corresponding dendogram was obtained using Ward´s Minimum Variance grouping method. To improve the graphics presentation, the determination coefficient was used as a scale of dissimilarity between clusters (Mohammadi and Prasanna, 2003).

Description of the Tuxpeño Race

To determine whether or not there was correspondence between the native maize populations and the Tuxpeño racial type, the references were the description made by Wellhausen *et al.* (1951) and the data collected on the Tuxpeño racial type (Vera-39) evaluated in the experiment According to the previously mentioned authors, the Tuxpeño race has the following characteristics: plant height, 2.7 m; number of leaves above the ear, six; total tassel length, 42.6 cm; peduncle length 5.1 cm; length of the branched segment, 14.4 cm; number of primary branches, 22. The ear is cylindrical, with a length of 19.7 cm, diameter of 4.4 cm, with 12.6 rows on average and 16 bracts. The grain has a width of 9.3 mm, thickness of 3.7 mm and length of 12.8 mm. Days to 50 % silking are 118.

RESULTS AND DISCUSSION

Collection composition

From the 93 accessions evaluated (Table 1), 65 had white grain (69.9 %), 13 blue (13.9 %), nine yellow (9.7 %) and six red (6.5 %). Gil *et al.* (2004) reported that in a collection of 198 native populations conducted in 1997 in the Ayotoxco region, a predominance of white grain populations (83.8 %), was found, they were followed by yellow materials (10.2 %) and at a lesser extent, the blue (2.0 %) and pinto ones (4.0 %). This can be considered as an indicator that the pattern of variation -at least in terms of grain color- has remained relatively constant for eleven years.

Earliness variation

In relation to the earliness levels, Gil *et al.* (2004) reported that the materials they evaluated reached 50 % silking between 60 and 77 days, hence they were classified as ‘ultra-early’. For the populations evaluated in the present study, an average of 126 to 131 days to silking (DFF) was determined, which according to the scale used by the above mentioned authors, classifies them as late materials. This can be explained by the fact that in 1997, the materials were evaluated in the spring-summer season, while in the present work the populations were studied during the fall-winter season, period in which the lower temperatures cause a delay in flowering (Pérez *et al.*, 2002). For this variable, the data are consistent with the results of Wellhausen *et al.* (1951), so it is inferred that they may have been evaluated in the same growing season.

Analysis of Variance

The combined analysis of variance (Table 2) revealed that among localities (LOC) there were highly significant differences ($P\leq 0.01$) in 37 variables, significant ($P\leq 0.05$) in one (number of ear bracts) and non-significant in two. Among populations (POBL) there were highly significant differences ($P\leq 0.01$) in 39 variables and significant ($P\leq 0.05$) in only one (dry matter percentage). The interaction localities × populations (LOC × POBL) was highly significant ($P\leq 0.01$) for seven variables and significant in ten ($P\leq 0.05$); for all other variables (23), the interaction was non-significant.

Table 2. Mean squares of the combined analysis of variance for 100 maize populations, evaluated in four municipalities of the humid tropics of Puebla.

VARIABLES	Mean squares				CV
	LOC	POBL	LOC*POBL	ERROR	%
Vegetative characteristics					
Plant height (ALTPL) (cm)	39209.34**	3493.53**	415.81ns	340.20	6.95
Ear height (ALTMZ) (cm)	5019.44**	2394.07**	300.26 *	222.55	9.71
Leaves above the ear (NHARR) (number)	1.45**	0.51**	0.10ns	0.09	5.28
Leaves below the ear (NHABA) (number)	16.43**	3.30**	0.49ns	0.40	7.81
Plant score (ASPL)	6.13**	0.82**	0.30ns	0.23	12.73
Ear score (CALMZ)	111.16**	0.52**	0.29 *	0.21	13.83
Phenological characteristics					
Days to anthesis (DFM)	11662.24**	117.593**	25.436**	16.499	3.35
Days to silking (DFF)	18037.34**	178.545**	34.979**	22.455	3.74
Floral asynchrony (ASFLO) (days)	799.00**	11.985**	5.708ns	5.022	39.34
Grain weight (PESGRANO) (g)	403829.34**	804.104**	400.016ns	323.894	25.25
Shelling factor (FACDESGA)	0.13**	0.007**	0.001 *	0.001	4.02
Dry matter percentage (MASECA)	0.08**	0.004 *	0.003ns	0.002	5.48
Yield (RENDIMIE) (Kg ha^{-1})	240441231**	1613238**	855925**	304562	43.87
Tassel characteristics					
Total length (LONTOL) (cm)	5068.44**	65.66**	30.85ns	23.78	7.83
Length of the branched segment (LOPARA) (cm)	1890.05**	47.67**	19.84ns	17.24	9.80
Length of the central branch (LORACE)(cm)	956.83**	24.93**	13.87 *	10.15	11.41
Peduncle length (LONPED) (cm)	186.64**	10.94**	6.18 *	4.42	10.06
Number of branches (NUMR)	155.84**	25.11**	8.21ns	6.49	13.78
Cornhusk characteristics					
Ear coverage (CUBERTO) (cm)	60.79**	5.15**	2.57ns	2.75	18.18
Length of the bracts (LONHOJA) (cm)	1295.97**	14.28**	4.15ns	3.13	6.74
Number of bracts (NUMBRAC)	13.12 *	6.34**	3.07ns	2.81	14.77
Bracts weight (PESOTOT) (g)	34687.56**	155.70**	76.21 *	53.25	25.95
Ear characteristics					
Ear length (LONMAZ) (cm)	1683.97**	5.82**	2.69 *	1.94	9.57
Ear diameter (DIAMAZ) (cm)	45.22**	0.20**	0.06ns	0.07	7.25
Rows per ear (HILMAZ) (number)	63.02**	3.54**	1.38 *	1.01	8.61
Grains from opposite rows (NGRAO)	8174.84**	74.09**	23.67ns	21.34	15.79
Grain characteristics					
10 grains thickness (GROSOR) (cm)	0.0005ns	0.010**	0.002ns	0.002	12.89
10 grains length (LARGO) (cm)	2.1299**	0.016**	0.007**	0.004	6.43
10 grains width (ANCHO) (cm)	0.2095**	0.006**	0.003**	0.001	4.92
Weight of 100 grains (PDEGRANO) (g)	11400.9231**	35.138**	21.492**	13.427	15.44
Indexes					
Plant height/ear height (PLANMAZ)	0.187368**	0.118**	0.013**	0.009	5.40
Leaves below/leaves above (ABAJARRI)	1.051883**	0.089**	0.024 *	0.017	9.17
Ear length/ear diameter (LONGDIA)	15.285470**	0.647**	0.206ns	0.213	11.59
Grain length/grain width(LONANCH)	1.115469**	0.035**	0.006ns	0.006	6.51
Plant height/tassel (PLANESPI)	3.941439**	0.803**	0.234ns	0.204	10.54
Peduncle length/tassel (PEDULO)	0.034022**	0.002**	0.001ns	0.001	11.45
Length of the central branch/ tassel (RACEL)	0.000002ns	0.003**	0.002ns	0.001	9.39
Grain length/grain thickness (LONGROS)	12.223776**	0.739**	0.140ns	0.141	13.80
Grain thickness/ grain width (GROSANC)	0.058289**	0.017**	0.004ns	0.004	13.68
Grain length x width x thickness (LOANGRO)	0.460724**	0.007**	0.003 *	0.002	14.05

* and ** = Significant differences at $P \leq 0.05$ and $P \leq 0.01$, respectively; ns = No significant differences; LOC = localities; POBL = populations; LOC*POBL = localities × populations; CV = coefficient of variation.

In all cases, degrees of freedom for localities and populations were 1 and 99, respectively. Degrees of freedom for the LOC*POBL interaction were 99 for DFM ASPL ALTPL ALTMZ NHARR NHABA PLANMAZ ABAJARRI LONTOL LOPARA PEDULON LORACE RACELON LONPED NUMR PLANESPI; 98 for DFF ASFLO FDESG MASECA RENDIMIE; 97 for LONH NBRAC COBTOT PTOT LONMAZ DIAMAZ NHIL GROGR LONGR ANCGR PESGRANO LONDIA LONANCH LONGRO GROANC LAANGRO NGRAO P100GR and 96 for CALMZ.

The results indicate the existence of genetic diversity in the evaluated varieties, and that they respond differently, in some variables, when the environment changes, as denoted by the 'localities × populations' interaction. According to Hortelano *et al.* (2008), the fact that the interaction 'localities × populations' was not significant in a large proportion of the variables they studied, implies that even though the environments of evaluation showed variations in the level of expression of the characteristics studied, those variations were constant and similar between populations, regardless of the localities. Data on Table 2 are consistent with the reports of Hortelano *et al.* (2008) as to the percentage of significant differences found in the sources 'localities' and 'populations'; this was not the case for the interaction 'localities × populations'. The percentages for the sources 'localities' and 'localities × populations', are similar to those found in native maize from Molcaxac, Puebla, reported by Ángeles *et al.* (2010). The results from the combined analysis of variance suggest that in the humid tropics of Puebla, and especially in the region of study, there is a considerable genetic diversity (expressed through morphological attributes) among the native maize populations there cultivated.

Principal component analysis

The principal component analysis revealed that with the first three components, 54 % of the original phenotypic variation observed in the populations was explained. The individual contribution of each component was: 22.9 % for the first component (CP1), 19.8 % for the second (CP2) and 11.2 % for the third (CP3). According to the magnitude of the eigenvectors (Table 3), the original variables with greater weight on CP1 were: days to silking, length of the branched segment and number of branches of the tassel, and the indexes plant height divided by ear height, leaves below the ear divided by leaves above the ear and ear length divided by its own diameter. In CP2, the original variables with greater weight were number of grains in opposite rows and shelling factor, along with the indexes grain thickness divided by grain width and grain volume. The CP3 was largely influenced by ear diameter and grain width.

Of the 12 most important original variables found in this study (Table 3), eight of them are directly or indirectly related with Hortelano's *et al.* (2008) findings, when they studied the morphological diversity of native maize from the Puebla valley. These variables are: days to silking, ear diameter, number of grains and grain width. The similitude was also present for some indexes, for example: plant height divided by ear height, ear length divided by ear diameter, grain thickness divided by grain width and grain volume. There was also coincidence with two of the most important original variables reported by López *et al.* (2005): plant height and days to silking, and with three reported by Ángeles *et al.*, (2010): days to silking, ear diameter and grain width. The most relevant original variables found in the four studies (including this) are: days to silking, ear diameter and grain width.

According to Herrera *et al.* (2000) the most important variables to determine differences among maize populations from the same region are the morphological characters, such as the vegetative and agronomic attributes, those of the tassel and grain, and specifically those from the ear, since they are the least affected by the environmental factors. On this matter, Goodman and Paterniani (1969) stated that the characters chosen for numerical taxonomic studies must be those that are less affected by the environment.

Cluster Analysis

To define similarity relationships between native maize and controls, a dendrogram was generated in which, at a cutoff distance of 0.4 units; six groups (Figure 2) were identified. These are described next.

Group I comprised the six improved materials included in the experiment; it is evident that they conformed a group scarcely related with the native populations from the region. According to data from Table 4, this group was characterized by being the earliest (107 days), having a short length of the branched segment, and few branches on the tassel. It also had the lowest values for the following indexes: 'leaves below the ear / leaves above the ear', 'ear length / ear diameter' and grain volume, and had the highest value for 'plant height/ ear height'. In general terms, this indicates that they were plants with little growth, compared with the other materials. Group II consisted of 15 populations from the four municipalities studied: Ayotoxco (1), Hueytamalco (5), San José Acateno (7) and Tenampulco (2); most of them came from the central region; 86.8 % had white grain. The group was characterized by presenting the highest values for number of branches in the tassel, grains in opposite rows (33 grains), and shelling factor (86.6 %); nevertheless, it had one of the lowest values in grain volume. Group III was one of the largest; it consisted of 21 populations from the four municipalities: Tenampulco (1), Ayotoxco (9), Hueytamalco (5) and San José Acateno (6). In this group, white populations predominated (57.1 %), followed by the yellow ones (19 %), blue (14.2 %) and red (9.7 %). In general, this group showed intermediate values for all variables, except for the index 'grain thickness / grain width', which was the lowest, and grain volume, which also had low values. Group IV included six materials; three of them were blue grained. Five of these populations were collected in the municipality of Tenampulco and one in Hueytamalco, in communities located in the northern part of the studied region. This group was characterized by having the narrowest ears and the lowest number of grains per row, as well as the lowest shelling factor. The index 'grain thickness / grain width' was the highest among the six groups and its grain volume was one of the highest. Group V grouped 21 populations, three from Ayotoxco, one from San José Acateno, four from Hueytamalco and 13 from Tenampulco, 95 % were white-grained. In general, this group was the latest in terms of maturity (Table 4) and was among the groups with the largest values for tassel characteristics, it had the widest grains, a high index of 'leaves below / leaves above the ear' and a low index for 'plant height / ear height'. Finally, the group VI was the largest, with 30 native maize populations which came from the four municipalities: Tenampulco (1), Ayotoxco (16), Hueytamalco (6) and San José Acateno (7). Distribution in terms of grain color was: white 63.3 %, blue 20 %, yellow 10 % and red 6.7 %. In this group the Tuxpeño racial control was included, suggesting that this group of native maize was the one with the highest similarity to this race.

Table 4 shows the average values for the racial control (Tuxpe). When compared with the characteristics reported by Wellhausen *et al.* (1951) for the Tuxpeño race, it turns out that those of the former are of a lesser magnitude. Out of the six groups identified, Group VI was the one that had the highest correspondence with the Tuxpeño racial control (Vera-39), since they coincided in most of the characteristics evaluated. Materials of the Group VI are concentrated in a warm and humid area, with rains throughout all the year (García, 1973), and altitudes from 147 to 352 m, thus suggesting that this is the area where the native maize populations more related to that race can be found.

Table 3. Eigenvectors for the first three main components (CP) and 21 original variables used for the characterization of 100 maize varieties from the humid tropics of Puebla.

Original Variables	**CP1 [4.82]**[1] **(22.9%)**[2]	**CP2 [4.17]**[1] **(19.8%)**[2]	**CP3 [2.36]**[1] **(11.2%)**[2]
Days to silking (DFF)	**0.352**	0.020	-0.114
Leaves above ear (NHARR)	0.214	0.164	0.270
Length of the branched segment (LOPARA)	**0.303**	0.139	-0.205
Peduncle length (LONPED)	-0.005	-0.065	-0.216
Number of branches (NUMR)	**0.310**	-0.049	0.091
Ear coverage (CUBRTO)	0.102	0.249	-0.032
Number of bracts (NUMBRAC)	0.099	0.035	0.056
Ear diameter (DIAMAZ)	-0.139	-0.053	**0.481**
Rows per ear (HILMAZ)	-0.213	0.011	0.084
Grains from opposite rows (NGRAO)	0.122	**-0.418**	0.028
Grain width (ANCHO)	0.082	0.067	**0.502**
Shelling factor (FDESG)	0.035	**-0.446**	-0.029
Plant height/ ear height (PLANMAZ)	**-0.384**	0.125	-0.080
Leaves below/ leaves above (ABAJARRI)	**0.323**	-0.155	-0.099
Ear length/ ear diameter (LONDIA)	**0.335**	0.110	-0.164
Grain length/ grain width (LONANCH)	0.059	-0.280	-0.256
Plant height/total tassel length (PLANESPI)	0.271	-0.178	0.284
Peduncle length/total tassel length (PEDULON)	-0.238	-0.140	0.006
Length of the central branch/total tassel length (RACELON)	-0.101	0.170	-0.188
Grain thickness/ grain width (GROSANC)	0.031	**0.417**	-0.201
Grain length x width x thickness (LOANGRO)	0.159	**0.345**	0.234

[1]Characteristic value of the component, [2] percentage of variation explained by each component, [3]in bold, original variables with the greatest explanatory power for each principal component.

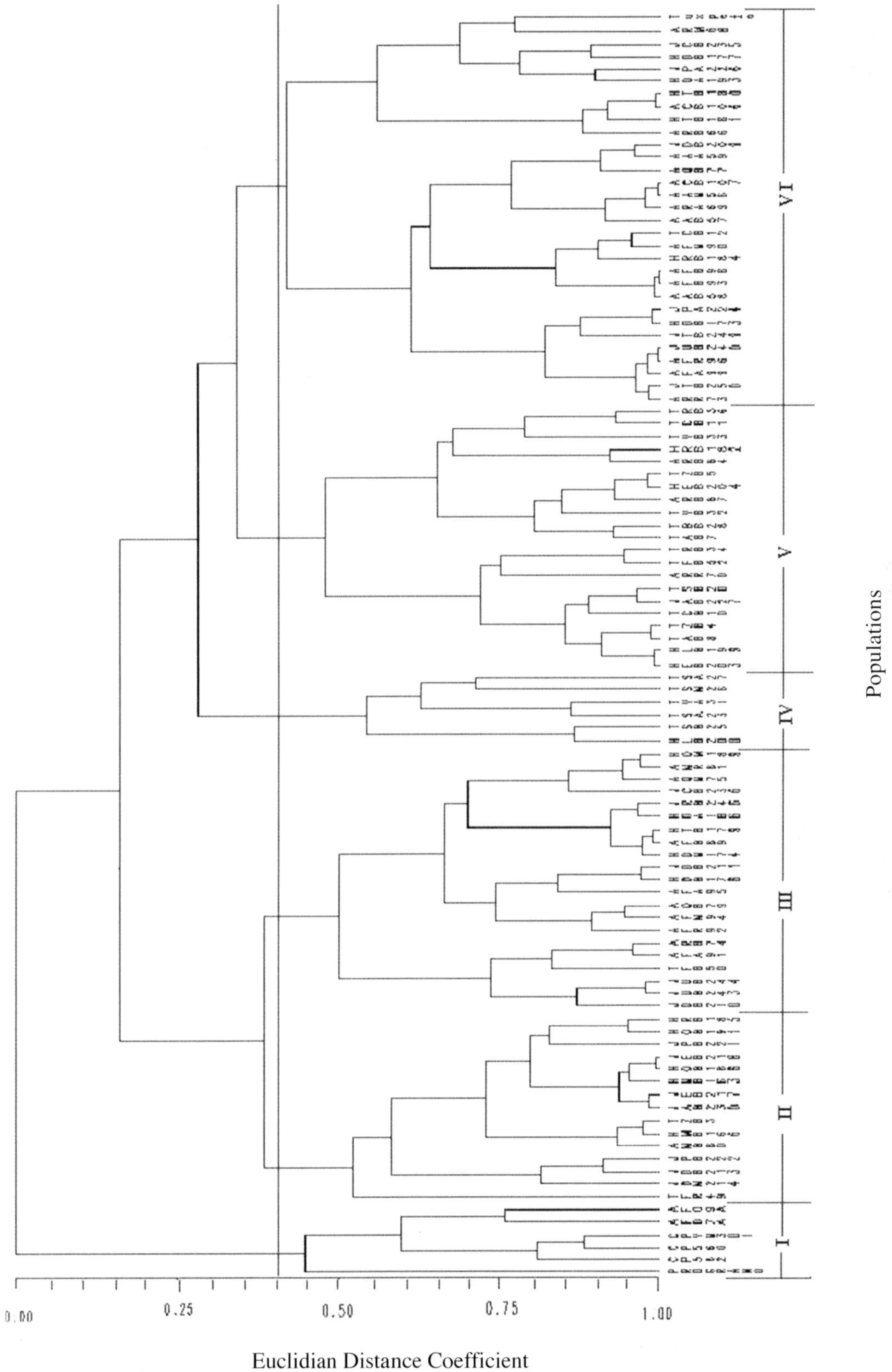

Figure 2. Dendrogram of 100 maize populations from four municipalities in the humid tropics of Puebla.

Figure 3 shows the location of the maize populations in a three-dimensional plane, determined by the first three main components. The first aspect that stands out is that Group I was separated from the rest, situation that is explained by the fact that it included all the introduced materials, thereby evidencing the little morphological similitude with the materials from the region. Therefore, the most relevant analysis derives from the inspection of the clustering patters of the native populations and the racial control. It was observed that the accession representative of 'Tuxpeño' was part of Group VI, which indicates that the materials on this group were the ones that kept, at a larger extent, the distinctive traits of such race. It is noteworthy that around Group VI, the Goups II to V were located, denoting that these sets of native populations maintain similarity with Tuxpeño. This is consistent with the concept of race, which according to Anderson and Cutler (1942) is defined as a group of related individuals with enough characteristics in common that enable its differentiation from other groups and that, according to Muñoz (2005), pools a group of varieties relatively similar, adapted to a particular ecological region. Thus, and supported on the results herein obtained (Figures 1, 2 and Table 4), it can be affirmed that even though on the native maize populations from the region, natural and artificial selection processes have been taking place, which have led to a certain level of morphological differentiation, in general terms they have kept the attributes proper of the Tuxpeño race.

When the collection sites of the 93 native populations were considered, it was found that most of the populations from Groups II and III were distributed along the warm humid climate with rains all year [Af(m)], the same climatic type where Group VI is found, but with slight differences. On the other hand, Groups IV and V are more associated to the warm humid climate with abundant rainfalls in summer [(A)C(fm)], most of their populations belonged to the municipality of Tenampulco and had morphological characteristics more similar to the accessions included in Group VI, where the Tuxpeño racial control was found.

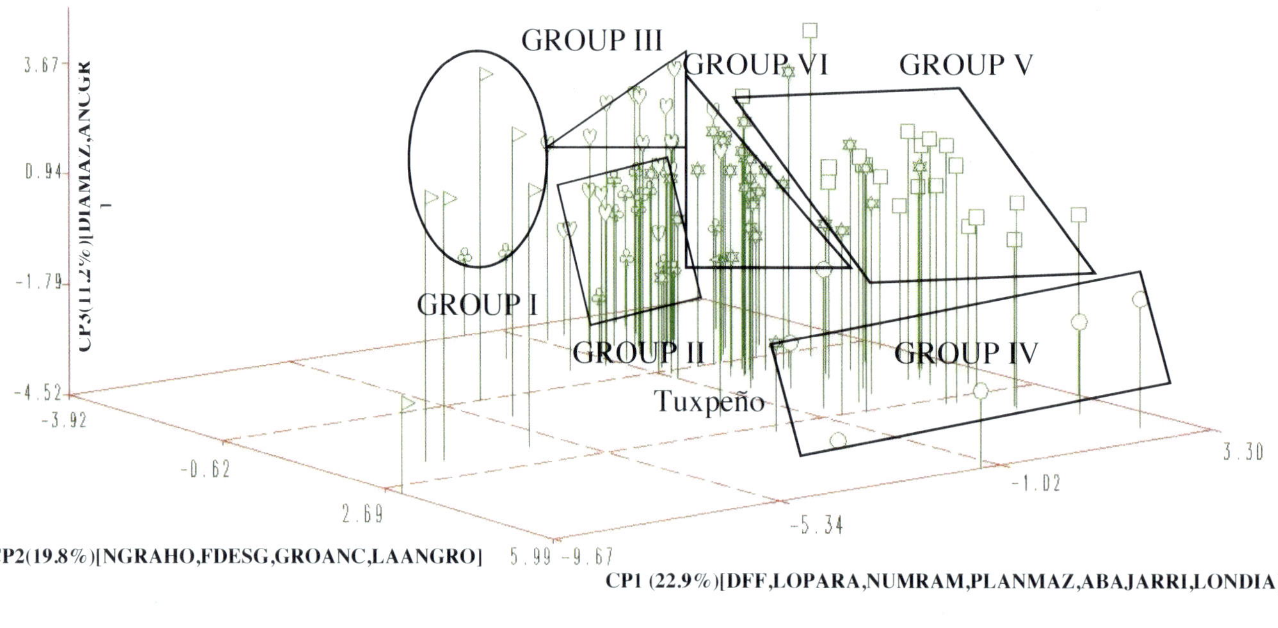

Figure 3. Dispersion of 100 populations evaluated in the humid tropics of Puebla in the space defined by the first three main components (CP).

Table 4. Means for the 12 most important original variables for the five groups identified in the dendrogram of 100 maize populations from the humid tropics of Puebla.

GROUP	DFF (days)	LOPARA (cm)	NUMR	DIAMAZ (cm)	NGRAO	ANCHO (cm)	FDESG (%)	PLANMAZ	ABAJARRI	LONGDIA	GROSANC	LOANGRO (cm^3)
I	107.0c	35.0c	13.8b	4.0a	23.6bc	0.84ab	0.78cd	2.23a	1.07c	3.0b	0.48bcd	0.33c
II	127.3ab	41.9ab	18.3a	3.5bc	33.0a	0.79b	0.86a	1.71c	1.51a	3.9a	0.42cd	0.29c
III	125.4ab	39.5bc	18.1a	3.6b	31.8a	0.86ab	0.85ab	1.73c	1.47a	3.9a	0.41d	0.33bc
IV	127.8ab	44.6a	16.7ab	3.2c	21.1c	0.81ab	0.74d	1.79bc	1.45a	4.4a	0.64a	0.40a
V	132.3a	44.6a	19.9a	3.6ab	27.1ab	0.87[a]	0.79bcd	1.68c	1.50a	4.2a	0.50bc	0.39ab
VI	126.8ab	43.9ab	19.0a	3.7ab	29.8a	0.84ab	0.82abc	1.73c	1.42ab	3.9a	0.47bcd	0.35abc
Tuxpe	124.0b	42.0ab	20.4a	3.6b	22.8bc	0.80b	0.83abc	1.99b	1.20bc	3.2b	0.53b	0.32c
LSD	8.16	4.81	4.22	0.37	5.85	0.06	0.05	0.23	0.22	0.59	0.08	0.06

DFF = Days to silking, LOPARA = Length of the branched segment, NUMR = Number of branches, DIAMAZ = Ear diameter, NGRAO = Number of grains in opposite rows, ANCHO = Grain width, FDESG = Shelling factor, PLANMAZ = Plant height / Ear height, ABAJARRI = Leaves below / leaves above, LONGDIA = Ear length/ Ear diameter, GROSANC = Grain thickness/ grain width, LOANGRO = Grain volume. Tuxpe = Values for the Tuxpeño racial control evaluated in the present study. LSD= Least significant difference.

CONCLUSIONS

In the humid tropics of Puebla there is a considerable morphological diversity among the native maize populations there grown, such variation was grouped in six groups distinguished basically by differences in days to silking, length of the branched segment of the tassel, number of branches of the tassel, ear diameter, number of grains in opposite rows, grain width, shelling factor and in the following indexes: plant height divided by ear height, leaves below divided by leaves above the ear, ear length divided by ear diameter, grain thickness divided by grain width, and grain volume.

Maize landraces grown in the humid tropics of Puebla can be regarded as part of the Tuxpeño race. Among them, the populations collected in the altitudinal strata from 147 to 352 masl, in a warm humid climate, with rains all the year, are the ones that keep the highest similarity with the aforementioned race.

Acknowledgements

We thank the financial support from Fondos Mixtos CONACYT – Gobierno del Estado de Puebla, (Mix Research Funds from both The National Council of Science and Technology (CONACYT) and The Puebla State Government), during the years 2009 – 2010, through the project PUE-2007-01-76993 "Genetic diversity, conservation and improvement of local maize (*Zea mays* L.) populations in major producing regions of Puebla", from which the present study was derived. Likewise, for the financial support from the Consejo Estatal de Ciencia y Tecnología del Estado de Puebla (Council of Science and Technology from Puebla State, CONCYTEP) via a scholarship (Beca-Tesis 2011).

REFERENCES

Ángeles, G. E., Ortiz, T. E., López, P. A., López, R. G. 2010. Caracterización y rendimiento de poblaciones de maíz nativas de Molcaxac, Puebla. Revista Fitotecnia Mexicana 33 (4): 287-296.

Anderson, E. and Cutler, H. C. 1942. Races of *Zea Mays*: I. Their Recognition and Classification. Annals of the Missouri Botanical Garden 29: 69–88.

Bellón, M. R., Adato, M., Becerril, J., Mindek, D. 2005. Impact of Improved Germplasm on Poverty Alleviation: The Case of Tuxpeño-Derived Materials in México. México, D. F. CIMMYT.

Bellón, M. R., Barrientos, P., A. F., Colunga, G. P., Perales, H., Reyes, A., J. A., Rosales, S. R., Zizumbo, V. D. 2009. Diversidad y conservación de recursos genéticos en plantas cultivadas. *En*: Capital natural de México, Volumen II. Estado de conservación y tendencias de cambio. CONABIO, México. pp. 355-382.

Brown, W. L. and Goodman, M. M. 1988. Races of corn. *En*: Sprague, G. F. and Dudley J. W. (ed.). Corn and Corn Improvement. 3° Ed. American Society of Agronomy. Madison, Wisconsin. pp. 33-79.

Caballero, H. F. y Cervantes, S. T. 1990. Estudio genético y taxonómico de poblaciones de maíz de la raza Tuxpeño. Agrociencia serie Fitociencia 1 (2): 43-64.

Franco, T. L. e Hidalgo, R. (eds.). 2003. Análisis Estadístico de Datos de Caracterización Morfológica de Recursos Fitogenéticos. Boletín técnico no. 8, Instituto Internacional de Recursos Fitogenéticos (IPGRI), Cali, Colombia.

García, E. 1973. Modificaciones al sistema de clasificación climática de Köppen. Instituto de Geografía. UNAM. México.

Gil, M. A., López, P. A., Muñoz, O. A., López, S. H. 2004. Variedades criollas de maíz (*Zea mays* L.) en el Estado de Puebla, México: diversidad y utilización. *En*: Chávez-Servia, J. L., Tuxill, J. y Jarvis, D. I. (eds.). Manejo de la diversidad de los cultivos en los agroecosistemas tradicionales. Instituto Internacional de Recursos Fitogenéticos (IPGRI), Cali, Colombia. pp. 18-25.

Goodman, M. M. and Paterniani, E. 1969. The races of maize III: Choices of Appropriate Characters for Racial Classification. Economic Botany 23:265-273.

Herrera, C., B. E., Castillo, G. F., Sánchez, G., J. J., Ortega, P. R., Goodman, M. M. 2000. Caracteres morfológicos para valorar la diversidad entre poblaciones de maíz en una región: caso la raza Chalqueño. Revista Fitotecnia Mexicana. 23: 335-354.

Hortelano, S., R. R., Gil, M. A., Santacruz, V. A., Miranda, C. S., Córdova, T. L. 2008. Diversidad morfológica de maíces nativos en el Valle de Puebla. Agricultura Técnica en México. 34 (2): 189-200.

Instituto Nacional de Estadística, Geografía e Informática (INEGI). 2009. Prontuario de información geográfica municipal de los Estados Unidos Mexicanos. Puebla. Disponible en Internet:

<http://mapserver.inegi.org.mx/dsist/prontuario/index2.cfm> Consultado el 11 de abril de 2012.

Instituto Nacional de Estadística, Geografía e Informática (INEGI). 2010. Anuario estadístico de Puebla. Tomo I. INEGI-Gobierno del Estado de Puebla. INEGI. México.

Jarvis, D. I., Myer, L., Klemick, H., Guarino, L., Smale, M., Brown, A., H. D., Sadiki, M., Sthapit, B. Hodgkin, T. 2006.Guía de capacitación para la conservación *in situ* en fincas. Versión 1. Instituto Internacional de Recursos Fitogenéticos (IPGRI), Roma, Italia.

López, R. G., Santacruz, V. A., Muñoz, O. A., Castillo, G. F., Córdova, T. L.Vaquera, H. H. 2005. Caracterización morfológica de poblaciones nativas de maíz del istmo de Tehuantepec, México. Interciencia 30:284-290.

Kato, Y., T. A., Mapes, S. C., Mera, O., L. M., Serratos H., J. L., Bye B., R. A. 2009. Origen y Diversificación del Maíz: Una Revisión Analítica. Universidad Nacional Autónoma de México, Comisión Nacional para el Conocimiento y Uso de la Biodiversidad. México.

Martínez, G. A. 1988. Diseños experimentales. Métodos y elementos de teoría. Trillas. México.

Mohammadi, S. A. and Prasanna, B. M. 2003. Analysis of genetic diversity in crop plants-salient statistical tools and considerations. Crop Science 43:1235–1248

Muñoz, O. A. 2005. Centli-Maíz. Prehistoria e Historia, Diversidad, Potencial, Origen Genético y Geográfico, Glosario Centli-Maíz (2° edición). Colegio de Postgraduados. Montecillo, México.

Pérez, C. A., Molina, G., J. D., Martínez, G. A. 2002. Adaptación a clima templado de razas tropicales y subtropicales de maíz de México por selección masal visual. Rendimiento, altura de planta y precocidad. Revista Fitotecnia Mexicana. 25 (4): 435-441.

Sánchez, G., J. J. and Goodman, M. M. 1992. Relationships among the Mexican races of maize. Economic Botany. 46: 72-85.

Sánchez, G., J. J., Goodman, M. M, Stuber, C. W. 2000. Isozimatic and morphological diversity in the races of maize of México. Economic Botany. 54: 43-59.

Statistical Analysis System (S.A.S.) Institute. 2002. SAS User´s Guide Statistics. Release 9.0. SAS Institute. USA.

Serratos, H., J. A 2009. El Origen y la Diversidad del Maíz en el Continente Americano. Greenpeace. México. Disponible en Internet: <http://www.greenpeace.org/mexico/Global/mexico/report/2009/3/el-origen-y-la-diversidad-del.pdf> Consultado el 25 de julio de 2010.

Servicio de Información Agroalimentaria y Pesquera (SIAP) 2010 SAGARPA-Secretaría de Agricultura Ganadería Desarrollo Rural Pesca y Alimentación. México. Cierre de la producción agrícola por estado. Disponible en Internet: <http://www.siap.gob.mx/index.php?option=com_wrapper&view=wrapper&Itemid=350> Consultado el 06 de septiembre de 2011.

Wellhausen, J. E., Roberts, M. L., Hernández, X. E., en colaboración con Mangelsdorf, P. C. 1951. Razas de maíz de México, su origen, características y distribución. Xolocotzia. Tomo II. Revista de Geografía Agrícola. pp. 609-732.

Wen, W., Franco, J., Chavez, T., V. H., Yan, J., Taba, S. 2012. Genetic characterization of a core set of a tropical maize race Tuxpeño for further use in maize improvement. PLoS ONE 7(3): e32626. Disponible en Internet: <doi:10.1371/journal.pone.0032626> Consultado el 02 de abril de 2012.

Yanes, G. G. 2011. Diversidad del ecosistema. *En*: Coordinador Nacional Comisión Nacional para el Conocimiento y Uso de la Biodiversidad (CONABIO). La Biodiversidad en Puebla: Estudio de Estado. México. Comisión Nacional para el Conocimiento y Uso de la Biodiversidad, Gobierno del Estado de Puebla, Benemérita Universidad Autónoma de Puebla. pp. 71-90.

3

Activity of Leaf-Cutting Ant Atta *Sexdens Piriventris* Submited to High Dilution Homeopathic Preparations

Alexandre Giesel[1], Mari. I. C. Boff[2], Paulo. A. S. Gonçalves[3], and Pedro Boff[4*]

[1] UDESC-Santa Catarina State University, Centro de Ciências Agroveterinárias, 88520-000. Lages, SC. Brazil. E-mail: alexandregiesel@gmail.com;

[2] UDESC-Santa Catarina State University, Centro de Ciências Agroveterinárias, 88520-000. Lages, SC. Brazil.E-mail: a2micb@cav.udesc.br;

[3] EPAGRI-Agricultural Research and Rural Extension Agency of Santa Catarina State, Cx. Postal 121, CEP 88400-000. Ituporanga, SC, Brazil. E-mail: pasg@epagri.sc.gov.br;

[4] EPAGRI-Agricultural Research and Rural Extension Agency of Santa Catarina State, Cx. Postal 191, CEP 88502-970 Lages, SC, Brazil. E-mail: pboff@epagri.sc.gov.br; pedro.boff@pq.cnpq.br

**Corresponding Author*

SUMMARY

The effect of high dilution preparations on the movement and foraging activities of *Atta sexdens piriventris* was evaluated. Five colonies of ants were located on each of the five experimental areas using a randomized complete block design. Three main forage trails from each colony were selected from where evaluations were made. Ten mL of high dilution preparation of *Atta sexdens piriventris* nosodes and *Belladonna* homeopathy solution were sprayed over 0.5 m of each selected trail, 1.0 m far from the nest. The controls were pure water and non treated trails. Applications were made daily during 10 days. The total number of ants moving on each trail one meter away from the nest, carrying or not plant fragments, were assessed before the daily application. Dilution preparations at 30CH (thirtieth centesimal Hahnemannian dilution) of *A. sexdens piriventris* nosodes and *Belladonna* reduced the activities of ants from the fifth day after the first application. The treatment effect lasted more than 20 days after the last application. The use of preparation at 30CH dilution order to reduce the foraging activity of leaf-cutting ants is a potential non residual method to manage leaf-cutting ants.

Keywords: *Atta*; foraging; non residual preparation.

INTRODUCTION

Leaf-cutting ants of the genus *Atta* are important generalist herbivores largely occurring throughout South America, especially in tropical regions (Fowler *et al.*, 1989; Fernández and Ospina, 2003). Species of *Atta*, together with those of *Acromyrmex* genus represent the true leaf-cutting ants overspread in the Neotropics (Fernández, 2003). However, *Atta* ants are highly polyphagous and they defoliate not only native forest species but also many agricultural and horticultural crops (Correa *et al.*, 2005). *Atta sexdens piriventris* Santschi [1919] is one of the predominant species in Southern Brazil agricultural areas (Link and Link, 2007; Grürzmacher *et al.*, 2002). Leaf-cutting ants cut young plant parts such as sprouts, buds, leaves, and flowers and they use it as substrate for their symbiotic fungus *Leucoagaricus gongilophorus* inside their nest (Roces, 1990). The fungus is the main food source for the colony, mostly required by larvae, young ants, and the queen (Della Lucia and Moreira, 1993; Jolivet, 1998). The population of leaf cutting ants have been increased due to deforestation, large area cattle pasture and crop monoculture. Also, degraded areas maintained by intensive agricultural systems and the use of pesticides have reduced natural enemies and facilitated some insect species to assume the status of pest as do the leaf-cutting ants (Cherrett *et al.*, 1986). Foraging activities of ants can seriously affect agriculture by reducing yields in annual and perennial crops (Grürzmacher *et al.*, 2002). Most modern conventional control methods of leaf-cutting ants are based on the application of large amounts of synthetic insecticides, which are costly, non selective, and hazardous to the environment (Della Lucia and Vilela, 1993; Jaccoud *et al.*, 1999). On the other hand, leaf-cutting ants in natural ecosystems assume important position in the food chain by recycling organic matter, breaking down the dormancy of some seeds, and restoring the soil structure (Jolivet, 1998; Kaspari, 2003; Delabie *et al.*, 2003).

Ecologically based technologies must be developed in such a way that management strategies of pests and diseases should aim to keep down its population levels and improve resilience to the crop, instead of total elimination of the causal agent. Moreover, as stated by García-Hernández *et al.* (2009) the increase of organic food demand has required several methods of pest management mainly at large plantations, which restrictions or opportunities are regulated by certificated agencies (Brasil, 2008). One possibility that has recently emerged is the use of high dilution preparation to treat plants (Betti *et al.*, 2003). High dilution preparations consist of successive one hundredth dilution started from mother tincture followed by 100 sucussions that are vertical 90° angular movements (Brasil, 1997). Intervention with high dilution preparations can restore health to the organism and its entire habitat without substantial changes in the ecological niche (Boff, 2008). This is a suitable technology aimed to design sustainable agriculture systems, which may consider all the complex interaction present in the agroecosystem. Despite the fact that very little is known about the mechanism of action of high dilutions preparations, its effects can be measured throughout experimental studies (Brizzi *et al.*, 2005). According to Bonato (2004), high dilution preparations can be used for restoring the dynamic balance of agroecosystems by targeting a particular point of disturbance; such as pests and diseases, or to improve the resilience throughout inducing resistance/tolerance to the crop. Boff (2008) pointed out that high dilution preparations obtained from local resources such as nosodes, can provide the farmers a simple, versatile, and cheap tool to quickly overcome local problems.

The use of nosodes in agroecosystems could be appropriate to small family farms which represents more than 80% of the farmers in Southern Brazil. Nosodes are bio-therapeutics preparations that are made from insect/pests present in the plant, or parts of the actual diseased plant. Another source of high dilution preparations is the normal homeopathic remedies found in the manipulation pharmacies that are steadily increasing in Brazil (Boff, 2008). In both cases, the high dilution preparations are prepared by combined successive dilution and sucussion (sequence of rhythmically vertical 90° angular movements) until the desirable potency is reached (Andrade, 2001). The laboratory protocols for obtaining homeopathic preparations are well documented in the Farmacopéia Homeopática Brasileira (Brasil, 1997), that follows similar procedures done in Mexico and USA (Tichavský, 2009). In Brazil, homeopathy became a recognized specialization by the Medicine, Veterinary, Pharmacist, and Agronomist Federal Council that are supported by law.

The efficacy of high dilution preparations in agriculture has been recently demonstrated by several works. However, it is still a new area for research in the field of Agroecology. Almeida (2003) observed a significant reduction in the incidence of corn caterpillar *Spodoptera frugiperda* by using high dilution preparations made from larvae of *S. frugiperda*, also called nosodes. Indication of a suitable high dilution preparation can also be found throughout the Homeopathic Matéria Médica, a compendium describing the therapeutic properties of homeopathy remedies for human treatments that were collected from studies made with formal protocols (Brasil, 1997). According to Andrade (2001) we can make analogies with the Matéria Médica and carry out further studies to evaluate its effect on agriculture problems. This is possible because the principle of similitude in homeopathy is concerning to all living systems. However, the cure properties pointed out in Homeopathic Matéria Médica must be tested in agricultural systems, since living organisms might respond differently among kingdoms (Bonato, 2004).

The present study was carried out in 2007 during four field-work months, was restricted to *Atta sexdens piriventris* colonies in Planalto Serrano Catarinense region. Its objective was to study the effect of high dilution preparations of *Atta sexdens piriventris* nosodes and *Belladonna* homeopathy solution on the movement and foraging activities of the leaf cutter ant.

MATERIAL AND METHODS

Study area

The selected nests for study were located in highland fields comprising about 2,500 m^2 of the Planalto Serrano Catarinense region, Santa Catarina State, Brazil. This region is part of the Atlantic Forest, at an altitude of 600 to 800 m above the sea level and situated between 27° 52' and 27° 43' W; 50° 28' and 50° 31' S. The field areas consisted of highland fields originally covered by native grass associated with *Araucaria angustifolia* forest ecosystem. Nowadays, this region has been modified by agriculture and re-forestation with exotic pine species. The research was carried out during the summer season (December-March).

Experimental set up

Data were collected from 25 colonies of *A. sexdens piriventris* located in the field and randomly submitted to high dilution treatments, with five repetitions. The ants of *Atta* genus were selected by the presence of three spine pairs on the dorsal thorax of worker ants and by the presence of crescent-shaped mounds of soil on the soil surface. The selected colonies were generally more than 50 m away from the closest nest and not more than 100 m away from one another. The distance of 100 m was also kept far from pinus re-forestation and the surrounding area of nests was predominantly covered by tough grasses. A sample of worker ants was collected and their species was confirmed at the laboratory following the key of Della Lucia and Moreira (1993). Treatments were randomized in five blocks each one in different areas. Each ant colony was one experimental plot receiving a single treatment. In each experimental plot, the three main foraging trails were labeled at one meter from the nest, from where applications of high dilution preparations and evaluations of ant activity were made. The ant nests area was calculating by measuring the minimum and maximal diameter and estimated according to Zanetti *et al.* (2002).

High dilution preparation treatments

The treatments were: a) ant nosodes prepared from maceration adults *A. sexdens piriventris*, b) ant nosodes prepared from trituration of adults *A. sexdens piriventris*, c) *Belladonna* homeopathy solution (*Atropa belladonna*) obtained from a homeopathic pharmacy, d) pure water, and e) non-treated ant colony. Both, the last two as control plots. All the following homeopathic dilutions were prepared in the Homeopathy and Plant Health Laboratory of Lages Experimental Station in EPAGRI (Agriculture Research and Extension Service Agency of Santa Catarina State, Brazil), according to Farmacopéia Homeopática Brasileira (Brasil, 1997), which describes standard procedures about homeopathic pharmacotechniques in Brazil and elsewhere (Brizzi *et al.*, 2005).

The macerated and triturated ant nosodes were obtained from a sample of 50 worker ants collected randomly from foraging trails of well-established colonies in the research area. For the trituration method, one gram of pre-ground ants was mixed with 99 g of lactose into a porcelain grail. The mixture was ground during 1(one) hour in the sequential period of time, being 10 min in circular movements and 5 min recollecting the material from the grail wall to the centre in vertical movements. This procedure produced the 1CH trit. (first centesimal Hahnemannian dilution of trituration) ant nosode. Again, by taking one part of 1CH trit. and adding 99 parts of lactose into a porcelain grail and ground it during one hour as described above, it was reached the 2CH trit. This procedure was repeated until the 3CH trit. was obtained. After that, each potency level until the 28CH (twentieth centesimal Hahnemannian dilution - stock) was prepared taking one part from the previous potency into 99 parts of alcohol 70%, and followed by 100 sucussion (sequence of rhythmically vertical 90° angular movements) using a mechanical dynamizer (Autic®, Mod. Denise 10-20).

The macerated preparation was originated from the mother tincture of nosode, which was obtained by the mixture of 1 gram of active worker ants mixed with 45 mL of water+alcohol+glycerine (1:1:1) solution and kept all into an amber glass bottle at room temperature for 48 h (Brasil, 1997). The homeopathic dinamization protocol was then followed: 0.2 mL of mother tincture and 19.8 mL of alcohol were placed in a 30 mL amber bottle. The bottle was placed in a mechanic dinamizer (Autic®, Mod. Denise 10-20) followed by 100 sucussions in order to obtain the 1CH (first centesimal Hahnemannian dilution). The same procedure was repeated to obtain the 2CH and the successive potencies until 28CH (the stock homeopathic matrix).

The *Belladonna* homeopathy at the level of third centesimal Hahnemannian dilution (3CH) was obtained from a drug store and prepared until 28CH potency, using alcohol at 70%.

The high dilution preparations were turned to 30CH potency only on the day of application, just before going to the field, using water instead of alcohol. The water treatment was obtained throughout the same procedure, starting with 100 sucussions of pure water until 30CH was reached.

Treatment applications

Application of treatments was performed as double-blind, where neither the applicator nor the evaluator knew the identity of the treatment. The treatments were individualized and identified by codes. Each high dilution preparation of ant nosodes, *Belladonna* and pure water were applied with an individual garden sprayer (Brudden®). Applications consisted of spraying 10 mL of homeopathic preparation over 0.5 m on each of the three main foraging trails of each of the five colonies per treatment. The application sites were located 1.0 m away from the ant nest. Daily applications were made in the morning, from 7:00 to 9:00 during the first 5 days and in the afternoon, from 17:00 to 19:00 during the last 5 days. This schedule was followed due to practical reasons about the travel and facilities.

Evaluation and data analysis

The foraging activity and total movement were recorded just before the applications of high dilution preparations. Foraging activity was considered when ants moved carrying green plant fragments. Total movement was considered when all ants were moving with and without green plant fragments. Both activities were estimated separately by counting the ants passing throughout a given point of the forage trail located 1.0 m away from the ant nest in one min. Evaluations were made just before each daily application during the 10 day treatment period. The last evaluation was made 20 days after the last application.

Statistical analysis was made using the statistical software SAEG®. Data analysis was performed considering the sum of the three foraging trails of each ant colony nest. Analysis of variance was performed to analyse the interaction between application days and homeopathic preparations, and the significant reduction of activities due to treatments. Total movement and foraging activity reduction were calculated by comparing the ninth or 29th day to the first day just before starting high dilution applications. These reductions were compared by Tukey test ($p = 0.05$).

RESULTS

The species of ants in all studied ant nests was confirmed to be *Atta sexdens piriventris* Santschi, 1919. The ant nests in our study had an area ranging from 111 to 234 m^2. No significant difference (F test; df=4, p=0.98) was detected among treatments after randomizing them.

No significant interaction between "treatment" and "application time" factors was obtained for total movement (F test; df=40, p=0.33) and foraging activity (F test; df=40, p= 0.98). However, the total movement was highly correlated with foraging activity for all treatments, thus *Belladonna* (r=0.99, p<0.001), triturate of *Atta* (r=0.99, p<0.001), macerate of *Atta* (r=0.98, p<0.001), pure water (r=0.97, p<0.001) and no application (r=0.95, p<0.001). High dilution preparation of *Belladonna,* ant macerated, and ant triturated reduce steadily the total movement and foraging activities in comparison to control with pure water or no application from the fifth application onwards, and maintained this effect for more than 20 days (Figure 1 and 2).

The total movement and foraging activity of ants were significantly reduced at the ninth and at the 29th day after application in comparison to the first application day when ants were submitted to the macerated or triturated ant nosodes and homeopathy solution of *Belladonna* (Table 1). *Belladonna* at 30CH showed to be the most effective treatment, by consistently reducing total ant foraging activities. Pure water did not reduce total movement or the foraging activity when compared with the non intervention treatment.

The daily rate of movement and forage activity indicated that after 5 days of applications, the reduction was maintained in the ant nests treated by nosodes and *Belladonna* homeopathic treatment (Figures 3 and 4). Control plots (pure water and no application) tended to keep the reduction rate close to zero, while high dilution treatments of triturated and macerated *Atta* reduced in more than 0.1 the daily rate of ant foraging activity after seven days of applications (Figure 4).

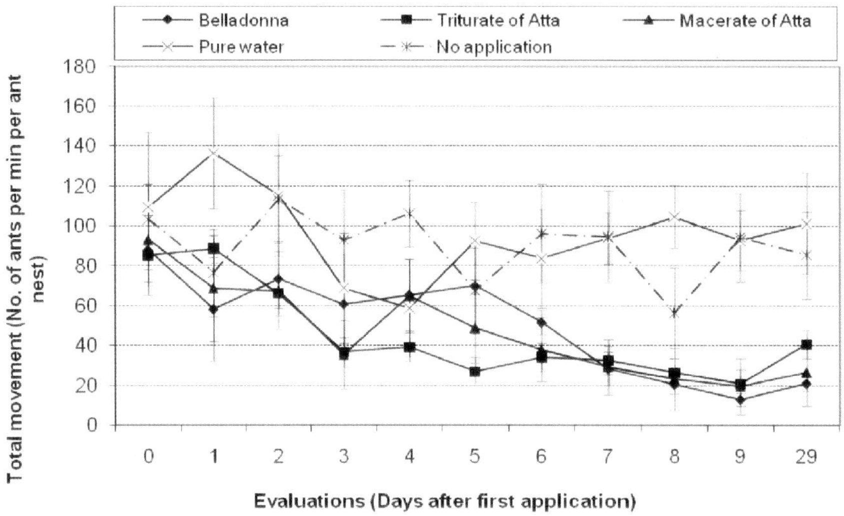

Figure 1. Ant movement of *Atta sexdens piriventris* submitted to high dilution preparations at 30CH (thirtieth centesimal Hahnemannian dilution). Santa Catarina State, Brazil, 2007. Bars are standard error of mean from five replicates.

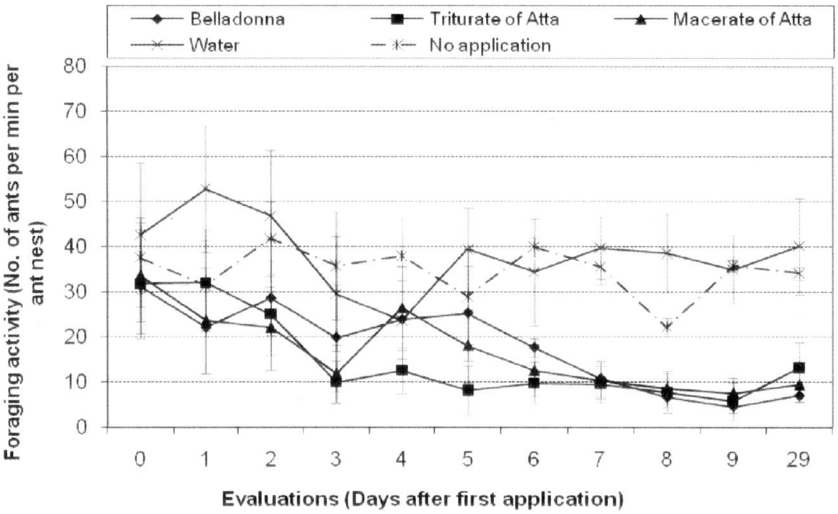

Figure 2. Foraging activity of *Atta sexdens piriventris* submitted to high dilution preparations at 30CH (thirtieth centesimal Hahnemannian dilution). Santa Catarina State, Brazil, 2007. Bars are standard error of mean from five repetitions.

DISCUSSION

High dilution preparations made from *Atta sexdens piriventris*, as macerated or triturated nosodes and *Belladonna* homeopathy could significantly reduce the total ant movement and foraging activities of the specie in the main ant trails. Due to possible yield losses caused by ant foraging activities, a variety of control methods, including use of synthetic insecticides, thermal fogging, poison baits, and explosives to blow up the colonies are usually recommended (Link and Link, 2007). Such conventional practices aim to eliminate the entire colony. Despite the damage that leaf-cutting ants may cause, the conventional control methods do not consider the important role in nutrient recycling and soil aeration of tropical ecosystems that ants may play (Zeh *et al.* 1999; Farji-Brener and Medina, 2000). Reducing the activities of leaf cutter ants without causing the colony extinction is an important strategy to have a low side effect and allow maintaining the ant's colonies playing their role in the food-chain. A ten-day application period of high dilution preparations could reduce up to 85% of foraging activities and the method reported here carries a minimum risk to biological diversity, frequently associated with the synthetic pesticides. The reduction of foraging activities without leading the colonies to extinction allows farmers to consider ants as part of the agroecosystem and accept their presence without unnecessary alarm (Fernández, 2003). Farmers in Panama have protected plants from the attack of *Atta cephalote* by using a refused plant material of this leaf cutting species colony dumped as a deterrent while ant colonies still are in the field (Zeh *et al.*, 1999). It is also important to maintain the biodiversity which improves the resilience property in the agroecosystem as a whole. According to Boff (2008) resilience is a function of complexity of a given ecosystems and the complexity is built up with diversity.

Farmers could use the high dilution method to reduce or suppress ant foraging activity during the most critical period for crop losses. Further studies are needed to see the consistence to manage ant foraging by high dilution homeopathic preparations under farmer conditions. Adopting this method, farmers could be able to establish an acquaintanceship with the leaf cutter ants in their agricultural systems. From our results we can state that 30 days is a particularly good period to protect plants from foraging attacks by leaf cutter ants in annual or perennial crops since sprouting and flowering occur in a short period of time. It is just necessary that farmers are able to identify critical periods in crop cycles when high dilution applications are needed. In case of changing ant paths or nests we should also consider a new situation and design another strategy. Also, it would be expected to protect the flowering period of most crops and, if necessary, indicate a new set of ten day applications. Nevertheless, this must be tested for different production system conditions (Zanetti *et al.*, 2002).

The frequency of applications is also an important factor to obtain an increasing effect of high dilution preparations. Significant reductions of ants' activities were observed after five days applications. It means that a single dose of a high dilution preparation, as sometimes recommended to animal and human treatments, probably will not work on the management of leaf cutting ants. In the case of *A. sexdens piriventris*, more than one application was necessary most probably for reaching the minimum number of worker ants that could cause changes in behaviour of the colony. Considering the ant nest as a society, we may also argue that a high dilution preparation treatment could take a while to have an effect, but when it does so the ant colony changes as a whole. In our study, it was observed a high correlation between reduction of foraging activities and reduction on total activities. This data means that the high dilution preparation acts on the behaviour of the entire ant colony with direct impact to ant movement and foraging activities. However, Della Lucia and Moreira (1993) pointed out that frequency must be also adjusted according to the size of ant nest. The nests we have studied had an area ranging from 111 to 234 m^2. This can give an idea to the farmer in how to manage ant leaf-cutting in colonies of similar size and also to take into account the residual effect. Moreover, we did not find new ant colonies surrounding those we have treated by high dilution preparations until 30 days after starting the applications. This means that high dilution preparations did not perturb the ant colonies at a level to stimulate them to move into new areas. Bonato (2004) reported that when high dilution preparations in agriculture have a prolonged effect, it could be an indication that such intervention approached to the *simillimum* of the agricultural system. This is particularly important for perennial crops, such as orchards because it gives a longer effect in the same treated area. Organisms sensitive to high dilution treatments became more easygoing and less exciting which may also reduce the feeding activities.

Table 1. Reduction of foraging and total movement of *Atta sexdens piriventris* observed at the 9th and 29th days compared to the day zero (0) after starting applications of high dilution preparations. Santa Catarina State, Brazil, 2007.

Treatments	9 days applications*		After 29 days start application*	
	Movement reduction (%)	Foraging reduction (%)	Movement reduction (%)	Foraging reduction (%)
Belladonna 30CH	85.19 a	85.25 a	76.08 a	77.56 a
Triturated *Atta* 30CH	75.52 a	81.76 a	52.23 a	58.49 a
Macerated *Atta* 30CH	79.14 a	77.23 a	71.39 a	72.02 a
Water 30CH	15.35 b	17.84 b	7.49 b	5.63 b
No application	8.89 b	4.78 b	17.40 b	9.04 b
C.V.(%)	49	50	60	59

*Columns followed by the same letter for the same activity (total movement or foraging) did not differ significantly (Tukey, $p = 0.05$). Mean of five replications.

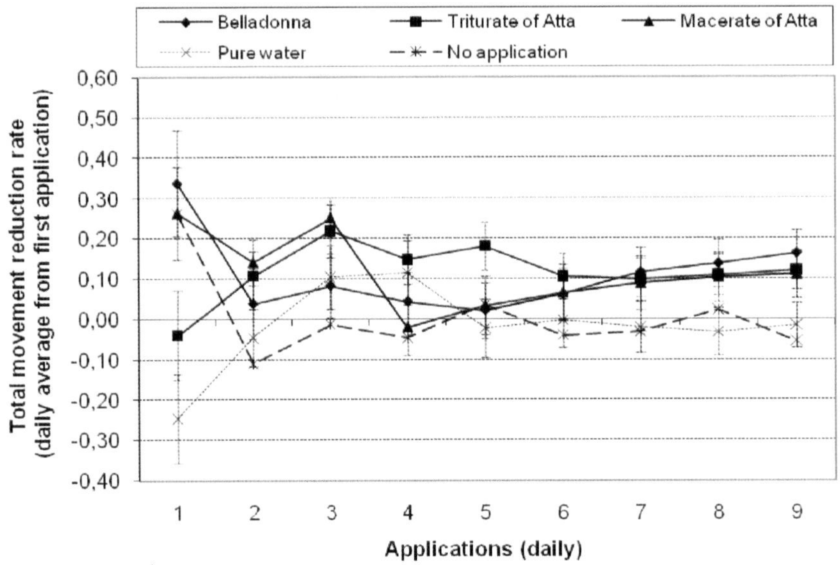

Figure 3. Reduction in daily total movement rate of ant *Atta sexdens piriventris*, considering the time from the first application. Lages, Santa Catarina State, Brazil, 2007. Bars are standard error of mean from five repetitions.

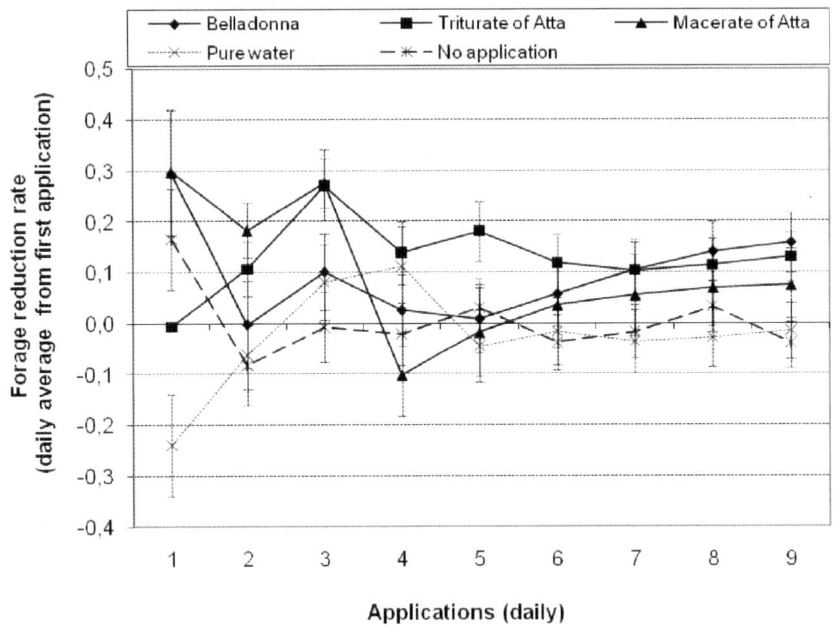

Figure 4. Reduction of daily foraging activity rate of ant *Atta sexdens piriventris*, considering the time from the first application. Lages, Santa Catarina State, Brazil, 2007. Bars are standard error of mean from five repetitions.

Daily applications for such period of time showed to be a practical way to apply homeopathic preparation to manage leaf cutting ants in family farms and organic systems. This is feasible because daily inspections around the orchards and garden are common in small farms and it may be included other routine checking ant foraging activities to do the spraying.

Concerning the method to obtain the mother tincture and get ant nosodes, both procedures, maceration and trituration, were similar on their effect on ant activities. *Belladonna* at 30CH was also as good as ant nosodes at 30CH for reducing foraging activities. Since *Belladonna* homeopathy can be found in pharmacies that manipulate other remedies, it would be a good option for farmers that are not trained for managing high dilutions, or in cases which local conditions are not suitable to prepare the high dilution preparation according to standard protocols.

The homeopathic therapy in human treatments is used everywhere in the world despite criticisms and the pretext of placebo effect (Brizzi *et al.*, 2005). The non effect of pure water at 30CH and similar results in the non intervention plot refused the placebo effect in our studies. In Brazil, there are several successful examples where animal production systems and companion animals are treated with high dilution preparations only (Boff, 2008; Sampaio, 1995). Nevertheless, this is not normal for plant health or plant protection strategies. The lack of research in crop systems with high dilution preparations can be the main cause of the scarce use of agro-homeopathy. However, homeopathy as an applied science implies a change in our perception of life, where co-existence and mutual aid would be the most important principle to develop sustainable living systems. In this sense, it is believed that high dilution preparations applied to agriculture is more feasible to implement under organic crop systems. In addition to that, high dilution preparations are an unprotected right medicine that means they are under public domain. Agro-ecology scientists must take the challenge to improve agro-ecosystems with technologies such high dilution preparations with no side effect and easy to implement.

CONCLUSIONS

High dilution preparations obtained from worker ants (nosodes) and *Belladonna* at 30CH dinamization, reduce the foraging activity of *A. sexdens piriventris* and their effects last more than 20 days. This is a potential strategy to manage leaf-cutting ants on several agricultural and horticultural crops considering further research showing consistent results on its efficacy. High dilution preparations are potential measures to manage leaf-cutting ants without colonies extermination, as do conventional strategies. Additional studies are necessary to check whether the *Atta sexdens periventris* nosodes and *Belladonna* homeopathy could be useful for controlling this insect species in large community areas.

ACKNOWLEDGEMENTS

The authors thank CNPq (Conselho Nacional de Desenvolvimento Científico e Tecnológico) and FAPESC (Fundação de Apoio a Pesquisa Cientifica e Tecnológica do Estado de Santa Catarina) for providing financial support for this research through the Guarani-Serra Geral Consortium Network, Project FAPEU/FAPESC n. 16261/10-2. They are also grateful to the farmers, students from UDESC, and technicians form EPAGRI for their valuable help during the study. The last author is PQ-CNPq researcher.

REFERENCES

Almeida, A.A. 2003. Preparados homeopáticos no controle de *Spodoptera frugiperda* (Lepidoptera: Noctuidae) em milho. M. Sc. dissertation. Federal University of Viçosa, MG, Brazil. 54 p.

Andrade, M.C. 2001. Homeopatia e as plantas medicinais. In: Seminário Brasileiro sobre Homeopatia na Agropecuária Orgânica. Pinhal (SP)/Brasil, p. 37.

Betti, L., Borghini, F., Nani, D. 2003. Plant models for fundamental research in homeopathy. Homeopathy, 92: 129-130.

Boff, P. 2008. Agropecuária saudável: da prevenção de doenças, pragas e parasitas a terapêutica não residual. Epagri/ Udesc, Florianópolis, Brasil. 80p.

Bonato, C.M. 2004. Mecanismo de atuação da homeopatia em plantas. In: Seminário Brasileiro sobre Homeopatia na Agropecuária Orgânica. Toledo (PR), Brasil. pp:17-44.

Brasil. 2008. Instrução Normativa N. 64 de 18 de dezembro de 2008. Regulamento técnico para sistemas orgânicos de produção animal e vegetal. Diário da Republica Federativa do Brasil, DF, 19 de dezembro de 2008-Seção I, 21p.

Brasil. 1997. Farmacopeia Homeopática Brasileira. 2nd. Atheneu: São Paulo/Brasil. 36p.

Brizzi, M., Lazzaroto, L., Nani, D., Borghini, F., Peruzzi, M., Betti, L. 2005. A biostatistical insight into the $As_{(2)}$ $O_{(3)}$ high dilution effects on the rate and variability of wheat seedling growth. Forsch Komplementarmed Klass Naturheilkd, 12: 277-283.

Cherrett, L.M., Peregrine, D.J., Etheridge, P., Mudd, A., Phillips, F.T. 1986. Some aspects of the development of toxic baits for the control of leaf-cutting ants. In: Proceedings of the International Congress-Iussi (7). London. pp. 69-75.

Correa, M.M., Bieber, A.G.D., Wirth, R., Leal, I.R. 2005. Ocurrence of *Atta cephalotes* (L.) (Hymenoptera: Formicidae) in Alagoas, Northeaster Brazil. Neotropical Entomology, 34: 695-698.

Delabie, J.H., Ospina, M., Zabala, G. 2003. Relaciones entre hormigas y plantas: una introducción.

In: Fernández, F. (ed). Introducción a las hormigas de la región Neotropical. Instituto de Investigación de Recursos Biológicos Alexander von Humboldt, Colombia. pp: 167-180.

Della Lucia, T.M.C., Moreira. D.O. 1993. Caracterização dos ninhos. In: Della Lucia, T.M.C. (ed). As Formigas Cortadeiras. Ed. Folha Nova de Viçosa, Brasil, pp. 32-40.

Della Lucia, T.M.C., Vilela, E.F. 1993. Métodos atuais de controle e perspectivas. In: Della Lucia, T.M.C. (ed). As Formigas Cortadeiras. Ed. Folha Nova de Viçosa, Brasil. pp: 163-190.

Farji-Brener, A.G., Medina, C.A. 2000.The importance of where to dump the refuse: seed Banks and fine roots in the nests of the leaf-cutting ants *Atta cephalotes* and *A. colombica*. Biotropica, 32: 120-126.

Fernández, F. 2003. Subfamilia Myrmicinae. In: Fernández, F. (de) Introducción a las hormigas de la región Neotropical. Instituto de Investigación de Recursos Biológicos Alexander von Humboldt. Colombia. pp: 307-330.

Fernández, F., Ospina, M. 2003. Sinopsis de las hormigas de la región Neotropical. In: Fernández, F. (ed.) Introducción a las hormigas de la región Neotropical. Instituto de Investigación de Recursos Biológicos Alexander von Humboldt. Colombia. pp: 49-64.

Fowler, H.G., Pagani, M.I., Silva, A.O., Forti, L.C., Pereira da Silva, V., Vasconcelos, H.L. 1989. A pest is a pest? The dilemma of neotropical leaf-cutting ants: Keystone taxa of natural ecosystems. Environmental Management, 13: 671-675.

García-Hernández, J.L., Valdez Cepeda, R.D., Servín-Villagas, R., Murillo-Amador, B., Rueda-Puente, E.O., Salazar-Sosa, E., Vásquez-Vásquez, C., Troyo-Diéguez, E. 2009. Manejo de plagas en la producción de hortalizas orgânicas. Tropical and Subtropical Agroecosystems, 10: 15-28.

Grürzmacher, D.D., Loeck, A.E., Medeiros, A.H. 2002. Ocorrência de formigas cortadeiras na região da Depressão Central do Estado do Rio Grande do Sul. Ciência Rural, 32: 185-190.

Jaccoud, D.B., Hughes, W.O.H., Jackson, C.W. 1999. The epizootiology of a *Metarhizium* infection in mini-nests of the leaf-cutting ant *Atta sexdens rubropilosa*. Entomologia Experimentalis et Applicata, 93: 51-61.

Jolivet, P. 1998. Interrelationship between insects and plants. CCR Press. Florida, EUA. 309p.

Kaspari, M. 2003. Introducción a la ecología de las hormigas. In: Fernández, F. Introducción a las hormigas de la región Neotropical. Instituto de Investigación de Recursos Biológicos Alexander von Humboldt. Colombia. pp: 97-112.

Link, D., Link, F.M. 2007. Biologia e controle da formiga saúva limão sulina, *Atta sexdens piriventris*. Biológico, 69: 195-196.

Roces, F. 1990. Olfactory conditions during the recruitment process in a leaf-cutting ant. Oecologia, 83: 261-262.

Sampaio, A. 1995. Homeopatia em medicina veterinaria. El Erial LTDA: Curitiba, BR, 226p.

Tichavský, R. 2009. Homeopatía para las plantas. Fujimoto Promociones: Monterrey, MX, 236p.

Zanetti, R., Carvalho, G.A., Santos, A, Souza-Silva, A., Godoy, M.S. 2002. Manejo Integrado de Formigas Cortadeiras. UFLA (MG), Brasil. pp: 355-363.

Zeh, J.A., Zeh, A.D., Zeh, D.W. 1999. Dump material as an effective small-scale deterrent to herbivory by *Atta caphalotes*. Biotropica, 31(2): 368-371.

4

Allelic and Genotypic Frequency of Kappa Casein Gene in Double Purpose Cattle

N.G. Cortés-López[1], S. del Moral [1], José A. Rueda [2], C. Luna-Palomera [3], C.A. Meza-Herrera[4] and J. Abad-Zavaleta[1*]

[1] *Universidad del Papaloapan. Instituto de Biotecnología. Calle Circuito Central No. 200, Col. Parque Industrial. CP. 68301. Tuxtepec, Oaxaca, México. Tel. (287) 87 59 2 40, ext. 220.*E-mail: jabad@unpa.edu.mx.*

[2] *Universidad del Papaloapan. Licenciatura en Zootecnia. Av. Ferrocarril S/N, Col. Ciudad Universitaria, C.P. 68400. Loma Bonita, Oaxaca, México. Tel. (281) 872-22-37, ext. 220.*

[3] *Universidad Juárez Autónoma de Tabasco. División Académica de Ciencias Agropecuarias. Av. Universidad S/N, Col. Magisterial, Villahermosa, Tabasco, México. Tel. (993) 35-8-15-85.*

[4] *Universidad Autónoma Chapingo. Unidad Regional Universitaria de Zonas Áridas. Bermejillo, Durango, México. Tel. (872) 77-6-01-60.*

**Corresponding Author*

SUMMARY

In this work, allelic frequencies (A and B) of *CASκ* gene were determined as selection criteria on milk quality in double purpose cattle. Blood samples were taken at 200 bovine females and were placed in tubes containing EDTA. Genetic marker MB002 was amplified from extracted genetic material. RFLP testing was performed using restriction enzyme *Hinf I* for the diagnosis of A and B alleles on *CASκ*. Genotype frequencies obtained corresponded to 0.34, 0.01 and 0.65 for AA, BB and AB, respectively. Allelic frequency were 0.67 and 0.33 for A and B alleles, respectively. Also, there was a heterocigocity average of 0.6481. The study population is not in Hardy Weinberg equilibrium, the value of $\chi2 = 14.8$ with 2 degrees of freedom ($P < 0.005$). Based on allelic frequency of *CASκ* B (0.34) observed in this study under theoretical assumptions, double-purpose cattle may be a feasible option to increase milk quality when BB sires are used for cross breeding. In this way, B alleles associated to higher milk quality can be enriched in few generations.

Keywords: double purpose bovine, allelic frequency, *CASκ* gene.

INTRODUCTION

Milk contains carbohydrates, lipids, proteins, essential amino acids, fatty acids, vitamins, inorganic elements and water. Milk proteins are divided in soluble and insoluble fractions. In the soluble fraction there are serum proteins: α-lactoalbumin (α-La) and β-lactoglobulin (β-Lg) while the insoluble fraction contains caseins: α-s1 (αs1-Cn), α-s2 (s2-Cn), (β-Cn) and κ-casein (κ-CN). Casein function is to retain fatty acids in milk to form micelles and forming an emulsion (Thompson *et al.*, 2009; Formaggioni *et al.*, 1999). All caseins have a genetic polymorphism involving the replacement of one or two amino acids and occasionally deletion of a segment. The presence of certain genetic variants in milk has an important effect upon coagulation properties of milk (Corral *et al.*, 2006; Gutiérrez *et al.*, 2000; Fox *et al.*, 2000; Formaggioni *et al.*, 1999; Freyer *et al.*, 1999).

In cheese production, k-CN is very important due to its involvement in stabilization and formation of micelles, preventing milk caseins precipitation. Caseins are encoded in a group of autosomal genes located on chromosome 6 at position 6q31-33 spanning 250 kb (Requena *et al.*, 2007; Corral *et al.*, 2006; Gutiérrez *et al.*, 2000; Fox *et al.*, 2000; Formaggioni *et al.*, 1999; Freyer *et al.*, 1999). κ-CN gene is comprised of 5 exons and 4 introns (Cervantes *et al.*, 2007; Requena *et al.*, 2007; López and Vasquez, 2004; Medrano and Cordova, 1990) (figure 1 and 2). Main allelic variants of κ-CN are A and B. These variants are present in all breeds with variable frequencies; however, other variants such as C, E, F o G have been also reported (Requena *et al.*, 2007; Barroso *et al.*, 1998; Horne and Muir, 1994).

Milk derived from animals with B variant has a higher proportion of κ-CN, smaller micelles, higher protein content, increased stability to heat and freezing, reduced coagulation time, a more consistent curd as well as increased cheese yield (5-10%) when compared to animals showing *CASκ* A (Requena *et al.*, 2007; Barroso *et al.*, 1998; López, 1998; Horne y Muir, 1994). These characteristics are important to dairy industry. High frequency of B allele is a common feature in most populations of dairy cattle (Osta, 1994; Zardowny and Kühnlein 1990) excepting Holstein cattle, where A allele is more common (Viana *et al.*, 2001; Bonvillani *et al.*, 2000).

Molecular marker MB002 (also known as JK5-JK3) has been used in several studies to identify allelic variants A and B (Veli *et al.*, 1994). This molecular marker covers exon 4 and intron 4-5 with a size of 350bp (figure 1 and 2); Requena *et al.*, 2007; López and Vásquez, 2004; Medrano and Córdova, 1990). Differences between A and B alleles can be defined using PCR-RFLP technique by restriction with *Hinf I* (Cervantes *et al.*, 2007).

In tropical areas, double-purpose cattle is a term commonly used to describe crossbreeding of Zebu, Creole and European cattle in Mexico, double-purpose cattle has been used for both meat and milk production. At the Papaloapan region, crossbreeding of cattle has been done indiscriminately due to a lack of professional orientation. Occasionally, genetically superior sires have been used, however, most of the time only individuals with a desirable phenotype to the producer are used. Double-purpose cattle is a genetic group well adapted to harsh environmental conditions like high temperature and moisture, low quality of pastures, and high incidence of parasites commonly found in tropical and subtropical areas. Productivity of this system can be considered lower when compared to intensive systems specialized in milk and meat production. However, double-purpose genotypes have advantages with respect to specialized genetic groups, mainly when low-input production systems are considered. These marginal production systems under tropical and subtropical ecosystems are usually managed under low economic and biotic input schemes and with minimum technology (Ortega and Ward, 2005; Baez 2000). The objective of the present work was determine allelic frequencies A and B to *CASκ* in double purpose cattle as well as to discuss some implications to enhance possible strategies designed to promote genetic improvement for milk quality traits.

MATERIAL AND METHODS

Location and environmental conditions

This work was done in the Papaloapan region of Oaxaca This region is located northwest of the Oaxaca Mountains, at a north latitude of 18° 05' west long 96° 08', and 20 m altitude (OEIDRUS-OAXACA, 2005). Papaloapan region has a warm subtropical climate with an annual average temperature of 33 ° C and average annual rainfall ranging from 1,300 to 3,000 mm.

Samples and animals

The double-purpose cattle are crossbreeds of *Bos taurus, Bos indicus* and creole, intended for both meat and milk production. *Bos taurus* breeds admixed are American and European Brown Swiss, Holstein and Simental, *Bos indicus* breeds are Brahman, Gyr, Sardo Negro, Indobrasil, Nelore and Guzerat. However, admixture level is unknown given unsystematically and undocumented crossbreeding.

Blood samples were randomly taken from 200 females of different age, physical condition and physiological state, under lactation and representing 10% of total sampled population. Sampling was done in 25 different ranches.

Whole blood was placed in tubes containing EDTA (1.8 mg/ml final concentration) as anticoagulant.

Blood was ice transported and stored in a - 20 ° C freezer until processed.

Genotyping

DNA extractions were carried out using Blood DNA Extraction kit GF-1 (Vivantis) following manufacturer's recommendations. Polymerase chain reaction was carried out from a starting template of approximately 100 ng of genomic DNA in a 10 µL volume. Accu Prime Taq DNA polymerase System (Invitrogen) was used for amplification. Primers MB002F (ATCATTTATGGCCATTCCACCAAAG) and MB002R (GCCCATTTCGCCTTCTCTGTAACAGA) flanking MB002 molecular marker (fig. 1) were used at 10 pM. Reaction was conducted in a MAXYGENE termocycler (AXYGEN Scientific®). Temperature conditions were 94 °C for 5 min; 30 cycles of 94 °C for 1 min, 56.1 °C for 30 s and 72 °C for 30 s; final extension step for 10 min. PCR products were separated on 2% agarose gels stained with ethidium bromure 0.5% (v/v). Gel was examined with UV light via a photodocumentation system (INGENIUS SYNGENNE®).

PCR-RFLP (Restriction fragment length polymorphism) was conducted using amplification of MB002 molecular marker and restriction enzyme *Hinf I* (Fermentas®). Restriction digest was carried out in a final volume of 30 µL containing 10 µL of PCR product and 1 U of restriction enzyme on R buffer 10X for 2 h at 37°C (figure 2). Digested fragments were analyzed on 3% agarose gels. Band patterns were observed via photo documentation system (Ingenius SYNGENNE ®).

Statistical analysis

Allelic and genotypic frequencies were obtained by means of χ^2 test performed on statistical program SAS.

RESULTS

MB002 marker was PCR amplified from genomic DNA with oligonucleotides MB002F and MB002R, obtaining a 344bp fragment. From 200 samples tested, only 108 were PCR positive.

For those samples where marker MB002 was amplified, *Hinf I* was used (figure 2) to restrict PCR product and allow allele discrimination. Homozygous AA individuals rendered two distinct bands representing three restriction fragments, two fragments of similar length (131 and 132bp) and a smaller 81pb fragment. Homozygous BB individuals showed a clear two band (263 and 81bp) restriction pattern (figure 2 and 3). Heterozygous AB individuals (figure 3) showed a three fragment restriction pattern (263, 131/132 and 81bp).

Table 1 contains genotypic and allelic frequencies found and expected from the study population of dual purpose cattle. Genotype frequencies obtained corresponded to 0.34 for homozygous AA, 0.65 for heterozygous AB and 0.01 for homozygous BB. Allele frequencies of 0.67 for A allele and 0.34 for B allele were observed. Table 1 also shows average heterozygosity of 0.6481. The study population is not in Hardy Weinberg equilibrium ($P < 0.05$). A allele was more prevalent than B allele, showing a frequency of 0.67 and 0.34, respectively.

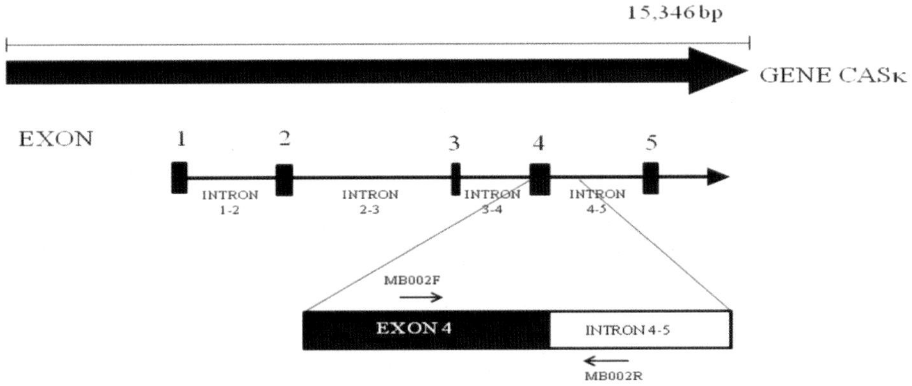

Figure 1. Exons and introns of *CASκ* gene from bovine.

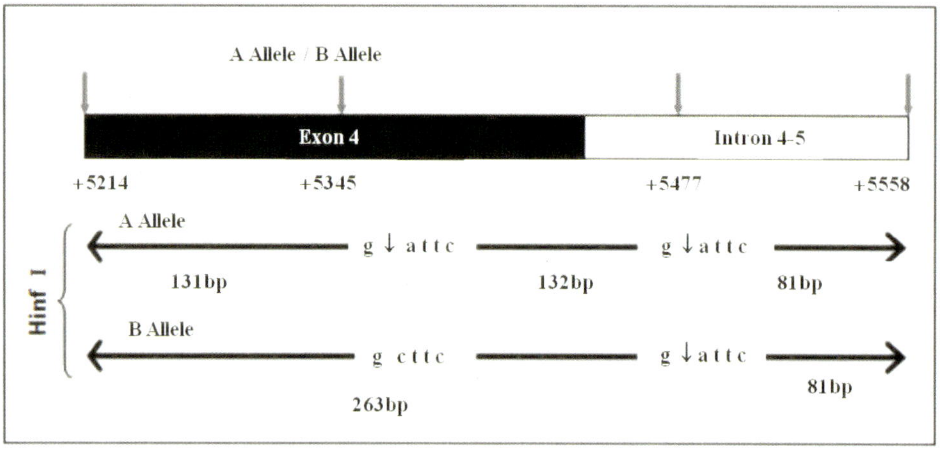

Figure 2. *Hinf I* restriction pattern of MB002 marker, using 344 bp PCR product illustrated in Fig 1 (adapted from Lopez and Vasquez, 2004).

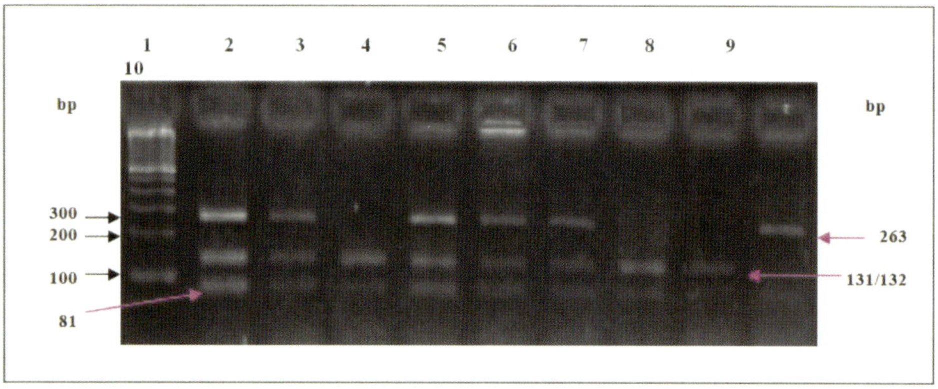

Figure 3. Representative agarose gel showing *CASκ* genotyping performed in this research in double purpose cattle. MWM: molecular weight marker, AA: homozygous A, BB: homozygous B, AB heterozygous.

Table 1. Genotypic and allelic frequencies observed and expected in double purpose cattle.

Genotype	Observed		Calculated		Allele	Allelic frequency
	Number of individuals	Genotypic frequency	Number of individuals	Genotypic frequency		
AA	37	0.34	48.00	0.44	A	0.67
AB	70	0.65	48.00	0.45	B	0.33
BB	1	0.01	12.00	0.11		
χ^2	14. 8	Degrees of freedom	2	(P < 0.005).		

DISCUSSION

From 200 samples tested, only 108 were PCR positive for marker MB002. As shown in figures 1 and 2, MB002 genetic marker is located in exon 4 and intron 4-5. Factors impeding amplification on all samples may be that intronic regions are poorly conserved among allelic variants (Cervantes *et al.*, 2007; Requena *et al.*, 2007; Lopez and Vasquez, 2004; Medrano and Cordova, 1990) and high variability in exon 4. This shows that genetic diversity of *CASκ* gene in exon IV has been underestimated in domestic livestock, and therefore it is necessary to devote more research efforts characterizing this genomic region (Chen *et al.*, 2008). Because of exon and intron variability, it is possible that MB002F and MB002R primers here used have not found a complementary region and therefore marker amplification was not positive for those 92 samples. Therefore, it is possible that negative samples contain another allelic variant than those expected in this work. We are currently looking for allelic variants in the negative samples.

Several studies evaluating genetic variability of *CASκ* gene in different breeds of dairy cattle have been done (Viana *et al.*, 2001; Kemenes *et al.*, 1999; Grosclaude, 1998; Van Eenennaam and Medrano, 1991, McLean *et al.*, 1984). Holstein breed has been highlighted due to its high milk yield; although, it has been low ranked because of its decreased levels of total solids, fat and protein (Benavides, 2003). In contrast, Jersey breed depicts a decreased milk yield regarding to Holstein breed, nonetheless, quality of its milk is far superior compared to Holstein and other dairy breeds (Grosclaude, 1998; Van Eenennaam and Medrano, 1991, McLean *et al.*, 1984). The main drawback of Jersey breed, however, is that its production is limited to certain environmental conditions and therefore it has not been commonly spread around the world. The latter has generated expectations about other alternatives in terms of milk production, considering both quality and volume, by other cattle breeds (Munizaga *et al.*, 2004). Under such scenario, characterization of double-purpose cattle acquire relevance given its widespread use as genotypes that produce both milk and meat, mainly under tropical and subtropical regions, despite of adverse environmental conditions involving high temperature and high humidity. Studies on double-purpose cattle are scarce (Cervantes *et al.*, 2007). The study here presented is the first of its kind in the Papaloapan region.

Several studies characterizing cattle breeds have shown that B allele has higher frequencies in *Bos taurus* with respect to *Bos indicus* (Golijow *et al.*, 1999; Kemenes *et al.*, 1999; Tambasco, 1998, Del Lama and Zago, 1996; Backer and Manwell, 1980). Likewise, in double-purpose cattle evaluated in this study, A allele was predominant over B allele (0.67 and 0.34, respectively), most likely due to high input from A allele races like Zebu, Holstein, Nelore and Guzerat, Gyr (0.74, 0.90, 0.91 y 0.92, 0.93, respectively).

Frequencies obtained for allelic variant B (0.34), surpassed those found in other breeds, such as Guernsey breed (Van Eenennaam and Medrano, 1991, Voelker *et al.*, 1997), Shorthorn breed (Van Eenennaam and Medrano, 1991), Holstein (Viana *et al.*, 2001), Zebu (Cervantes *et al.*, 2007), Nelore, Guzerat, and Gyr (Kemenes *et al.*, 1999) with values of 0.27, 0.11, 0.10, 0.26, 0.09, 0.08 and 0.07, respectively. In the Papaloapan region, commonly used breeds like Jersey, Brown Swiss and Normand have higher B alleles values (0.77, 0.57 y 0.56, respectively).

On the other hand, in several breeds (Holstein-Friesian Irish, Dutch-Friesian, Limousin, Montbeliarde, Charolais, Normande, Norwegian Red and Kerry) has been reported that A allele is dominant in all groups (dairy, beef and double purpose) with an allelic frequency of 0.89 in high-yielding dairy animals and beef animals 0.75. AB haplotype is comparatively rare in dairy cattle (0.11) compared with both beef and double purpose animals. BB haplotype, though rare overall (0.09), is higher in double-purpose animals (0.18) than dairy (0.056) animals (Keating *et al.*, 2007).

Van Eenennaam and Medrano (1991) demonstrated that in mammary glands, allelic variant B has a higher level of expression with respect to allelic variant A and therefore, the proportion of κ-CN in milk depends on the animal genotype in the following order BB> AB> AA, reaching a protein difference of 3% in animals with allelic variant B against animals with allelic variant A. We found only one individual with BB genotype, because of high heterozygous frequency found.

Increasing BB frequency through targeted crossbreeding is an interesting strategy to increase cheese production. If a breed that shows an allelic frequency around 0.25 for kappa-casein B allele is crossed with selected BB stallions, in few generations BB frequency can be dramatically increased (Prinzenberg *et al.*, 1996). Double-purpose cattle in the Papaloapan region can be subject of this strategy. It can be directed towards milk volume production given its high A allele frequency, or towards milk quality by increasing B allele frequency which is already high (Table 1).

Double purpose cattle is the main traditional alternative for milk production, considering an animal production model managed and operated under marginal conditions ,because its excellent adaptation to harsh environmental conditions in the tropics, where specialized breeds cannot develop to full potential.

As shown in table 1, genotype frequencies (AA, AB and BB) observed and expected under the equilibrium condition were not equal. Therefore, the population is not in H-W equilibrium and random mating for this *locus* did not occur in the last generations, according to χ^2 test. Heterozygosity measured (0.6481) suggests a highly hybrid poheterogeneity population regarding this locus, in the studied population, probably asa result of indiscriminate crossbreeding and lack of orientation toward a defined production goal. Finally, using PCR-RFLP -a simple and low cost tool- those animals with allelic variant B could be identified. Therefore, making genetic improvement programs through targeted crosses, with the goal of obtaining animals with better milk quality, are feasible. This will contribute to economic development in the region. Moreover, the short time from collection of whole blood samples to genotype identification by PCR-RFLP offers great advantages in contrast to the time required to evaluate an animal by conventional methods. Recent reports indicate that improvements in this technique allow discriminating between 9 allelic variants in a basic molecular biology laboratory (Pacheco *et al.*, 2011).

CONCLUSION

Through amplification and genotyping of marker MB002 is possible to identify allelic variants A and B of *CASκ* in double-purpose cattle from the Papaloapan region. In this study, only 108 out of 200 samples amplified marker MB002. Therefore, the remaining 92 animals should contain a different allelic variant than those studied in this work. In the population under study, A allele predominated over B allele, with a frequency of 0.67 over 0.34, respectively. High population heterogeneity makes feasible a genetic program to increase milk quality in double-purpose cattle. High frequency of heterozygotes AB is in contrast to almost absent frequency of homozygotes BB. This seems to be caused by indiscriminate crosses combined with minimal guidance to get a defined production goal. Present results will be used for genotypic and phenotypic indicators, in order to select those individuals with homozygous BB and heterozygous AB alleles to produce sires with BB genotype. Furthermore, using homozygous BB sires for crossbreeding will increase the frequency of B allele in the studied dual purpose cattle population and therefore it will improve milk quality, as well as cheese yield .

ACKNOWLEDGMENTS

This research was supported by a grant from PROMEP awarded to J.A-Z. (103.5/07/2740) and a student scholarship awarded. to N. G. C - L. We thank Lice Jennie Renn and PhD Julian Peña-Castro for their collaboration reviewing grammar of this manuscript.

REFERENCES

Barker, A.C.M., Manwell, C. 1980. Chemical classification of cattle. I breed groups. Animal Blood Groups Biochemistry Genetic. 11:127-150.

Barroso, A., Dunner, S., Cañón, J. 1998. Technical note: Detection of Bovine Kappa-casein variants A, B, C, y E by Means of Polymerase Chain Reaction-single strand Conformation Polymorphism (PCR-SSCP). Journal Animal Science. 76:1535-1538.

Benavides Castro, T.A. 2003. Efecto de las Variantes Genéticas A y B de κ-Caseína y β-lactoglobulina sobre las Propiedades de Coagulación de la Leche. Tesis de Licenciatura. Escuela de Ingeniería en Alimentos. Facultad de Ciencias Agrarias. Universidad de Chile. Valdivia Chile.

Bonvillani, A.G., Di Renzo, M.A., Tiranti, I.N. 2000. Genetic polymorphism of milk protein loci in Argentinian Holstein cattle. Genetics and Molecular Biology. 23:819-823.

Cervantes, P., Luna, M., Hernández. M., Perez-Gil, F., Ponce, P., Uffo O. 2007. Polimorfismo genético de en el *locus* de la kappa-caseína, en vacas de diferentes razas y cruces en el trópico mexicano. Revista Salud Animal. 29:78-84.

Chen, S.Y., Costa, V., Azevedo, M., Baig, M., Malmakov, N., Luikart, G., Erhards, G., Beja-Pereira, A. 2008. New alleles of the bovine k-casein gene revealed by resequencing and haplotype inference analysis. Journal Dairy Science. 91:3682-3686.

Corral, J. M., Izquierdo, M., Mateos, S., Parejo, J.C., Salazar, J., Rabasco, A., Martínez-Trancón, M., Sansinforiano, M. E., Hernández, F. I., González-Crespo, J., Jiménez-Hernando, M.A, Portilla, F. J., Padilla, J.A. 2006. Asociación del polimorfismo del gen de la k-caseína (CSN3) con la producción y composición de la leche en ovino merino. Disponible en: www.aida-itea.org/jornada38/genetica/qtls_y_genes_candidatos_ii/qg c9_corral.pdf.

Del Lama, S.N., Zago, M.A. 1996. Identification of κ-casein and β-lactoglobulin genotypes in Brazilian *Bos indicus* and *Bubalus bubalis* population. Brazilian Journal of Genetics. 19:73-77.

Formaggioni, P., Summer, A., Malacarne, M., Mariani, P. 1999. Milk protein polymorphism: detection and diffusion of the genetic variants in bos genus. Univ degli Studi di Parma An della Facoltà di Med Vet. Disponible en: www.unipr.it/arpa/facvet/annali/1999/index2.htm.

Fox, P.F., Guinee, T.P., Cogan, T.M., McSweeney, P.L. 2000. Fundamentals of Cheese Science. Aspen Publishers, Inc. Gaithersburg, Maryland. 32-34.

Freyer, G., Liu, Z., Erhardt, G., Panicke, L. 1999. Casein polymorphism and relation between milk production traits. Journal of Animal Breeding and Genetics. 116:87-97.

Golijow, C.D., Giovambattista, G., Ripoli, M.V., Dulout, F.N. Lojo, M.M. 1999. Genetic variability and population structure in loci related to milk production traits in native Argentine Creole and commercial Argentine Holstein cattle. Brazilian Journal of Genetics. 22:395-398.

Grosclaude F. 1988. Le polymorphisme génétique des principales lactoproteines bovines. Relation avec la quantité, la composition et les aptitudes fromagères dulait. Inra Productions Animales. 1:5-17.

Gutierrez, A., Pintado, B., de la Fuente, J. 2000. Uso del polimorfismo genético de las lactoproteínas para mejorar la calidad de la producción lechera. Boletín Anembe. Disponible en: www.anembe.com/revista/indice.cgi?folder=mar_abr_00.

Horne, D.S., Muir, D.D. 1994. Genetic polymorphism of κ-casein and rennet coagulation time. Effects of serum phase components. Milchwissenschaft. 49:446-449.

Keating, A.F., Davoren, P., Sminth, T.J., Ross, R.P., Cairs, M.T. 2007. Bovine k-Casein Gene promoter haplotypes with potential implications for milk protein expression. Journal of Dairy Science. 90:4092-4099.

Kememes P.A., Reginato, L.C.A., Rosa, A.J.M., Parker, I.V., Razook, G.A., Figuereido, L.A., Silva, N.A., Etchegaray M.A.L., Coutinho, L.L. 1999. κ-casein, β-Lactoglobulin and growth hormone allele frecuencies and genetic distances in Nelore, Gyr, Guzerá, Camcu, Charolais, Canchin and Santa Gertrudis cattle. Genetics and Molecular Biology. 22:539-541.

López Benavides, M. 1998. Correlación entre la constitución genética kappa-caseína y características de producción en vacas Holstein de primer parto en la sabana Bogotá. Primer Simposio de Investigación en la Universidad de la Salle. Colombia.

López, R.E., Vásquez, N. 2004. Determinación del sexo y genotipificación del gen de la κ-caseína en embriones bovinos. Revista Colombiana de Ciencias Pecuarias. 17:231-240.

McLean, D. M.; Graham, E. R. B.; Ponzoni, R. W. and McKenzie, H. 1984. Effects of milk protein genetic variants on milk yield and composition. Journal Dairy Res. 51:531-546.

Medrano, J. F. y E. A. Córdova. 1990. Genotyping of bovine kappa-casein loci following DNA sequence amplification. *Biotechnology* 8:144-146.

Munizaga, F. G.; Barrales, V. L. y Valenzuela, A. L. 2004. Desarrollo de un modelo para evaluar la factibilidad productiva y económica de la progenie resultante del cruzamiento de vacas Holstein Friesian con toros de las razas francesas Montbeliarde y Normando. Departamento de Ciencias Animales, pontificia Universidad Católica de Chile.

Ortega, L. y Ward R. 2005. El sistema de Ganadería de Doble Propósito: Un sistema eficiente. Facultad de Agronomía. Universidad de Zulia. Maracaibo-Venezuela, 22-26.

Osta R. 1994. Caracterización genética de proteínas lácteas y sexaje de embriones en ganado vacuno mediante la aplicación de la biotecnología al análisis del DNA. Tesis Doctoral, Facultad de Veterinaria, Universidad de Zaragoza, España.

Pacheco-Contreras, V., D.L. Lourenco-Jaramillo, G.M. Parra Bracamonte, J.C. Martínez González y A.M. Sifuentes-Rincón. 2011. Convenient genotyping of nine bovine K-casein variants. Electronic Journal of Biotechnology. 14:1-6.

Prinzenberg, E., S. Hiendleder, T. Ikonen and G. Erhardt. 1996. Molecular genetic characterization of new bovine kappa-casein alleles CSN3F and CSN3G and genotyping by PCR-RFLP. Animimal Genetic. 27:347-349.

Requena, F.D., E.I. Agüera y F. Requena. 2007. Genética de la caseína de la leche en el bovino Frisón (Milk of casein of genetic in the Frison bovine). REDVET. Revista electrónica de Veterinaria. 8, 1, 1695-7504.

Tambasco M.D. 1998. Detecao de polymorphism dos genes de κ-casina, β-lactoglobulina em animais da raca Jersey. Monografia: Universidad Federal de Sao Carlos. S.P.

Thompson A., Boland M., Singh H. 2009. Milk Proteins from Expression Food. Food Science and Tecnology, International Series. ELSEVIER. New Zeland. 1-19.

Van-Eenennaam A.L. y J.E. Medrano. 1991. Differences on allelic protein expression in the milk of heterozygous k-casein cow´s. Journal Dairy Science. 74:1491-1496.

Veli E., E. Rivas, V. Rivas, M. Verastegui & S. Pastor. 2004. Evaluación de la variabilidad de genes de kappa caseína en poblaciones de bovinos criollos de ticllos y huaschao, región ancash. Disponible en: www.inia.gob.pe/genetica/zoogeneticos/ArticuloV Congreso.pdf.

Viana, J. L.; Fernández, A.; Iglesias, A.; Sánchez, L.; Becerra, J. 2001. Análisis de los genotipos más frecuentes de la kappa caseína en la raza vacuna rubia gallega mediante PCR/RFLPs. Archivos de Zootecnia 50:91-96.

Voelker, G. R., G. T. Bleck y M. B. Wheeler. 1997. Single-base Polymorphisms within the flanking region of the bovine α-lactalbumin gene. En: Journal Dairy Science. 80:194-197.

Zardowny D y U. Kühnlein. 1990. The identification of the kappa-casein genotype in Holstein dairy cattle using the polymerase chain reaction. Theory Applied Genetic. 80:631-634.

5

Fertility Status of Eutric Nitisol and Fertilizer Recommendation using Numass in the Selected Areas of Jimma Zone, Southwestern Ethiopia

Abebe Nigussie*, Gebremedihin Ambaw and Endalkachew Kissi

Jimma University College of Agriculture and Veterinary Medicine

Department of Natural Resource Management, P.O. Box 307, Jimma, Ethiopia

E-mail address: abenigussie@yahoo.com

**Corresponding author*

SUMMARY

Nutrient management decision support system (NuMaSS) is a tool that diagnoses soil constraints and selects the appropriate management practices, based on agronomic, economic and environmental criteria, for location-specific conditions. However, its validation in different soil orders and agroecosystems should be tested. Thus, the present study was conducted to assess the fertility status of Eutric Nitisol and suggests the application of NuMaSS as a diagnostic tool to recommend supplementary N and P fertilizers for maize across different agroecology zones. Accordingly, fifty soil samples were collected from five locations and analyzed for the major soil physical and chemical properties. The soil pH-H_2O was below the critical level for all soil samples. Nitrogen, Cu and Zn were deficit in all soil samples. Deficiency of available phosphorous was also observed in 88% of the soil samples. High probabilities of nitrogen and phosphorous deficiency were also observed in the study areas using NuMaSS. Moreover, 24 kg/ha P and 73 kg/ha N for Dedo; 24kg/ha P and 82 kg/ha N for Agaro; 20 kg/ha P and 100 kg/ha N for Serbo; 27 kg/ha P and 94 kg/ha N for Dimtu; and 4.9kg/ha P and 97 kg/ha N for Asendabo were recommended for maize using NuMaSS.

Keywords: NuMaSS; Soil fertility; Critical level; maize.

INTRODUCTION

Declining of soil fertility is a fundamental impediment to crops production and a major reason for slow growth in food production in Sub-Saharan Africa (SSA) (Sánchez *et al*., 1995; Quinones et al., 1998; Shapiro and Sanders, 1998). In Ethiopia, soil degradation and nutrient depletion have gradually increased in area and magnitude and have become serious threats to agricultural productivity (Hurni, 1988; Campbell et. al, 1996; Bojo and Cassels, 1995; Fasil and Charles, 2009). Nutrient balance calculations for countries by Stoorvogel and Smaling (1990) also showed that Ethiopia was among the countries with the highest rates of net nutrient losses. The annual nutrient deficit is estimated at -41 kg N, -6 kg P and -26 kg K by ha. Thus, there is an urgent need to improve soil fertility and nutrient management in the country.

Depletion of soil nutrient is a reversible constraint as long as soil test based fertilizer application is in place (Fasil and Charles, 2009). However, assessing soil fertility status is difficult because most soil chemical properties either change very slowly or have large seasonal fluctuations; in both cases, it requires long-term research commitment (Taye and Yifru, 2010). Moreover, the knowledge requirements to properly diagnose and prescribe best management alternatives for location-specific nutrient problems throughout the world exceed the capacity of any human expert. Scarcity and cost of experts; however, can be alleviated by adopting a new system called nutrient management decision support system (NuMaSS). NuMaSS is a tool that diagnoses soil constraints and selects the appropriate management practices, based on agronomic, economic and environmental criteria, for location-specific conditions.

In Jimma zone, Nitosols are the dominant soil types and they are much utilized for crop production. There are about 6827 km^2 of Nitisols in the zone covering about 35% of the district's landmass (BPEDORS, 2000). According to Weigel, (1986) and Elias (2002), Nitisols are inherently fertile, but large Ethiopian areas have now been depleted due to continuous cultivation, leaching and erosion. However, the fertility status of Eutric Nitisol that dominates the study area is not quantified. Furthermore, farmers usually apply blanket recommendation of fertilizer in the study area irrespective of soil orders and agroecosystems. Therefore the objective of the study is to evaluate the fertility status of Eutric Nitisol in Jimma zone and to determine the recommended amount of N and P fertilizer for maize across different agroecosystems using NuMaSS.

MATERIALS AND METHODS

Description of the Study Areas

The study was conducted in the Jimma Zone, southwestern Ethiopia, which is located between $7^0 13'$ – $8^o 56'$N latitude and $35^0 49$ -$38^0 38'$E longitude with an estimated area of 19,506.24 km^2 and elevation between 1000-3500 m.a.s.l. The zone constitute three major climates 78%, 12% and 10% belongs to subtropical, temperate and tropical or thermal zones, respectively. The major soil orders of the zone are Nitosol 6827 km^2 (35%), Vertisol, 2925.9km^2 (15 %) and others 9553 km ^{2} (50%). Almost 49.6% of the zone total area devote to cultivation. The remaining 50.4% of the total area of land were under vegetation cover (22.8% under forest, 18% under woodland, and 9.6% under grassland (BPEDORS, 2000). For the present study, five study areas which have good agricultural potential and dominated by Eutric Nitisol were selected from different agroecosystems (Table 1).

Soil Sampling and Analysis

Prior to soil sampling, soil profile studies were carried out in order to characterize and classify the dominant soil groups in the study areas using WRB (2006). Accordingly, ten soil samples from each site were collected randomly from different farmlands at the depth of 25cm that make a total of fifty soil samples. The soil samples were air dried, mixed well, ground, and passed through 2 mm sieve and the laboratory analysis were conducted using the following standard procedures. Soil texture was determined by hydrometer. Soil pH-H2O and pH-KCl were measured using a pH meter in a 1:2.5 soil: water and soil: KCl ratios, respectively. Soil organic carbon was determined by the Walkley-Black oxidation method, total nitrogen (micro-Kjeldahl method), and available phosphorus was determined using Bray I extraction method as described by Van Reeuwijk (1992). Total exchangeable bases were determined after leaching the soils with ammonium acetate. Amounts of Ca^{2+} and Mg^{2+} in the leachate were analyzed by atomic absorption spectrophotometer and K^+ and Na^+ were analyzed flame photometrically. Cation exchange capacity was determined at soil pH level of 7 after displacement by using 1N ammonium acetate method in which it was, thereafter, estimated titrimetrically by distillation of ammonium that was displaced by sodium (Chapman, 1965). Percent base saturation was calculated by dividing the sum of the base forming cations (Ca, Mg, Na and K) by the CEC of the soil and multiplying by 100. Available micronutrients (Fe, Mn, Zn, and Cu) were extracted with DTPA as described by Lindsay and Norvell (1978).

Table 1: Detail description of the study sites.

Location	Altitude (m)	Coordinates	Mean Annual Rainfall (mm)	Mean Annual Temperature (^{0}C)
Dedo	1500-2000	7° 31' 0" N, 36° 52' 0" E	1500.00	20-25
Agaro	1716-1836	7°49'N 36°39'E	1663.00	19.9
Serbo	1640	7°35' N, 36°46'-37°14'E	1150.00	18-22
Dimtu	2166	7° 58' 0" N, 37° 11' 0" E	1168.00	27.6
Asendabo	1750	7° 53' 0" N, 35° 25' 0" E	1131.08	27.6

Source: BPEDORS, 2000

Statistical Analysis

The data were analyzed using descriptive statistics and analysis of variance using SAS 9.2. Duncan's multiple range was also employed to test the significance difference between locations. Correlation analysis was also performed in order to determine the relationship between the selected soil properties. The recommended amounts of nitrogen and phosphorous fertilizers for maize across different agroecologocal zones were determined using Nutrient Management Support System (NuMaSS) version 2.2.

RESULTS AND DISCUSSION

Soil texture

The textural class of Eutric Nitisol across all locations was clay; however, there was high variation in soil separates between the sampled soils. The coefficients of variation between the soil samples were 19.70%, 21.06% and 40.49% for clay, silt and sand contents, respectively. Significance difference ($P<0.05$) in the percentage of soil separates across locations was also observed (Table 2). All of the soil samples have clay content more than 30%, which is the marginal ranges of total clay requirement for Nitisol (WRB, 2006). The highest percentages of clay, silt and sand were observed at Dimtu, Sebo and Dedo, respectively while the lowest values were recorded at Dedo, Dimtu and Asendabo, respectively (Table 2).

Soil pH, EC and exchangeable acidity

The statistical analysis showed that soil pH-H_2O varied significantly ($P<0.01$) across locations (Table 2). Soil pH-H_2O ranged from 5.23 at Agaro to 5.60 at Dimtu. All of the soil samples had pH-H_2O less the critical level (6.5-8.5) given by Landon (1991). From the sampled soils, 16% rated as strongly acidic while the rest rated as moderately acidic. According to Landon (1991) and Tisdale et al. (1993), most nutrients for field crops are available at pH value of above 5.5; however, 67% of the sampled soils have pH less than 5.5. The lowest values of soil pH at the study sites could be because of loss of base forming cations down the soil profiles even beyond sampling depth through leaching and drain to streams in runoff generated from accelerated erosion. This enhances the activity of Al^{3+} and H^{+} in the soil solution, which reduces soil pH and thereby increases soil acidity. The depletion of basic cations in crop harvest, as indicated in their significant reduction is the other cause for the fall in soil pH. The second reason is continuous use of ammonium based fertilizers such as diammonium phosphate, $(NH4)_2HPO_4$, in such cereal based cultivated fields, which upon its oxidation by soil microbes produces strong inorganic acids. These strong acids in turn provide H^{+} ions to the soil solution that in turn lower soil pH. Acidic nature of Nitisol was also reported by Yihenew (2002). Thus, it is pertinent to raise the soil pH through liming to increase crop productivity of the study areas.

Due to high amount of rainfall (>1000mm) that results leaching of cations, all of the soil samples had also very low electrical conductivity. The values of exchangeable acidity were also rated as low for most of the sampled soils. A significance ($P<0.01$) variation in exchangeable acidity was also observed between locations. Exchangeable acidity ranged from 0.15meq/100g at Asendabo to 0.80meq/100g at Serbo (Table 2). The observed small values of exchangeable acidity indicate the easiness to manage the acidity problem of the study areas.

Table 2: Mean values of soil separate, pH, EC, exchangeable acidity, OC, TN and available phosphorous across different agroecological zones in Jimma.

Location	Dedo	Agaro	Serbo	Dimtu	Asendabo	Critical level	P-value
Clay (%)	41.34^{b}	44.67^{b}	50.08^{ab}	63.33^{a}	58.67^{a}	-	*
Silt (%)	34.33^{ab}	32.00^{ab}	37.92^{a}	23.67^{c}	28.33^{bc}	-	*
Sand (%)	24.33^{a}	23.33^{a}	12.00^{b}	13.00^{b}	13.00^{b}	-	*
Textural class	Clay	Clay	Clay	Clay	Clay	--	-
pH-H_2O	5.48^{ab}	5.23^{c}	5.35^{bc}	5.60^{a}	5.57^{a}	6.5-8.5	**
pH-KCl	4.37^{ab}	4.28^{b}	4.04^{c}	4.39^{ab}	4.45^{a}	-	**
EC (mmhos)	0.043^{c}	0.057^{b}	0.046^{c}	0.050^{bc}	0.068^{a}	-	**
Ex.Ac (meq/100g)	0.25^{b}	0.30^{b}	0.80^{a}	0.20^{b}	0.15^{b}	-	**
OC (%)	2.71^{a}	2.39^{ab}	1.50^{d}	2.20^{bc}	1.88^{c}	0.5-1.0	**
TN (%)	0.25^{a}	0.22^{b}	0.16^{c}	0.18^{c}	0.17^{c}	<1.00	**
C/N	11.17	10.96	9.72	12.10	10.83	10-20	NS
AVP (ppm)	6.97^{b}	6.69^{b}	7.33^{ab}	6.63^{b}	8.96^{a}	10-15	*

Means in the same row followed by the same letters are not significantly different at 5% level of significance; *significant at P<0.05; **significant at P<0.01; NS = non significance; EC = electrical conductivity; Ex.Ac = Exchangeable acidity; OC = organic carbon; TN = total nitrogen; C/N = carbon to nitrogen ratio; AVP = available phosphorus

Organic carbon, total nitrogen and available phosphorous

The mean values of organic carbon, total nitrogen and available phosphorous across different locations is given in Table 2. Organic carbon and total nitrogen varied significantly (P<0.01) across locations. Organic carbon also showed high variability in the investigated soil samples (CV=26.57%). It ranged from 1.50% at Serbo to 2.71% at Dedo (Table 2). According to Landon (1991) and Sanchez et al, (1982), all of the sampled soils had organic carbon greater than the critical level (0.5-1%). High organic carbon in Nitisol was also reported by Shimeles *et al* (2006), Wakene and Heluf (2002), Mesfin (1998), and Eylachew (1999). The surface soils of Nitisol may contain several percent of organic carbon (WRB, 2006). The highest values of organic carbon could be because of poor drained condition of the soil and high amount of rainfall that reduces the rate of organic matter decomposition in the study sites.

Total nitrogen also showed high variability (CV = 22.87%) among the soil samples. Total nitrogen ranged from 0.16% at Serbo to 0.25% at Dedo. Based on the classification of Landon (1991) and Sanchez et al, (1982), total nitrogen was found as one of the limited plant nutrient in the study sites. The values of total nitrogen in all soil samples were below the critical level (<1%) (Table 2). The observed nitrogen deficiency in all soil samples could be because of low input of plant residues, nitrogen rich organic materials like manure and compost in cereal based farming systems. As the area receives high rainfall, the nitrogen leaching problem can be another reason for the decline of total nitrogen in cropped fields. Moreover, farmers of the study area do not integrate leguminous plants on their farmlands. Shimeles *et al* (2006) also reported nitrogen deficiency in the cultivated Nitisol.

A significance (P<0.05) difference in available phosphorous was also observed between locations. High variability (CV = 75.31%) in available phosphorus was also observed among the sampled soils. Available phosphorous ranged from 7.19 ppm at Agaro to 12.16 ppm at Asendabo (Table 2). From the soil samples, 88% had available phosphorous below the critical level (10-15 ppm) given by Landon (1991) and Sanchez et al, (1982). The low level of available phosphorous in the study area might be due to its fixation by Al and Fe, as their presence is expected at the pH values of the soils of the study areas (Tisdale et al, 1993). High phosphorous sorption capacity of Nitisol was also reported by WRB (2006). The correlation matrix (Table 5) also showed a positive and significance (r = 0.37*) relationship between soil pH and available phosphorous. Unlike total nitrogen and available phosphorus, the carbon to nitrogen ratio; however, was with in the critical level except at Serbo which

Nitrogen and Phosphorous Fertilizer Recommendation using NuMaSS

The recommended amounts of nitrogen and phosphorous for maize across different agroecosystem using NuMaSS tool are presented in Table 6. According to the tool, high probabilities of phosphorous and nitrogen deficiencies were observed in all study sites. This finding is also in line with the field data recorded during the survey (Table 2). The recommended amount of phosphorous for maize ranged from 4.97Pkg/ha at Asendabo to 27kg P/ha at Dimtu while the recommended amount of N ranged from 73 kg/ha at Dedo to 100kg/ha at Serbo. The smallest phosphorous recommendation at Asendabo could be because of the residual effect of fertilization. Due to its proximity to the city, the intensive agricultural practices demand excess application of DAP at Asendabo. In line with NuMaSS fertilizer recommendation, the blanket fertilizer recommendation in the study areas for maize is 100kg DAP and 100kg urea which is equivalent to 20.08 kg/ha of phosphorous and 64 kg/ha of nitrogen. Moreover, 110 kg N and 20 kg P was recommended for maize in Nitisol (Wakene et al., 2001). Therefore, NuMaSS could be used as a tool to diagnose and recommend the amount of fertilizer required for maize in different agroecosystems of the study area.

Table 6: Nitrogen (N) and phosphorous (P) fertilizers recommendation for maize across different agroecological zones using NuMaSS.

	Maize			
	Prob. Level (%)	P (kg/ha)	Prob. Level (%)	N (kg/ha)
Dedo	91	24	87	73
Agaro	91	24	87	82
Serbo	90	20	87	100
Dimtu	92	27	87	94
Asendabo	85	4.9	87	97

CONCLUSION

The study revealed that the fertility status of Eutric Nitisol highly varied across location and it has serious nutrient limitations. All of the sampled soils have pH-H_2O below the critical limit for plant growth. Nitrogen, Cu and Zn deficiency was also observed in all soil samples. Similarly, 88% of the sampled soils have available phosphorous below the critical level. Therefore, management practices such as increase input of plant residues, nitrogen rich organic materials like manure and compost are required. Moreover, application of lime, micronutrients based fertilizers should be introduced in order to boost the crop productivity of the study area. Since the fertilizer recommendation from field experiment is in harmony with NuMaSS fertilizer recommendation, NuMaSS can be used as a tool in order to diagnose nutrient deficiency and recommended fertilizers for maize in different agroecological zones. Finally, for validation, it is recommended that the study should be repeated in different soils and agroecological zones in Ethiopia and elsewhere.

Acknowledgement

The authors would like to acknowledge Jimma University College of agriculture and veterinary medicine for funding this research. The authors would like also acknowledge technicians of the college for their assistance during the field work. Moreover, we thank the late Mr. Nigussie Nigatu for his encouragement.

REFERENCES

Balesh, T., Jens A.B., Fred, J.H., Bernand, V. 2008. The prospects of reduced tillage in tef (*Eragrostis tef* Zucca) in Gare Arera, West Shawa Zone of Oromiya, Ethiopia. Soil Tillage Research 99: 58–65.

BPEDORS. 2000. Physical and Socio-Economical Profile of 180 District of Oromia Region, Ethiopia.

Bojo, J., Cassels, D. 1995. Land Degradation and Rehabilitation in Ethiopia: A Reassessment. World Bank Africa Technical Department Working Paper 17. Washington, D.C.

Campbell, C.A., McConkey, B.G., Zentner, R.P., Sells, F., Curtin, D. 1996. Long-term effects of tillage and crop rotations on soil organic carbon and total nitrogen in a clay soil in south western Saskatchewan. Canadian Journal of Soil Sciences. 76: 395-340.

Chapman, H.D., 1965. Cation Exchange Capacity. In: C.A. Black, *et al.* (Eds.), Methods of Soil Analysis. Agronomy 9, American Society of Agronomy. Inc., Madison, Wisconsin, pp. 891–901.

Chi, M., Richard, A.E. 1999. Cation Exchange Capacity of Kaolinite. Clays and Clay Minerals' 47 (2): 174-180.

Elias, E. 2002. Farmers Perception of Soil Fertility Change and Management. SOS-Sahel and Institute for suatainable Development, Addis Ababa, Ethiopia, pp: 252

Eylachew, Z. 1999. Selected physical, chemical and mineralogical characteristics of major soils occurring in Chercher highlands, eastern Ethiopia. Ethiopian Journal of Natural Resource,1(2): 173-185.

Fassil, K., Charles, Y. 2009. Soil Fertility Status and Numass Fertilizer Recommendation of Typic Hapluusterts in the Northern Highlands of Ethiopia. World Applied Sciences Journal 6 (11): 1473-1480.

Hurni, H. 1988. Degradation and conservation of the resources in the Ethiopian highlands. Mountain Research and Development, 8 (2/3): 123-130.

Jones, J.B., 1972. Plants tissue analysis for micronutrients, In: "Micronutrients in Agriculture (Ed. Mortvedt J.J., Giodano P.M., Lindsay W.L.), Soil Science Society of America, Madison, Wisconsin. pp. 319.

Jones, J., Eck, H.V. 1973. Plant analysis as an aid in fertilizing corn and grain sorghum. In Soil Testing and Plant Analysis and Revised L.M. Walsh and J.D. Beaton (eds.), Soil Science Society of American, pp: 349-364.

Landon, J.R., 1991. Booker Tropical Soil Manual, A handbook of soil Survey and Agricultural Land evaluation in the tropics and subtropics. Longman publication. London. pp 500.

Lindsay, W.L., Norvell, W.A. 1978. Development of a DTPA soil test for zinc, iron, manganese and copper. Soil Science Society of American Journal. 42:421-428.

Louis, E.A. 2010. Physicochemical characteristics of the rhizosphere soils of some cereal crops in Ambo Woreda, West Shoa, Ethiopia. Maejo International Journal of Science and Technology. 4(1), 93-100.

Mesfin, A. 1998. Nature and management of Ethiopian soils. Alemaya University of Agriculture, Ethiopia. pp 272.

Quinones, M.A., Borlaug, N.E., Dowswell, C.R. 1998. A fertilizer-based green revolution in Africa. In: RJ Buresh, PA Sanchez & F Calhoun (eds), Replenishing soil fertility in Africa. SSSA, Special Publication No.51, Madison, WI, USA. pp 81-95.

Sánchez, P.A., Valencia I., Izac, A.M., Pieri, C. 1995. Soil fertility replenishment in Africa. Nairobi, Kenya: ICRAF.

Sanchez, P.A., Couto, W., Boul, S.W. 1982. The fertility capability soil classification system: interpretation, applicability and modification. Geoderma, 27: 83-309.

Shapiro, B.I., Sanders, J.H. 1998. Fertilizer use in semiarid West-Africa. Profitability and supporting policy. Agricultural Systems, 56 (5): 467–482.

Shimeles, D., Mohammed, A., Abayneh, E. 2006. Characterization and Classification of the Soils of Tenocha-Wenchancher Micro Catchment, Southwest Shewa. M.Sc thesis, Alemaya University, Ethiopia. pp 111.

Stoorvogel, J.J., Smaling, E.M.A., 1990. Assessment of soil nutrient depletion in Sub-Saharan Africa: 1983-2000. Report 28, DLO Winand Staring Center for Integrated Land, Soil and Water Research (SC-DLO), Wageningen, Netherlands.

Taye B., Yifru, A. 2010. Assessment of Soil Fertility Status with Depth in Wheat Growing Highlands of Southeast Ethiopia. World Journal of Agricultural Sciences 6 (5): 525-531.

Teklu, B., Amnat, S., Yongyuth, O., Ed, S. 2007. Status of Mn, Fe, Cu, Zn, B and Mo in Rift Valley Soils of Ethiopia: Laboratory Assessment. Kasetsart Journal (Nat. Sci.) 41: 84 – 95.

Tisdale, S.L., Nelson, W.L., Beaton, J.D., Havlin, J.L. 1993. Soil fertility and fertilizers. 5th edition. Macmillan Publishing Co., NY, pp 634.

Van Reewijk, L.P. 1992. Procedures for Soil Analysis (3rd Ed.). International Soil Reference Center (ISRIC), Wageningen, The Netherlands, pp 371.

Vijay, M.V., Bruce, R.J., Alba, T. 1997. Copper solubility in Myersville B horizon soil in the presence of DTPA. Soil Science Society of American Journal. 61: 44-52.

Wakene, N., Heluf, G. 2001. Assessment of Important Physicochemical Properties of Dystric Udalf (Dystric Nitosols) Under different Management Systems in Boko Area, Western Ethiopia. MSc Thesis, Alemaya University, Ethiopa. pp 93.

Wakene, N., Tolera, A., Friesen, D.K., Abdenna, D., Berhanu, D. 2001. Evaluation of compost for maize production under farmers' condition. Seventh Eastern and Southern Africa Regional Maize Conference. 11th-15th February, pp 382-386.

Weigel, G. 1986. The soils of the Gununo area, Sidamo Research Unit, Ethiopia. Research report 8, Soil Conservation Research Project, Addis Ababa. pp 94.

World Reference Base (WRB). 2006. World Reference Base for Soil Resources: A framework for International Classification, Correlation and Communication. Food and Agriculture Organization of the United Nations. Rome, Italy. pp 145.

Yihenew, G., 2002. Selected physical and chemical characteristics of soils of Adet Research Center and its testing sites in northwestern Ethiopia. Ethiopian Journal of Natural Resources 4(2): 199-215.

6

Effect of Soybean Plant Populations on Yield and Productivity of Cassava and Soybean Grown in a Cassava-Based Intercropping System

E. U. Mbah[a*] and E. Ogidi[b]

[a]*Department of Crop Production Technology, Federal College of Agriculture, P.M.B. 7008, Ishiagu, Ebonyi State, Nigeria. E-mail: emmaumbah@yahoo.com*

[b]*Department of Biological Sciences, Michael Okpara University of Agriculture, Umudike, Abia State, Nigeria.*

[*]*Corresponding author*

SUMMARY

Intercropping cassava (*Manihot esculenta* Crantz), with varying soybean [*Glycine max* (L.) Merrill] plant populations may influence not only the performance of the component crops but also the residual nitrogen contribution to the cropping system. The treatment scheme was cassava fixed at 100% of its sole crop population of 10,000 plants ha^{-1} while soybean was varied at 25, 50, 75 and 100% of its sole crop population of 266,000 plants ha^{-1}, representing 66,500; 133,000; 199,500 and 266,000 plants ha^{-1}, respectively. The experiment was laid out in a randomized complete block design with three replicates. Fresh tuber yield in cassava increased as soybean population increased to the highest amount used in the study while number of pods plant^{-1} and seed weight plant^{-1} in soybean were depressed by intercropping in the two cropping seasons. However, soybean grain yield was significantly increased with increase in population due to higher aggregate plant populations per unit area in the intercrop. Averaged over the two cropping seasons, productivity showed that cassava/soybean intercrop, especially at the highest soybean population gave the highest yield advantage in terms of total land equivalent ratio, (2.00), land equivalent coefficient (2.41), crop yield equivalent for cassava (31.07 t ha^{-1}) and for soybean (2.25 t ha^{-1}), monetary equivalent ratio (1.56), total gross monetary returns (₦170,311.52) and total net profit (₦74,488.02). This implies that for maximum financial returns, additive intercropping of cassava and soybean at 100% cassava + 100% soybean is recommended because of improved productivity of the system as well as bonus yield and improved health of the farmers from the associated soybean component are achieved.

Keywords: Cassava; *Manihot esculenta* Crantz; soybean; *Glycine max;* intercropping; plant populations; productivity.

INTRODUCTION

Growing of two or more crops simultaneously on the same field in different but proximate stands is a common sustainable crop production system in the humid tropics (Fageria, 1992, Willey, 1979). It has been shown to be more efficient in resource utilization, and improves the overall ecology of the system (Babatunde, 2000) as well as increases monetary returns to the farmers (Mbah *et al.*, 2009). According to Olasantan and Lucas, (1992) the architecture and height of crop canopy as well as days to utilization of soil and aerial environment of the plants properly contribute to the competitiveness and performance of component crops in intercropping. Furthermore, the penetration of light into plant stand is diminished through interception and absorption by the leaves and other parts of the shoot systems.

Hence, in an intercropping system, light utilization by leaf and crop surfaces determine the potential shares of the light that are gained by the component crops. Introducing legumes, including, soybean into cassava-based cropping systems in the humid tropics of south east Nigeria is gaining increased attention because soybean fix atmospheric nitrogen and produce proteins, while cassava deplete the soil nitrogen and give carbohydrates.

Cassava and soybean mixtures improve the diet of farmers as well as the soils of their farms. This study was initiated purposely to evaluate the effects of increasing soybean plant density on growth, yield and biological productivity of the crop species in the mixture as well as to assess the appropriate soybean plant density in cassava/soybean mixture.

MATERIALS AND METHODS

Field experiments were conducted at the Teaching and Research Farm, Michael Okpara University of Agriculture, Umudike, Abia State. The study area lies at Longitude 07° 33' E, Latitude 05° 29' N, altitude 122 m in the humid low lands of southeastern Nigeria. The area has two distinct seasons – the dry season (November – April) and the wet season (May – October). The mean annual rainfall of the experimental site was 2,179.5 mm (2001/02) and 2,069.3 mm (2002/03) and displays a bimodal pattern that is characteristics of southern part of West Africa. The mean monthly temperature during the two cropping seasons ranged from a mean minimum of 25.0 °C to a mean maximum of 28.5 °C. The vegetation is rainforest, although much has been degraded to secondary forest and fallow. The top soil of the location is a sandy loam characterized as ultisol (paleustult). The pre-planting soil analysis (0 – 15 cm) of the experimental sites is presented in Table 1.

A low branching cassava variety (TMS 30572) was obtained from National Root Crops Research Institute (NRCRI), Umudike, Abia State, Nigeria, while a medium maturing soybean variety (TGX-1440-IE) obtained from International Institute of Tropical Agriculture (IITA) Ibadan, Oyo state, Nigeria were used for the study. There were six treatments fitted into a randomized complete block design with three replications. The experiment was carried out under the following treatment scheme:

1. 100 % sole cassava, (10,000 plants ha^{-1})
2. 100 % sole soybean, (266,000 Plants ha^{-1})
3. 100 % sole cassava + 25 % sole soybean
4. 100 % sole cassava + 50 % sole soybean
5. 100 % sole cassava + 75 % sole soybean
6. 100 % sole cassava + 100 % sole soybean

The sole of the component crops were established as control for assessing productivity of the system. The Plot size was 4 x 5 m (20 m^{-2}) with 0.50 m inter-plot and 1.00 m inter-block spaces. Cassava and soybean were planted the same day on 5 m long ridges made 1 m apart.

Table 1: Pre-planting soil analysis of experimental sites in 2001/02 and 2002/03 cropping seasons.

Soil Property	Values	
	2001/02	2002/03
Soil texture	Sandy loam	
pH 1:1 (soil:water)	5.20	4.50
% organic carbon	1.16	1.10
% Total N	0.10	0.13
% organic matter	2.16	2.32
Available P Cmol./kg	19.0	17.0
Exchangeable Ca^{2+} Cmol./kg	2.40	1.20
Exchangeable k Cmol./kg	0.17	0.20
Effective CEC Cmol./kg	4.62	6.24

Cassava cuttings were planted slanting (45°) on the crest of the ridges, one meter apart at the recommended sole plant population of 10,000 plants ha^{-1} while soybean seeds were sown, three seeds $stand^{-1}$ at the varying plant populations as shown in the treatment scheme. Missing stands of cassava were supplied while soybean seedlings were thinned to achieve one plant $stand^{-1}$ 2 weeks after planting (WAP). Manual hoe weeding and remoulding of ridges were carried out 3, 8 and 12 WAP. Slashing of the experimental plots was done 26 WAP. Compound fertilizer N:P:K:Mg 12:12:17:2 at the rate of 400 kg ha^{-1} (Enwezor *et al.* 1989) was applied in two split doses (200 kg ha^{-1}) at 3 WAP on soybean and the second half 8 WAP on cassava.

At the different sampling ages, three cassava and five soybean stands were randomly sampled from the inner ridges of each plot. Data on plant height and number of leaves $plant^{-1}$ were taken from the component crops. Plant height of cassava and soybean was measured with a meter rule, as the height from the base of the crop (ground level) to the tip of the plant while number of leaves $plant^{-1}$ was collected by counting.

Soybean was harvested at 4 months after planting (MAP) when 95 % of the pods had turned brown by pulling whole dry plants with the roots. Harvested soybean pods were sun-dried and threshed. Cassava was harvested at 12 MAP. The yield and yield components of cassava and soybean were taken from the net plots of three inner ridges (6 m^2) in sole and intercropped plots. Records were taken on total number of fresh tubers $plant^{-1}$ and weight of tubers ha^{-1} in cassava and number of pods $plant^{-1}$, seed weight $plant^{-1}$ and grain-yield (tons ha^{-1}) in soybean.

Biological and economic efficiency of the intercropping system was assessed by comparing the productivity of a given area of intercropping with that of sole crops using the following competition functions:

The land equivalent ratio (LER), $= \frac{Yci}{Ycs} + \frac{Ysi}{Yss}$

Where,
Yci = Yield of cassava in intercrop
Ycs = Yield of cassava in sole crop
Ysi = Yield of soybean in intercrop
Yss = Yield of soybean in sole crop, (Mead and Willey, 1980).

Land equivalent coefficient, (LEC) = LA x LB

Where,
LA = LER of main crop (Cassava)
LB = LER of intercrop, (Adetiloye *et al.*, 1983).

Crop yield equivalent (t ha^{-1}):

$$\frac{\text{Yield of crop 'a'}}{\text{Market price of crop 'a'}} \times \frac{\text{Market price of intercrop 'a + b'}}{1}$$

Where,
'a' = Yield of cassava or soybean in sole
'a + b' = Yield of cassava and soybean in intercrop, (Prasad and Srivastava, 1991).

Monetary equivalent ratio (MER), = $(r_1 + r_2)/R$

Where,
r_1 = Monetary return of cassava in the intercrop
r_2 = Monetary return of soybean in the intercrop
R = Higher sole crop monetary return compared with the others, (Adetiloye, 1989).

Gross monetary returns (GMR) (₦ ha^{-1}) was computed for each treatment based on the prevailing market prices of the crops in the study location while net profit (₦ ha^{-1}) was calculated as the difference between GMR and the variable costs of production (Land preparation, procurement of planting materials and planting, three farm weeding and one under-brushing, fertilizer procurement and application, harvesting and processing of soybean pods, harvesting of cassava fresh tubers and movement of farm produce to the nearest food market) of the component crops.

All data from each crop were separately subjected to analysis of variance relevant to RCB experiments as outlined by Gomez and Gomez (1984). The analysis was used to compare the variability in selected growth and yield parameters, due to the treatment applications. Comparison of treatment means were detected by standard error according to Obi (2001).

RESULTS AND DISCUSSION

Cassava: In both cropping seasons, intercropping increased the number of tubers $plant^{-1}$ and fresh tuber yield of cassava ha^{-1}, though not significantly (Table 2). This could be due to better photosynthetic efficiency by intercrops than sole cassava. Also, cassava tuber initiation and bulking were not subjected to any intercrop competition, having harvested the soybean earlier before tuberization process commenced in cassava. Total number of tubers $plant^{-1}$ increased with increase in soybean plant populations. However, fresh yield of cassava was not significantly affected by soybean plant populations in the two cropping seasons, though there were bigger individual sizes of root tubers in the intercrop as soybean populations increased. The big individual roots obtained could be attributed to more competition for light and other growth resources as soybean plant population increased under the additive mixtures. The results confirmed the observations of Chinaka and Obiefuna (2000) in their study on sweet potato/maize mixture as well as Ennin *et al* (2002) in soybean/maize intercrop. Furthermore, the varying maturity dates of the component crops led to a reduction in competitive effects of soybean on cassava fresh tuber yield. Soybean matured much earlier than cassava, an indication that cassava had enough time, space and assess to growth resources such as soybean crop left-over and fixed atmospheric nitrogen to compensate for losses suffered once the soybean competition was removed. The results corroborated studies by Cenpukdee and Fukai (1992) in cassava/legume intercrop in which they reported that 30 - 50 days after planting, canopy width of and light interception by cassava were very low such that the legume in the mixture intercepted 45 – 90 % solar radiation incident reaching the crops.

Soybean: Intercropping significantly depressed the number of pods $plant^{-1}$, seed weight $plant^{-1}$ and grain yield ha^{-1} in soybean in both seasons (Table 2). The highest grain yield was obtained in sole soybean because of reduced inter-specific competition for growth resources among the crops as well as higher aggregate population density per unit area observed in the intercropping situation. In the two cropping seasons, number of pods $plant^{-1}$ and weight of seeds $plant^{-1}$ were reduced as soybean plant density increased in the mixture due to competition for growth resources between the component crops, which reduced the rate of assimilated photosynthates in high density soybean planting.

Field grain yield ha^{-1} increased as soybean plant density increased up to 199,500 plants ha^{-1}.and then decreased with further increase in plant density. The increase in soybean grain yield was due to better utilization of environmental factors since there was less interference from the neighboring cassava plants with initial slow growing period. The results confirmed the observations of Udealor (2002) in cassava/vegetable cowpea mixture as well as Ebwongu *et al.* (2001) in maize/*Solanum* potato intercrop in which they reported improved performance of crops with low plant population due to species intercrop effects.

Assessment of the productivity of the system:

Biological: The land equivalent ratio (LER), land equivalent coefficient (LEC) and yield equivalent of cassava and soybean were used to assess the biological productivity of the component crops in the intercrop (Table 3). In the two cropping seasons, intercropping resulted in yield advantage. The total LER of cassava and soybean in the intercrop were all above unity, ranging from 1.53 to 1.99 (2001/02) and 1.53 to 2.01 (2002/03), an indication that higher productivity per unit area was achieved in intercropping than sole cropping. In particular, 100 % cassava + 100 % soybean combination gave the highest total LER of 1.99 (2001/02) and 2.01 (2002/03), implying that 99 % and 101 % in 2001/02 and 2002/03, respectively more land would be required as sole crop to produce the yield obtained under intercropping situations. Yield advantages resulted from intercropping because the component crops had different durations and growth patterns, hence, made major demands on resources at different times, which led to better temporal use of growth resources (Mbah and Muoneke, 2007).

Value obtained in LEC and yield equivalent of the component crops followed a trend similar to that of LER showing that intercropping cassava and soybean had strong yield advantage relative to sole cropping of the component crops in both cropping seasons.

These productivity results (LER, LEC and Yield equivalent) are consistent with the findings of Njoku *et al.* (2007) in sweet potato/okra, Muoneke *et al.* (2007) in maize/soybean, John and Mini (2005) in okra/cowpea, okra/cucumber and okra/amaranth intercrops as well as Mbah *et al* (2009) in cassava/soybean mixtures, which showed yield advantages in the systems.

Economic: Assessing the monetary equivalent ratio (MER) (Adetiloye, 1989), the results showed that in either of the two cropping seasons, increased soybean plant populations increased MER in the intercropping system (Table 4). In all the combinations, MER was above unity, an indication that it was more advantageous to plant the crops in mixture than sole cropping. Intercropping 100 % cassava with 100 % soybean gave the highest percent MER (63 %) in 2001/02 and (49 %) in 2002/03. In terms of gross monetary returns (GMR), increased soybean plant population increased partial GMR in cassava and soybean.

Table 2: Effect of soybean plant populations on yield indices of cassava and soybean in cassava-based intercropping system in 2001/02 and 2002/03 cropping seasons.

Crop	Yield index	100 % sole cassava, (10,000 plants ha^{-1})	100 % sole soybean, (266,000 Plants ha^{-1})	100 % sole cassava + 25 % sole soybean	100 % sole cassava + 50 % sole soybean	100 % sole cassava + 75 % sole soybean	100 % sole cassava + 100 % sole soybean	s.e.
2001/02								
Cassava	No. of tubers plant^{-1}	6.47	-	7.27	7.59	9.03	9.04	0.182
	Fresh root tuber Yield (t ha^{-1})	19.7	-	20.8	21.5	21.9	23.8	0.346
Soybean	No. of pods plant^{-1}	-	40.38	31.23	22.90	17.17	15.25	2.276
	Seed weight (g plant^{-1})	-	8.97	7.00	5.09	3.93	3.34	0.704
	*Grain yield (kg ha^{-1})	-	745.90	351.87	413.02	470.35	583.62	58.085
2002/03								
Cassava	No. of tubers plant^{-1}	6.65	-	6.81	7.93	8.58	10.01	0.491
	Fresh root tuber Yield (t ha^{-1})	21.1	-	21.3	21.8	22.0	24.1	0.961
Soybean	No. of pods plant^{-1}	-	33.26	26.26	20.92	16.40	13.70	2.273
	Seed weight (g plant^{-1})	-	5.20	4.05	3.22	2.12	1.78	0.416
	*Grain yield (kg ha^{-1})	-	540.19	254.24	303.39	367.87	437.53	55.733

s.e. = standard error of means; *Grain yield at 13% moisture content.

The trend was the same in both cropping seasons. The mean of total GMR obtained from the intercropping system was greater than that from the sole of the component crops (cassava and soybean) by 31 % and 62 %, respectively in 2001/02 and by 25 % and 73 %, respectively in 2002/03. Averaged over the two cropping seasons, the highest GMR was obtained under 100 % cassava intercropped with 100 % soybean (₦170,311.52). This implies that yield and monetary returns of cassava and soybean in the intercrop were optimized at this soybean plant population.

The highest total net returns was obtained from 100 % cassava intercropped with 100 % soybean (₦66,420.64) in 2001/02 and (₦82,555.40) in 2002/03. A plausible explanation is better utilization of the resources in the micro-climate of the intercrop and the economic values of the component crops used in the study.

CONCLUSION

On the strength of the present results, soybean intercropped with cassava, especially, at 100 % of its sole crop density can be advocated as a promising intercrop in cassava-based production systems. The intercrop system gave full benefit of the main crop (cassava) while soybean yield was added productive advantage. Also, farmers' health was enhanced

Table 3: Effect of soybean plant populations on land equivalent ratio, land equivalent coefficient and crop yield equivalent of cassava and soybean in cassava-based intercropping system in 2001/02 and 2002/03 cropping seasons.

Treatments	Land equivalent ratio †Partial Cassava	Land equivalent ratio †Partial Soybean	‡‡Total	Land equivalent coefficient	Crop yield equivalent (t ha^{-1}) Cassava	Crop yield equivalent (t ha^{-1}) Soybean
2001/02						
100 % sole cassava, (10,000 plants ha^{-1})	1.00	-	1.00	-	-	-
100 % sole soybean, (266,000 Plants ha^{-1})	-	1.00	1.00	-	-	-
100 % sole cassava + 25 % sole soybean	1.06	0.47	1.53	1.62	25.87	1.80
100 % sole cassava + 50 % sole soybean	1.09	0.55	1.64	1.79	27.45	1.91
100 % sole cassava + 75 % sole soybean	1.11	0.63	1.74	1.93	28.67	1.99
100 % sole cassava + 100 % sole soybean	1.21	0.78	1.99	2.41	32.20	2.24
2002/03						
100 % sole cassava, (10,000 plants ha^{-1})	1.00	-	1.00	-	-	-
100 % sole soybean, (266,000 Plants ha^{-1})	-	1.00	1.00	-	-	-
100 % sole cassava + 25 % sole soybean	1.06	0.47	1.53	1.62	24.69	1.85
100 % sole cassava + 50 % sole soybean	1.08	0.56	1.64	1.77	25.85	1.94
. 100 % sole cassava + 75 % sole soybean	1.09	0.68	1.77	1.93	26.90	2.02
100 % sole cassava + 100 % sole soybean	1.20	0.81	2.01	2.41	29.93	2.25

†Partial LER for cassava and soybean were obtained by dividing each intercrop yield by its corresponding sole crop yield.

‡Total LER was the sum of the partial LERs from cassava and soybean in the intercropping system.

Table 4: Effect of soybean plant density on crop equivalent coefficient, gross monetary returns and net profit of the component crops in cassava-based intercropping system in 2001/02 and 2002/03 farming seasons.

Treatment	*Monetary equivalent ratio	+Gross monetary return (₦ ha^{-1}) Partial Cassava	+Gross monetary return (₦ ha^{-1}) Partial Soybean	+Gross monetary return (₦ ha^{-1}) Total	Net return (₦ ha^{-1}) ‡‡partial Cassava	Net return (₦ ha^{-1}) ‡‡partial Soybean	Net return (₦ ha^{-1}) **Total
2001/02							
100 % sole cassava, (10,000 plants ha^{-1})	1.00	98,5000	-	98,500	29,200.00	-	29,200
100 % sole soybean, (266,000 Plants ha^{-1})	1.00	-	53,704	53,704	-	16,204	16,204
100 % sole cassava + 25 % sole soybean	1.31	104,000	25,334	129,334	43,600.00	1,934	45,534
100 % sole cassava + 50 % sole soybean	1.39	107,500	29,737	137,237	47,100.00	3,987	51,087
100 % sole cassava + 75 % sole soybean	1.46	109,500	33,865	143,365	49,100.00	5,995	55.095
100 % sole cassava + 100 % sole soybean	1.63	119,000	42,020	161,020	58,600.00	7,820	66,420
2002/03							
100 % sole cassava, (10,000 plants ha^{-1})	1.00	120,600	-	120,600	42,100	-	42,100
100 % sole soybean, (266,000 Plants ha^{-1})	1.00	-	43,215	43,215	-	5,765	5,765
100 % sole cassava + 25 % sole soybean	1.23	127,800	20,339	148,139	64.200	539	64,739
100 % sole cassava + 50 % sole soybean	1.29	130,800	24,271	155,071	67,200	1,071	68,271
100 % sole cassava + 75 % sole soybean	1.34	132,000	29,429	161,429	68,400	1,829	70,229
100 % sole cassava + 100 % sole soybean	1.49	144,600	35,002	179,602	81,000	1,555	82,555

*Monetary equivalent ratio (MER) of intercropped cassava and soybean was obtained by dividing the monetary returns of each intercrop by the highest sole crop monetary return.

+Cassava and soybean were sold at the prevailing market prices (₦ ton^{-1}) of ₦5,000 ton^{-1} and ₦72,000 ton^{-1}, respectively at the time of harvest in 2001/02 and at ₦6,000 ton^{-1} and ₦80,000 ton^{-1}, respectively at the time of harvest in 2002/03, cropping seasons. 1 US Dollar = ₦120:50 (Nigerian Naira) in 2001/02 and ₦128:50 (Nigerian Naira) in 2002/03.

‡‡Partial net profit was the difference between partial GMR and partial variable costs of production of cassava and soybean in the intercrop.

**Total net profit was the sum of the partial net profit from cassava and soybean in the intercrop.

REFERENCES

Adetiloye, P.O. 1989. A review of current competition indices and models for formulating component proportions in intercropping. In: Cassava Based Cropping systems Research, 11: 72 – 90. International Institute of Tropical Agriculture (IITA), Ibadan.

Adetiloye, P.O., Ezedinma, F.O.C., Okigbo, B.N. 1983. A land equivalent coefficient concept for the evaluation of competitive and productive interactions on simple complex crop mixtures. Ecological Modelling, 19: 27 - 39.

Babatunde, F.E. 2000. Effects of intercropping and cropping sequence on the marketable yield and productivity of Celosia and okra. Journal of Agriculture and Environment, 1: 56 – 61.

Cenpukdee, U., Fukai, S. 1992. Cassava/legume intercropping with contrasting cassava cultivars. 1. Competition between component crops under three intercropping conditions. Field Crops Research, 29: 113 – 133.

Chinaka, C.C., Obiefuna, J.C. 2000. Evaluation of optimum population and biological efficiency of sweet potato in sweet potato/maize intercropping system. Nigeria Agricultural Journal, 31:158-165.

Ebwaonga, M., Adipala, E., Ssekabembe, C.K., Kyamanywa, S., Bhagsaril, A.S. 2001. Effect of intercropping maize and solanum potato on yield of the component crops in central Uganda. African Crop Science Journal, 9, 83 – 96.

Ennin, S.A., Clegg, M.D., Francis, C.A. 2002. Resource utilization in soybean/maize intercrops. African Crop science Journal, 10: 251 – 261.

Enwezor, W.O.E., Udo, E.J, Usoroh, N.J., Ayotade, K.A., Adepetu, J.A., Chude, V.O., Udegbe, C I. 1989. Fertilizer use and Management Practices for Crops in Nigeria, Series No.2, Fertilizer procurement and distribution Division (FPDD): Federal Ministry of Agriculture, Water Resources and Rural Development, Lagos, Nigeria. Pp. 60 – 63.

Fageria, N. K. 1992. Maximizing crop yields. Marcel Dekker Inc., New York, USA. 279 pages.

Gomez, K.A, Gomez A.A. 1984. Statistical procedures for Agricultural Research 2nd ed. John Wiley and Sons Inc., New York, 680 pp.

John, S. A., Mini, C. 2005. Biological efficiency of intercropping in okra (*Abelmoschus esculentus* (L.) Moench). Journal of Tropical Agriculture, 43: 33 – 36.

Mbah, E. U. and Muoneke, C. O. 2007. Productivity of cassava/okra intercropping systema as influenced by okra planting density. African Journal of Agricultural Research, 2: 223 – 231.

Mbah, E.U., Nottidge, D.O., Keke, C.I. 2009. Growth and yield of cassava and okra as influenced by cassava varieties in cassava/okra intercrop on an acid ultisol. Proceedings of the 43rd Annual Conference of Agricultural Society of Nigeria, held at National Universities Commission and Raw Materials Research and Development Council, FCT, Abuja, Nigeria. 20–23 October, 2009. pp. 19 – 23.

Mead, R., Willey, R. W. 1980. The concept of a land equivalent ratio and advantages in yield from intercropping. Experimental Agriculture, 16: 217 - 218.

Muoneke, C.O., Ogwuche, M.A.O., Kalu, B.A. 2007. Effect of maize planting density on the performance of maize/soybean intercropping system in a guinea savanna agro-ecosystem. African Journal of Agricultural Research, 2: 667 - 677.

Njoku, S.C., Muoneke, C.O., Okpara, D.A., Agbo, F. M. 2007. Effect of intercropping varieties of sweet potato and okra in an ultisol of southeastern Nigeria. African Journal of Biotechnology, 6: 1650 – 1654.

Olasanta, .F.O. Lucas, E.O. 1992. Intercropping maize with crops of differing canopy heights and similar or different maturities using different spatial arrangements. Journal of Agricultural Science and Technology, 2: 13 – 22.

Obi, I.U. 2002. Statistical Methods of Detecting Differences Between Treatment Means and Methodology Issues in Laboratory and Field Experiments. AP Express Publishing Company, Limited, Nsukka, Nigeria. 116 pp.

Prasad, K. and Srivastava, R. C. 1991. Pigeon pea (*Cajanus cajan*) and soybean (*Glycine max*) intercropping system under rainfed situation. Indian Journal of Agricultural Science, 61: 243 – 244.

Udealor, A. 2002. Studies on Growth, Yield, Organic Matter Turn-over and Soil Nutrient Changes in Cassava (*Manihot esculenta* Crantz)/Vegetable Cowpea (*Vigna unguiculata* L. Walp). Mixtures. Ph. D. Thesis, Department of Crop Science, University of Nigeria, Nsukka. 169 p.

Willey, R.W. 1979. Intercropping: Its importance and research needs. Part 1. Competition and yield advantages. Field Crops Abstracts, 32: 1 – 10.

7

Effect of Mycorrhizal Inoculants in the Development of Mexican Landrace Avocado Rootstocks

Edgar Castro Alvarado[1], Ana Tztzqui Chávez Bárcenas[1], Pedro Antonio García Saucedo[1], Leovigilda Reyes Ramírez[1] and Ana Elizabeth Bárcenas Ortega[1*]

[1]*Facultad de Agrobiología "Presidente Juárez", Universidad Michoacana de San Nicolás de Hidalgo. Lázaro Cárdenas esq. Berlín S/N, CP 60190, Uruapan, Michoacán, México.E-mail: abarcenas@prodigy.net.mx*
**Corresponding author*

SUMMARY

The aim of this work was to assess the effect of two arbuscular mycorrhizal fungi (AMF) inoculants in the development of avocado rootstocks. Seeds of Mexican landrace avocado (*Persea americana* Mill. var. *drymifolia*) were used, with two commercial inoculants: T1 containing *Glomus fasciculatum, G. constrictum, G. tortuosum, G. geosporum,*and *Acaulospora scrobiculata,* and T2, containing *G. mosseae* and *G. cubense*. The plants inoculated with AMF showed more rapid growth than the no inoculant control as measured by plant height (50% and 54%), stem diameter (35% and 36%), leaf number (48% and 37%) and length (31% and 40%), and root fresh weight (85% and 59%); however, no significant differences were observed between T1 and T2. The chlorophyll concentration in the leaves from T1 was 16.4% and T2 was 19% higher than the control suggesting a higher photosynthetic capacity in T1 and T2. Finally the shoot/root ratio, as indicator of the potential development of plantations, was 79% and 50% higher in mycorrhizal plants than in the control. In conclusion both T1 and T2 inoculants improved growth rate and vigor of avocado nursery rootstocks producing higher quality plants.

Key words: Arbuscular Mycorrhizal; biofertilizers; *Persea Americana;* rootstocks.

INTRODUCTION

Avocado production is a major economic activity in Mexico, considered the world's leader in planted surface area, production, consumption and exportation of this fruit crop. More than 74% of the avocado's planted surface area in this country is located in the State of Michoacan, with an annual production of 1,117,338.5 tons (SIAP, 2012). This agronomic activity encompasses remarkable economic and social impact due to the volume of production and to the large amount of direct and indirect employment it generates (Sánchez, 2007).
The most popular cultivar in Michoacan is "Hass"

grown on the Mexican landrace, *Persea americana* Mill. var. *drymifolia,* as rootstock. Avocado has hairless mangrove-like roots (Salazar-García, 2002) that are highly dependent on mycorrhizal symbiosis (Alarcón *et al*., 2001). Arbuscular mycorrhizal fungi (AMF) establish symbiotic relationships with plant roots and perform important functions, contributing quite efficiently to the growth and survival of crops due to their capacity to increase water uptake (Augé, 2004; Ruiz-Lozano and Azcón, 2006) and the absorption of sparingly mobile nutrients such as P, Cu and Zn (Krishna *et al*., 2010; Smith and Smith, 2011). AMF also diminish the incidence of pathogenic diseases in plants (Pozo and Azcón-Aguilar, 2007) and contribute to ameliorate the soil structure (Cuenca *et al*., 2007).

The positive effect of mycorrhiza on avocado plants has been observed in seedlings and micropropagated plants. Menge *et al*. (1980) reported that *Glomus fasciculatus* increased the growth of 'Topa Topa' avocado seddlings, Rivera *et al.* (2011) observed a increases on the development of avocado when seedlings were inoculated with AMF, showing higher results with *G. hoi-like* than with *G. mosseae*, the survival and development of micropropagated avocado plants have shown to improved when they are inoculated with *Glomus* spp. during the *ex vitro* acclimatization (Azcón-Aguilar *et al*., 1992; Vidal *et al*., 1992), additionally mycorrhizal avocado plants are more tolerant to *Phytophthora cinnamomi* damage than the nonmycorrhizal plants (Davis *et al*., 1978).

Avocado seedlings are grown in local nurseries, on substrates treated with chemical insecticides and fungicides in order to keep them free from plant pathogens, treatments which can negatively affect beneficial AMF. Application of AMF biofertilizers (Gianinazzi and Vosátka, 2004) can reduce the negative effects of pesticide treatments and increase sustainability of the agricultural production systems. Due to the importance of this association and the benefits that these fungi provide, it is valuable to evaluate how best to use mycorrhizal inoculants at nurseries to prevent remediate the loss of microorganisms in the substrate, while fostering the rapid growth of seedlings, before they are planted in the field. Herein, we assess the effects of two commercial AMF inoculants on the development of Mexican landrace avocado rootstocks.

MATERIALS AND METHODS

Location

Studies were done in a greenhouse and laboratory facilities of the "Presidente Juárez" Agri-biology school of the Universidad Michoacana de San Nicolás de Hidalgo, at Uruapan, Michoacan, Mexico.

Biological Material

Since avocado is an allogamous plant with high genetic variability, the seeds to be used were obtained from one single tree. 90 fruits were collected from a Mexican landrace avocado tree (*P. americana* Mill. var. *drymifolia*) located at Tancitaro, municipality in Michoacan. The seeds were separated from the mesocarp of the fruit and 30 seeds with similar size and shape were selected to conduct the experiment.

Two commercial AMF inoculants were used: inoculant-1 is a consortia of five AMF species (*Glomus fasciculatum, G. constrictum, G. tortuosum, G. geosporum,* and *Acaulospora scrobiculata*) at a density of 20 spores g^{-1} (Mena *et al*., 2011) and inoculant-2, which contains 20 to 30 spores g^{-1} of *G. mosseae* and *G. cubense* species (Rodríguez *et al*., 2011).

Substrate

A blend of soil from local crop fields, sand and organic matter (pine tree bark) was used at a ratio of 3:1:1 (v:v:v), and was disinfected using wet heat in an autoclave at 121 °C and 15 psi for three cycles of 1 h every 24 h. The blend was air dried for 24 hours and bagged in black polyethylene 3-kg bags.

Treatments

Three conditions were evaluated: T1, avocado seeds coated with inoculant-1; T2, avocado seeds coated with inoculant-2; and T3, avocado seeds not inoculated (control). The experimental unit consisted in two avocado seeds, and five repetitions were considered per treatment, which were randomly distributed. The experiment was conducted during seven months after sprouting.

Inoculation and planting

The selected seeds were dried for 72 hours at room temperature, then the seed coat was removed and seeds were disinfected by submergence in 70% alcohol for 5 min; followed by soaking in 0.8 % sodium hypochlorite for 20 min, followed by a profuse rinse with sterile water and left to dry at room temperature. Subsequently disinfected seeds were covered with a fluid paste prepared with 50g of the corresponding, T1 or T2 inoculants and 30 ml of water, as described by Rivera *et al.* (2011), while T3 control seeds were left untreated. Afterwards the seeds were sown one per bag in the sterile substrate and placed in a greenhouse with partial shade and kept at a mean temperature of 19 °C.

Plant and mycorrhizal response measurements

Plant height, stem diameter, number and length of leaves were measured every month during six months, starting one month after sprouting. A graduated ruler and a Vernier gauge were used to measure plant height from the neck of the first root to the apex; the stem diameter at 5 cm from the soil and the leaf length from the point where the petiole joins the leaf blade.

Chlorophyll concentration was measured six months after inoculation and planting using the third leaf of each plant. Measurements were made with a Minolta Spad 502 meter taken on the same day at 10:00 am-11:00 am. The data were converted into the chlorophyll's concentration rate using the slope and intercept calculated from a standard curve (Figure 2) in which the Spad units of leaves with different green hues were correlated to chlorophyll concentrations measured using an acetone extraction method proposed by Porra *et al.* (1989) and the Arnon's formula (Arnon, 1949).

$$\text{Chlorophyll } (\mu g/g \text{ fresh tissue}) = \frac{(CCR) * (EV)}{\text{Fresh weight (g)}}$$

Where:
CCR= Chlorophyll's concentration rate (μg/ml)
EV= Extract's volume

Root fresh weight (RFW) was recorded as the weight in grams of freshly excised roots.

Foliar dry weight (FDW) and *root dry weight (RDW)* were determined by drying roots and shoots at 60 °C until constant weight was reached (72 hours). These weights were then used to provide the FDW/RDW or Shoot/Root ratio.

Mycorrhizal colonization. A sample of roots was taken from the plants and, each sample was stained as described by Chávez-Bárcenas *et al.* (2013). Stained roots were analyzed under a compounded microscope at 40X and 100X, quantifying the typical structure of mycorrhizal symbiosis present on the sample roots. The colonization percentage of individual root hair segments was calculated using Phillips and Hayman (1970) formula.

$$\text{Total colonization } (\%) = \frac{\text{N° colonized segments} * 100}{\text{N° total segments}}$$

Destructive analyses were performed seven months after sowing (six months after sprouting). At that point the roots of the avocado plants were carefully disaggregated from the soil and cut apart from the upper aerial part of the plant to achieve the determinations.

Statistical Analysis

Statistical differences in all the variables determined between the inoculated and not inoculated treatments were assessed by variance analysis and Tukey means comparison test ($p<0.05$), using the ANOVA module of the statistical package V 2.5 (Olivares, 1994).

RESULTS AND DISCUSSION

Mycorrhizal colonization

Arbuscular mycorrhizal structures were observed in the roots of plants inoculated with T1 and T2, while the roots of plants without AMF did not show any evidence of mycorrhizal colonization (Fig. 1). These results reveal that at least some AMF species from both inoculants are able to develop mycorrhiza with studied avocado roots.

The total colonization percentage in T1 roots was 6% higher than in T2 (Table 1). These results, together with the fact that inoculants in T1 include a wider AMF species diversity and lower spore density than T2, suggest that AMF species within T1 bear a greater capacity to establish symbiosis with avocado roots than AMF species from T2. Nevertheless, more acute and specific experiments should be design to ascertain the colonization capacity of the species within inoculants 1 and 2.

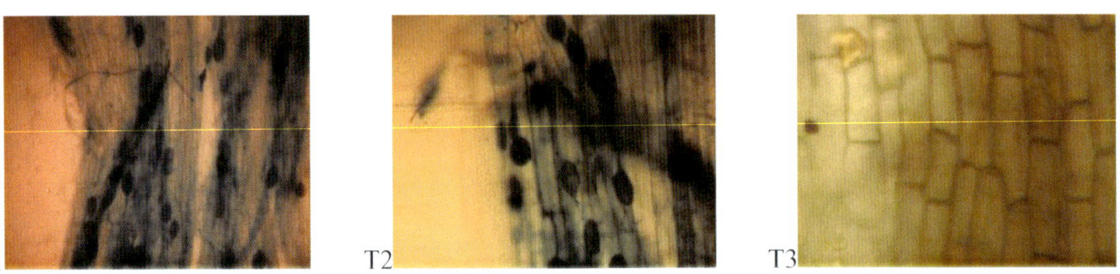

T1 T2 T3

Figure 1. Avocado plant roots colonized with AMF from T1 and T2, and non-colonized roots in T3 control.

Morphological variables

Both AMF treatments developed plants with better morphological traits than the non-inoculated control (Table 1). However, no statistically significant differences were observed between T1 and T2 plants, but T1and T2 differed significantly of control. Plant height was 50% to 54% higher in AMF treated plants than in the control. Similarly, higher values were registered in AMF inoculated plants as regards to stalk diameter (35% to 36%), leaf number (37% to 48%), leaf length (31% to 40%), and root fresh weight (59% to 85%), when compared to the not inoculated control. These results coincide with previous reports that showed positive effects of different AMF species on the growth and development of avocado plants when compared to a non-mycorrhizal control (Azcón-Aguilar *et al.*, 1992; Reyes *et al.*, 1997; Rivera *et al.*, 2011).

Shoot/root ratio

Shoot/root ratio was higher in T1 (0.74) than in T2 (0.63) and T3 (0.41); the difference between T1 and T3 was significant (Table 1). The shoot/root ratio is an important indicator of "quality" of nursery tree seedlings for achieving high survival and good take-off after transplanting (Larsen *et al.*, 1986; Zandstra and Lyptay, 1999; Dans *et al.*, 1999; Vaario *et al.*, 2009; Negreros-Castillo *et al.*, 2010), denoting equilibrium between the size of the plants aerial parts and their root size, and a balance status of carbon partitioning and allocation (Basil *et al.*, 2009; Landhäusser *et al.*, 2012). The shoot/root ratio is usually higher in more fertile soils and substrates (Lambers *et al.*, 1998; Greenwood *et al.*, 2007). Our results suggest that avocado nursery plants inoculated with AMF are closer to a shoot/root biomass balance, than the non-inoculated plants, this agreed with Navarro *et al.* (2011) who reported a higher balance of root/shoot radio in mycorrhizal plants of *Arbutus unedo* L. when compared with non-mycorrhizal during the nursery period, Zhang *et al.* (2011) also showed that AMF affect the interaction root vs. root allocation and thus allometric allocation and biomass-density relationships. Notwithstanding, our results also suggest that these allometric adjustments are also dependent on the AMF species, since both inoculants attained different shoot/root responses compared to the control.

Chlorophyll concentration

Chlorophyll concentrations in leaves of T1 and T2 plants were 16% and 19% higher, respectively, than those of the non-inoculated control (Table 2, Fig. 2). These results are consistent with Azcón-Aguilar *et al.* (1992) and Reyes *et al.* (1997) results, which showed that besides improving the uptake and utilization rates of nutrients in avocado plants, AMF also increased their photosynthetic capacity, promoted higher growth rates and enhanced plant health. This study also agrees with Alarcón and Ferrera-Cerrato (1999), who noted that the symbiotic association of mycorrhizal fungi with plant roots produces several changes and/or modifications at a physiological level; among which an increase in the photosynthetic capacity can be highlighted, due to a greater CO_2 fixation capacity, and the resulting increase in growth and biomass rates.

Moreover, the difference on the chlorophyll concentration between the inoculated treatments T1 and T2 (Table 2) are in agreement with previous studies that have shown a differential response of plants to AMF inoculants within treatments of diverse AMF composition, as well as when compared to non-inoculated, less developed plants (Janousková *et al.*, 2009; Mena *et al.*, 2011; Rivera *et al.*, 2011). Additionally, it is to be noticed that although T1 showed greater mycorrhizal colonization percentage (Table 1), T2 had higher chlorophyll content, suggesting that the mycorrhizal symbiosis with the AMF species in T2 might promote enhancement of the photosynthetic capacity of plants.

Table 1. Mean comparison of the variables measured in avocado nursery plants inoculated or not with AMF.

TREATMENTS	PH (cm)	SD (mm)	NL	LL (cm)	RFW (g)	S/RR	C%
T1	33.6a	8.79a	21.9a	13.3a	31.5a	0.75a	79.5a
T2	34.6a	8.87a	20.3a	14.2a	27.0a	0.63ab	75.0b
T3	22.4b	6.5b	14.8b	10.1b	17.0b	0.42b	00.0c

PH, Plant Height; SD, Stem Diameter; NL, Number of Leaves; LL, Leaf Length; RFW, Root Fresh Weight; S/RR, Shoot/Root ratio; C%, Total Colonization Percentage. The same letters in a column indicate statistically equal values ($p \leq 0.05$)

Table 2. Chlorophyll concentration in avocado nursery plants inoculated or not with AMF.

Treatments	Chlorophyll concentration (μg/g fresh tissue)
T1	1886.9b
T2	1929.1a
T3	1621.0c

The same letters in a column indicate statistically equal values ($p \leq 0.05$)

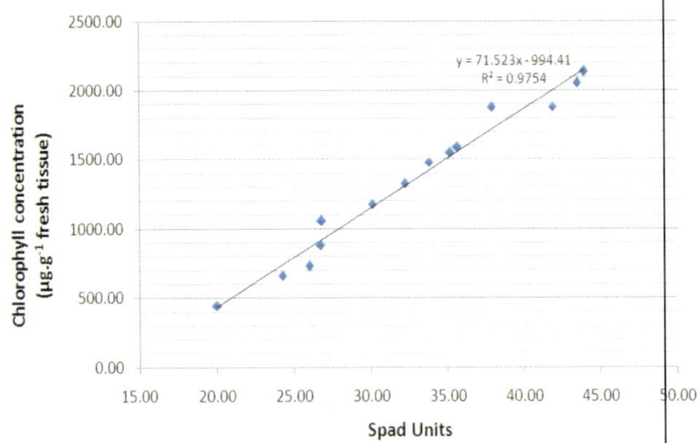

Figure 2. Standard curve of chlorophyll concentration in avocado leaves.

CONCLUSIONS

AMF from both inoculants used in this essay are able to establish a mycorrhizal symbiosis with Mexican landrace avocado; however inoculant T1 showed higher mycorrhizal colonization rates. Both inoculants promote positive developmental traits on avocado plants, with minor differences between them. Better plant fitness due to higher balance in shoot/root ratio was obtained with inoculant T1, while higher photosynthetic potential due to superior chlorophyll concentration is developed in leaves of plants was observed with inoculant T2. Together our results may have critical implications in the avocado production in Mexico.

ACKNOWLEDGEMENTS

We thank the Coordinación de la Investigación Científica, UMSNH for fundings CIC-UMSNH 15.2 and CIC-UMSNH 15.2 to support part of the present work. R.R.L. was a Masters fellow from the Consejo Nacional de Ciencia y Tecnología, and C.A.E. received an undergraduate scholarship from the CIC-UMSNH.

REFERENCES

Alarcón, A., Almaraz, J.J., Ferrera-Cerrato, R., González-Chávez, M.C.A., Lara H., M.E., Manjarrez M., M.J., Quintero, L.R., Santamaría, R.S. 2001. Manual: Tecnología de hongos micorrízicos en la producción de especies forestales en vivero. Colegio de Postgraduados-PRONARE-CONAFOR. México. pp. 98.

Alarcón, A., Ferrera-Cerrato, R. 1999. Manejo de la micorriza arbuscular en sistemas de propagación de plantas frutícolas. Revista Terra Latinoamericana. 17:178-191.

Arnon, D.I. 1949. Enzyemes in isolated chloroplasts. Polyphenol oxidase in Beta vulgaris. Plant Physiol. 24:1-15.

Augé, R. M. 2004. Arbusuclar mycorrhizae and soil/plant water relations. Canadian Journal of Soil Science 84:373-381.

Azcón-Aguilar, C., Barceló, A., Vidal, M. T., De la Viña, G. 1992. Further studies on the influence of mycorrhizae on growth and development of micropopagated avocado plants. Agronomie. 12:837-840.

Basil, G., Mazzarino, M.J., Roselli, L., Letourne, F. 2009. Efecto del compost de biosólidos en la producción de plantines de *Austrocedrus chilensis* (ciprés de la cordillera). Ciencias del suelo. Argentina. 27(1):49-55.

Cuenca, G., Cáceres, A., Oirdobro, G., Hasmy, Z., Urdaneta, C. 2007. Las micorrizas arbusculares como alternativa para una agricultura sustentable en áreas tropicales. Interciencia. 32:26-29.

Chávez-Bárcenas, A.T., Lua-Aldama, J., Salmeron-Santiago, I.A., Silva-Adame, M.B., García-Saucedo, P.A., Olalde-Portugal, V. 2013. A modified staining technique for the anatomical observation of mycorrhizal roots of woody trees. African Journal of Microbiology Research. 28:3589-3596.

Dans del Valle, F., Fernández de Ana M., F.J., Romero G., A. 1999. Manual de silvicultura del pino radiata en Galicia. Escuela Politécnica Superior de Lugo y Asociación Forestal de Galicia. pp. 199.

Davis, R.M., Menge, J.A., Zentmyer, G.A. 1978. Influence of vesicular-arbuscular mycorrhizae on *Phytophthora* root rot of three crops plants. Phytopathology 68:1614-1617.

Gianinazzi, S., Vosátka, M. 2004. Inoculum of arbuscular mycorrhiza fungi for production systems: science meets business. Canadian Journal of Botany 82:1264-1271.

Greenwood, D.J., Mckee, J.M.T., Fuller, D.P., Burns, I.G., Mulholland, B.J. 2007. A novel method of supplying permits predictable shoot growth and root : shoot ratios of pre-transplant bedding plants. Annals of Botany 99:171-182.

Janousková, M., Seddas, P., Mrnka, L., van Tuinen, D., Dvoracková, A., Tollot M, Gianinazzi-Pearson, V., Vosátka, M., Gollotte, A. 2009. Development and activity of *Glomus intraradices* as affected by co-existence with *Glomus claroideum* in one root system. Mycorrhiza 19:393-402.

Krishna, H., Das, G., Attri, B. L., Grover, M., Ahmed, N. 2010. Suppression of *Botryosphaeria* canker of apple by arbuscular mycorrhiza fungi. Crop Protection 29:1049-1054.

Lambers, H., Chapin III, F. S., Pons, T. L. 1998. Plant Physiological Ecology. Journal of Agronomy and Crop Science. 184:143-144.

Landhäusser, S.M., Pinno, B.D., Lieffers, V.J., Chow, P.S. 2012. Partitioning of carbon allocation to reserves or growth determines future performance of aspen seedlings. Forest Ecology and Management 275:43-51.

Larsen, H.S., South, D.B., Boyer, J. M. 1986. Root growth potential, seedling morphology and bud dormancy correlate with survival of loblolly pine seedlings planted in December in Alabama. Tree Physiology 1:253-263.

Mena, E.A., Fernández S.K., Jerez M.E., Olalde P.V., Serrato R. 2011. Influencia de la inoculación con *Glomus hoi-like* y un conglomerado de especies de HMA en el crecimiento de plantas de sorgo sometidas o no a estrés hídrico. Cultivos Tropicales 32:16-27.

Menge, J.A., LaRue, J., Labanauskas, C.K., Johnson, E.L.V. 1980. The effect of two mycorrhizal fungi upon growth and nutrition of avocado seedlings grown with six fertilizer treatments. Journal of the American Society for Horticultural Science 105:400-404.

Navarro, G.A., Bañón, A.S.P., Morte, A., Sánchez-Blanco, M.J. 2011. Effects of nursery preconditioning through mycorrhizal inoculation and drought in *Arbutus unedo* L. plants. Mycorrhiza 21:53-64.

Negreros-Castillo, P., Apodaca-Martínez, M., Mize, C. W. 2010. Efecto de sustrato y densidad en la calidad de plántulas de cedro, caoba y roble. Madera Bosques. 16:7-18.

Olivares S., E. 1994. Paquete de diseños experimentales FAUANL. Versión 2.5. Facultad de Agronomía. Universidad Autónoma de Nuevo León. Marín, Nuevo León. México.

Olivares S E (1994) Paquete de Diseños Experimentales FAUANL. Versión 2.5. Facultad de Agronomía de la UANL. Marín, N.L., México.

Porra, J. R., Thompson, W. A., Kriedemann, P. E. 1989. Determination of accurate extinction coefficients and simultaneous equations assaying chlorophylls a and *b* extracted with four different solvents: verification of the concentration of chlorophylls standards by atomic absorption spectroscopy. Biochimica et Biophysica Acta. 975:384-394.

Pozo, M., Azcón-Aguilar, C. 2007. Unraveling mycorrhiza-induced resistance. Current Opinion in Plant Biology 10:393-398.

Reyes A., J. C., Alarcón, A., Ferrera C., R. 1997. Aspectos relacionados sobre el uso de la endomicorríza arbuscular en aguacate (*Persea americana* Mill). In Memoria 1997. Fundación Salvador Sánchez Colín. CICTAMEX S.C. Coatepec Harinas, México. pp. 83-94.

Rivera E., R., Martín C., J. V., Calderón P., A., Torrez H., A. 2011. Utilización de cepas eficientes de hongos micorrízicos arbusculares en el desarrollo de portainjertos de aguacate en un sustrato suelo-cachaza. Cultivos Tropicales. 32:172-183.

Rodríguez, Y., Dalpé, Y., Séguin, S., Fernández, K., Fernández, F., Rivera, R. A. 2011. *Glomus cubense* sp. nov., an arbuscular mycorrhizal fungus from Cuba. Mycotaxon. 118:337-347.

Ruiz-Lozano, J.M., Azcón R. 2006. Hyphal contribution to water uptake in mycorrhizal plants as affected by the fungal species and water status. Physiologia Plantarum 95:472-478.

Salazar-García. S. 2002. Nutrición del Aguacate, Principios y Aplicaciones. INIFAP INPOFOS. México. pp. 165.

Sánchez R., G. 2007. El Cluster del aguacate de Michoacán, desarrollo regional para el liderazgo global. Fundación produce Michoacán. pp. 241.

SIAP. 2012. Servicio de Información Agroalimentaria y Pesquera. Producción anual, cierre de la producción agrícola por cultivo. Accessed 30 July 2013 In: http://www.siap.gob.mx/index.php?option=com_wrapper&view=wrapper&Itemid=350.

Smith, S.E, Smith, F.A. 2011. Roles of arbuscular mycorrhizas in plant nutrition and growth: new paradigms from cellular to ecosystem scales. Annual Review of Plant Biology 62:227-250.

Vaario, L.-M., Tervonen, A., Haukioja, M., Pennanen, T., Timonen, S. 2009. The effect of nursery substrate and fertilization on the growth and ectomycorrhizal status of containerized and outplanted seedlings of *Picea abies*. Canadian Journal of Forest Research 39:64-75

Vidal, M.T., Azcón-Aguilar C., Barea J.M. 1992. Mycorrhizal inoculation enhances growth and development of micropropagated plants of avocado. HortScience 27:785-787.

Zandstra, J.W., Liptay, A. 1999. Nutritional effects on transplant root and shoot growth – a review. Acta Horticulturae 504:23-31.

Zhang, Q., Zhang, L., Weiner, J., Tang, J., Chen, X. 2011. Arbuscular mycorrhizal fungi alter plant allometry and biomass-density relationships. Annals of Botany 107:407-413.

8

An Effective *In Vitro* Slow Growth Protocol for Conservation of the Orchid *Epidendrum Chlorocorymbos* Schltr

Guadalupe Lopez-Puc

Unidad Sureste de Centro de Investigación y Asistencia en Tecnología y Diseño del Estado de Jalisco A.C. Av. Normalistas # 800, Colinas de la Normal, Guadalajara Jalisco, México. C.P. 44270

SUMMARY

Efficient slow growth protocol for the orchid *E. chlorocorymbos* Schltr. was developed using *in vitro* conservation studies. Seedlings were placed in MS culture medium and a 2x3x3 factorial design applied to evaluate the effects of supplementing the medium with mannitol (0, 1, 2 and 3%), sucrose (0, 1, 2 and 3%) or sorbitol (0, 1, 2 and 3%). Experimental conditions were a 16:8 h photoperiod, 23 ± 2 °C temperature and 50-80% relative humidity. At 6 months, the best treatment was MS medium at half ionic strength with 1% sorbitol. The culture was also free of contamination. This resulted in slow growth and normal morphology during maintenance and successful growth afterwards. Shoots were subsequently recovered, multiplied and rooted on MS medium with sucrose 3% without addition of growth regulators.

Key words: *in vitro* storage; mannitol; sorbitol; *Epidendrum chlorocorymbos* Schltr.

INTRODUCTION

The family Orchidaceae is one of the groups most diverse of plants, consisting of ~35,000 species under 800 genera (Singh *et al.*, 2007). Orchids are distributed on all continents (except Antarctica), but its greater diversity is concentrated in tropical regions. Mexico has about 1260 species and 170 genera (Hágsater *et al.*, 2005). Among the genera of orchids, Epidendrum as the largest genus of orchids with over 1000 species, many of which occur in great abundance in Central America (Chen *et al.*, 2002). *Epidendrum chlorocorymbos* Schltr. is an orchid found throughout the Yucatan Peninsula (Carnevali *et al.*, 2001), has been reported as a medicinal plant use for cholesterol, sore ears (Burgos-Hernández, 2009). As occurs with orchids in many other locations in Mexico, this species experience heavy exploitation, leading to possible decreases in their native populations. In situ conservation of orchid species is very difficult because of the relatively slow growth of orchids and low germination rates which requires symbiotic relationships with mycorrhizal fungi in natural habitats (Godo *et al.*, 2010). One alternative for preventing irreparable loss of biodiversity is use of in vitro conservation of plant germplasm through using slow growth procedures or cryopreservation. Slow *in vitro* growth is usually achieved by using substances such as mannitol and sorbitol to reduce medium osmotic potential (Grout, 1991), a low concentration of nutrients (Engelmann, 1991). Slow growth storage via *in vitro* cultures has been reported in many species such as *Elettaria cardamomum* (Tyagi *et al.*, 2009), *Plumbago indica* Lim (Chaeoensub and Phansiri, 2004), garlic cultivars (Hassan *et al.*, 2007), *Cynara scolymus* (Bekhett and Usama, 2007), *Veronica multifida* spp capsellicarpa (Holobiuc *et al.*, 2008), *Drosophyllum lusitanicum* (Gonçalves and Romano, 2007); strawberry (Hassan and Bekheet, 2008), *Deutzia scabra* Thunb (Gabr and Sayed, 2010). Although none has yet been reported for *in vitro* conservation of *Epidendrum chlorocorymbos* Schltr. The present study objective was to evaluate in vitro technique for conservation this orchid specie for induce in slow-growth in an effort to develop genetic diversity conservation protocol.

MATERIAL AND METHODS

Vegetal material and disinfection.

Green capsules with seeds were collected from *Epidendrum chlorocorymbos* Schltr in the region of Escarcega, Campeche state, Mexico.

Capsule treatment and sterilization.

Capsules were placed in a drier at 23 ± 2 °C for 15 days to allow seeds to reach maturity. Mature capsules were washed with a detergent solution and rinsed with sterile distilled water. The capsules were then placed in a solution (1% commercial sodium hypochlorite and 1 drop of Tween 20) for 10 min, under constant agitation. The washed capsules were dipped in pure alcohol, quickly passed through a flame, the alcohol allowed to completely burn off and the capsules and tweezers left to cool. Once cooled, the sterile capsule was cut in half longitudinally with a sterile scalpel. The seeds extracted from the capsules were distributed over the media culture were added drops of water to separate and distribute in each container (Magenta boxes).

Culture medium and germination conditions.

Epidendrum chlorocorymbos Schltr seeds were germinated in semi-solid medium composed of MS medium (Murashige and Skoog, 1962), basal salt and vitamins supplemented with 30 g l^{-1} sucrose, 2.2 g l^{-1} gelrite without growth regulators. The seeds were placed in and incubated for 12 weeks. See culture conditions below.

Slow growth treatments.

At three months of age, *in vitro* germinated seedlings with two leaves, 1-2 roots and heights of (ca. 5 mm) *Epidendrum chlorocorymbos* Schltr were placed in the different slow growth treatments. A 2x3x3 factorial design was applied to evaluate the effect of each culture medium's ionic strength, carbon source (osmotic agent) and carbon source concentration on number shoots and number leaves formed after six months in vitro conservation. Two treatments were used as control (MS full ionic strength and half ionic strength without carbon source). All treatments are described in Table 1. For all treatments seedlings were placed in test tubes (20 X 150 mm) containing 20 ml medium. Each culture tube received one explant enclosed with polypropylene caps and covered with cling film.

Table 1.Treatments evaluated for slow growth in the orchid *Epidendrum chlorocorymbos* Schltr.

Treatment code	MS. Ionic Strength	Carbon Source		
		Sucrose (%)	Mannitol (%)	Sorbitol (%)
TFSA1	100%	1	0	0
TFSA2	100%	2	0	0
TFSA3	100%	3	0	0
THSA1	50%	1	0	0
THSA2	50%	2	0	0
THSA3	50%	3	0	0
TFM1	100%	0	1	0
TFM2	100%	0	2	0
TFM3	100%	0	3	0
THM1	50%	0	1	0
THM2	50%	0	2	0
THM3	50%	0	3	0
TFS1	100%	0	0	1
TFS2	100%	0	0	2
TFS3	100%	0	0	3
THS1	50%	0	0	1
THS2	50%	0	0	2
THS3	50%	0	0	3
TCON1	100%	0	0	0
TCON2	50%	0	0	0

Data recording and statistical analyses.

Each treatment had four replicates. Data were recorded after six months of conservation in slow growth treatments. Data were subjected to analyses of variance (ANOVA) and means were compared by Tukey test using Stat graphics plus version 5.1 program.

Table 2. Effect of treatments of slow growth on shoot and leaves formed in plantlets of *Epidendrum chlorocorymbos* Schltr.

Treatment code	Response variables Assessment at six months of culture Number Shoots	Number Leaves
TFSA1	13.25± 1.50^{ab}	48.00± 4.24^{b}
TFSA2	4.50±1.29^{cd}	13.50±3.87^{e}
TFSA3	12.00±1.63^{b}	36.00±4.89^{c}
THSA1	13.25±1.50^{ab}	39.70±4.50^{c}
THSA2	15.00±2.58^{ab}	37.75±14.56^{c}
THSA3	21.00±3.46^{a}	53.00±10.39^{a}
TFM1	0.00±0.00^{e}	0.00±0.00^{h}
TFM2	0.00±0.00^{e}	0.00±0.00 ^{h}
TFM3	0.00±0.00^{e}	0.25±0.50 ^{h}
THM1	0.00±0.00^{e}	0.75±0.50^{gh}
THM2	0.00±0.00^{e}	0.05±0.57 ^{gh}
THM3	0.00±0.00^{e}	0.05±0.57 ^{gh}
TFS1	2.50±0.50^{d}	9.00±1.63^{f}
TFS2	3.50±0.57^{d}	8.00±3.36^{fg}
TFS3	6.50±1.00^{c}	18.25±0.50 ^{de}
THS1	2.50±1.00^{d}	8.50±3.00^{fg}
THS2	7.25±1.50^{bc}	21.75±4.50^{d}
THS3	12.50±3.00^{b}	32.25±12.09^{c}
TCON1	0.00±0.00^{e}	0.00±0.00 ^{h}
TCON2	0.00±0.00^{e}	0.00±0.00 ^{h}

Multiplication after slow growth.

To test the viability at the end of the six-month experimental period in slow growth conditions, survival and growth were evaluated. shoots from selected treatment were removed from culture tube and washed under running tap water to remove the adhering gelling agent. Shoots were transferred to fresh MS medium with macro and micro elements and vitamins, containing 3% sucrose and solidified with 0.2% Gel-Rite.

Culture conditions of germination, slow growth and multiplication. All the chemicals used for preparing the media were of analytical grade (Sigma). In all cases, medium pH was adjusted to 5.7 with KOH 0.1 N and/or HCl 0.1 N before sterilizing in an autoclave at 121°C and 1.06 kg cm−2 pressure for 20 min. All experiments were incubated a 23 ± 2 °C, under a 16:8 (light:dark) photoperiod using Philips cool white fluorescent tubes of 1500 lux.

RESULTS AND DISCUSSION

Effect of treatments of slow growth on shoot and leaves formed in plantlets of *Epidendrum chlorocorymbos* Schltr was evaluated at six months.

Values are mean ± SE of 4 cultures for each treatment. Values superscripted with the same letter in each column are not significantly different on the basis of ANOVA Stat graphics plus.

Effect carbon source in slow growth

The beneficial effect of mannitol for in vitro conservation has been previously reported in *Discorea alata* (1.5%) Borges *et al.*, 2004; *Plumbago indica* (2%) (Chaeoensub and Phansiri, 2004); *Drosophyllum lusitanicum* (2%) Gonçalves and Romano, 2007; *Veronica multifida* ssp (3 and 6%) Holobiuc *et al.*, 2008. In this it work observed that mannitol treatments produced *Epidendrum chlorocorymbos* seedlings death, this explain since it has been reported the toxic effects of mannitol in vitro conservation, Lata *et al.*, 2010 reported that 2% manitol was not adequate for in vitro conservation of *Podophyllum peltatum*, Loureiro da Silva *et al.*, 2011 observed that 1-3 % mannitol negatively affect causing low rate of recovered shoots after storage of *Piper hispidernervum* and *Piper aduncum* shoots. However, the lethal concentration seems to be species dependent (Loureiro da Silva *et al.*, 2011), and storage time can influence, as reported Bekhett and Usama, 2007, in *Cynara scolymus* that the presence of mannitol in culture medium had a retardant effect on the growth and development of all cultures and data of reveal that high percentage of survival and healthy shoot bud cultures were recorded after the first three months of storage on medium containing 4% mannitol, but survival and healthy shoot cultures were sharply decreased after six months of storage.

The treatment with both sucrose and sorbitol resulted in the mass multiplication of shoots in treatments. Due to the formation of multiple shoots from the initial seedling it was not possible to measure the plant length variation and therefore this parameter was not considered. When analyzing the variables of shoots and leaves formed significant difference was found between the carbon sources, p-value are in Table 3 and 4 respectively. In Fig.1a was observed that sorbitol produced less shoots than sucrose, at 1 and 2% (Fig.1b). Similar results was observed on number leaves (Fig. 2). These results agree with those found for diminished growth reported in the presence of sorbitol. Although the protocol established for *E. chlorocorymbos* is less expensive since only use 1% while in *C. scolymus* was necessary at 4% (Bekhett and Usama, 2007), sorbitol 0.2 M has also been found to be suitable as an osmotic agent for the conservation of strawberry germplasm (Hasan and Bekheet, 2008), and 0.16 M sorbitol is reported to preserve African violet sprouts (Moges *et al.*, 2003). In *E. chlorocorymbos* was observed that lack of a carbon source produced abnormal morphology and plantlets did not develop in these treatments.

Effect ionic strength in slow growth

Lowering the content in mineral elements of culture medium resulted in terms of number of shoots and leaves no significant difference between the ionic strength of MS medium (Table 3 and 4 respectively), however, in the graphics of interaction is observed that sorbitol gives less number of shoots and leaves with 50% MS (Fig. 3). This agrees with the results reported in slow growth conservation of cardamom (Tyagi *et al.*, 2009), in vitro conservation coconut embryos (Sukendah and Cedo, 2005).

Table 3. Analysis of Variance for shoots - Sums square of Type III

Source	Sum of square	DF	Mean square	Ratio F	p-value
Main effects					
A:MS:Ionic strength	11.6806	1	11.6806	3.28	0.0755
B:Carbon source	2410.36	2	1205.18	338.00	0.0000
C: Concentration	93.5278	2	46.7639	13.12	0.0000
Interactions					
AB	341.861	2	170.931	47.94	0.0000
AC	28.5278	2	14.2639	4.00	0.0236
BC	107.889	4	26.9722	7.56	0.0001
Waste	206.806	58	3.56561		
Total (Corrected)	990.625	71			

F ratios are based on the residual mean square error

Table 4. Analysis of Variance for leaves - Sums square of Type III

Source	Sum of square	DF	Mean square	Ratio F	p-value
Main effects					
A:MS:Ionic strength	26.8889	1	26.8889	0.62	0.4340
B:Carbon source	21627.1	2	10813.6	249.59	0.0000
C: Concentration	1166.03	2	583.014	13.46	0.0000
Interactions					
AB	1955.44	2	977.722	22.575	0.0000
AC	397.028	2	198.514	4.58	0.0142
BC	1451.72	4	362.931	8.38	0.0000
Waste	2512.89	58	43.3257		
Total (Corrected)	29137.1	71			

F ratios are based on the residual mean square error

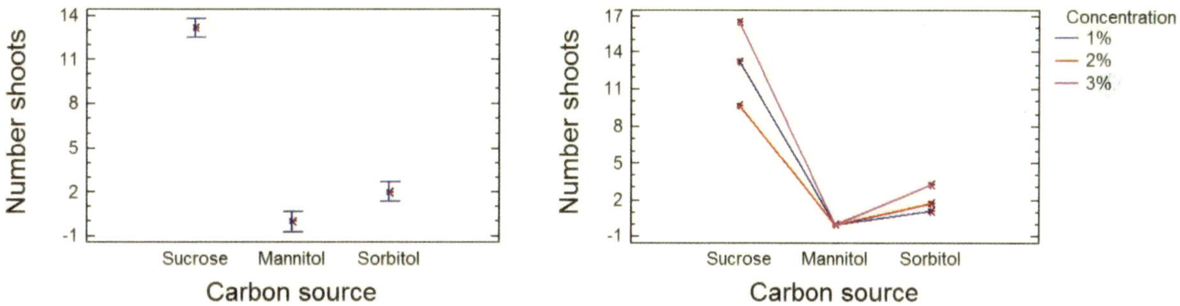

Figure 1.Graphics of number of shoots by *E. chlorocorymbos.* (a) Interval means and percentages Tukey of effect carbon source; (b) Interaction between factors: carbon source and concentration

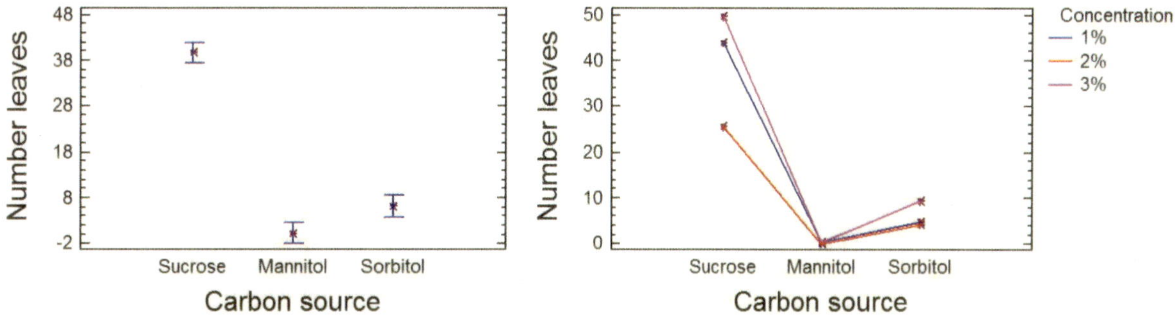

Figure 2.Graphics of number of leaves by *E. chlorocorymbos.* (a) Interval means and percentages Tukey of effect carbon source; (b) Interaction between factors: carbon source and concentration.

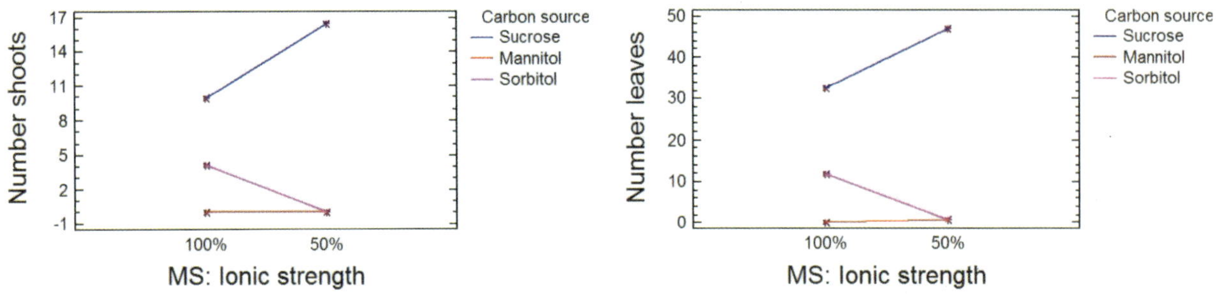

Figure 3.Graphics of Interaction between factors: Ionic strength and carbon source. Number of shoots. (b) Number leaves

Slow growth protocol for *E. chlorocorymbos*

A protocol *in vitro* slow growth for conservation of *E. chlorocorymbos* was established by use of osmotic regulators such as sorbitol, this regulator added to the culture medium reduce hydric potential, consequently restricting water availability to the plantlets and lowering mineral uptake (Shibli *et al.*, 2006). In contrast to most germplasm conservation protocols, low temperatures were unnecessary in the present study; cooling raises conservation costs also the conservation mediated by osmotics may be more economical than cryopreservation Tehrim and Sajid, 2011. At six months, it was clear that the treatments with the lowest shoots and leaves development were also those with the lowest plantlet development. These results of this study indicate that the optimum treatment is 50% of ionic strength MS supplemented with 1% sorbitol, as this gives the lowest amount of growth (Fig 4a) whilst the maximum growth occurred in MS with 3% sucrose (4b).

Multiplication after slow growth conservation

Survival after the conservation period was 100% in shoots of *E. chlorocorymbos* in the treatment THS1. Growth and rooting occurred in both cases and no abnormalities were observed (Figure 5).

CONCLUSIONS

Although research on the development of conservation techniques has been done for numerous plant species, this is a first report for in vitro conservation for *Epidendrum chlorocorymbos* Schltr germplasm. Growth reduction was attained by modifying culture medium composition through addition of the osmotic regulator sorbitol. Optimal conditions for slow growth were 1% sorbitol and 50% of ionic strength MS. Establishing the proper conditions for slow growth in orchids will aid in preserving their germplasm for the purpose of later reintroduction and sustainable use.

Figure 4. Shoots of *Epidendrum chlorocorymbos* Schltr after six month of slow growth conditions. (a) In treatment THS1 (MS 50%, sorbitol 1%). (b) In the treatment TFSA3 (MS 100% 3% sucrose).

Acknowledgment

The author is thankful for the financial support of Fomix Campeche-CONACYT.

Figure 5. Seddlings regenerated of *Epidendrum chlorocorymbos* Schltr after six months of conservation in slow growth conditions in treatment THS1

REFERENCES

Bekhett,S., Usama, I.A. 2007. In vitro conservation of globe artichoke (*Cynara scolymus* L.). International Journal of Agriculture Biology. 9 (3):404-407.

Borges, M., Ceiro, W., Meneses, S., Aguilera, N., Vázquez, J., Infante, Z., Fonseca, M. 2004. Regeneration and multiplication of *Dioscorea alata* germplasm maintained *in vitro*. Plant Cell, Tissue and Organ Culture. 76:87–90.

Burgos-Hernández, M. 2009. Flora vascular con características potenciales para el aprovechamiento y conservación de los fragmentos de selva en el municipio de Atzalan Veracruz. Tesis del Instituto de Ecología A.C. Xalapa Veracruz México. 1-143.

Carnevali Fernández-Concha, G., Tapia-Muñoz, J.L., Jiménez-Machorro, R., Sánchez-Saldaña, L., Ibarra-González, L., Ramírez, I., Gómez, M.P. 2001. Notes on the flora of the Yucatan peninsula II: A synopsis of the orchid flora of the Mexican Yucatan Peninsula and a tentative checklist of the Orchidaceae of the Yucatan peninsula biotic province. Harvard Papers in Botany. 5(2):383-466.

Chaeoensub, R., Phansiri, S. 2004. In vitro Conservation of Rose coloured leadwort: Effect of mannitol on Growth of plantlets. Kasetsart Journal Natural Science. 38:97-102.

Chen, L.R., Chen, J.T., Chang, W. 2002. Efficient production of protocorm-like bodies and plant regeneration from flower Stalk explants of the sympodial orchid *Epidendrum radicans*. In Vitro cellular & Developmental Biology - Plant. 38:441-445.

Engelmann, F. 1991. *In vitro* conservation of tropical plant germplasm-a review. Euphytica. 57:227-243.

Gabr, A.M.M., Sayed, S. 2010. Slow growth conservation and molecular characterization of *Deutzia scabra* Thunb. African Journal of Plant Science. 4(10):409-416.

Grout, W.V. 1991. Conservation *in vitro*. Acta Horticulturae. 289:171-178.

Godo, T., Komori, M., Nakaoki, E., Yukawa, T., Miyoshi, K. 2010. Germination of mature seeds of *Calanthe tricarinata* Lindl., an endangered terrestrial orchid, by asymbiotic culture *in vitro*. In Vitro cellular & Developmental Biology - Plant. 46:323–328.

Gonçalves, S., Romano, A. 2007. *In vitro* minimum growth for conservation of *Drosophyllum lusitanicum*. Biologia Plantarum. 51(4):795-798.

Hassan, N., El Halwagi, A., Gaber, A., El Awady, M., Khalaf, A. 2007. Slow-growth *in vitro* conservation of garlic cultivars grow in Egypt: Chemical characterization and molecular evaluation. Global Journal of Molecular Sciences. 2(2):67-75.

Hasan, N., Bekheet, S.A. 2008. Mid-term storage and genetic stability of Strawberry tissue cultures. Research Journal of Agriculture and Biological Sciences. 4(5):505-511.

Hágsater, E., Soto-Arenas, M.A., Salazar-Chávez, G.A., Jiménez-Machorro, R., López-Rosas, M.A., Dressler, R.L. 2005. Las orquídeas de México. Instituto Chinoín, México, D.F., Mexico.

Holobiuc, I., Blindu, R., Carasan, M., Helepciuc, F., Voichita, C., Negrean, G. 2008. *In vitro* conservation strategy in *Veronica multifida* spp capsellicarpa (Dubovik) A. Jelen. Romanian Journal of Biology-Plant Biology. 53(2):71-81.

Lata, H., Moraes, R.M., Bertoni, B., Pereira, A.M.S. 2010. *In vitro* germplasm conservation of *Podophyllum peltatum* L. under slow-growth conditions. In Vitro cellular & Developmental Biology - Plant. 46:22-27.

Loureiro da Silva, T., Scherwinski-Pereira, J.E. 2011. *In vitro* conservation of *Piper aduncum* and *Piper hispidinervum* under slow-growth conditions. Pesquisa Agropecuária Brasileira. 46(4):384-389.

Moges, A.D., Karam, N., Shibli, R.A. 2003. Slow growth *in vitro* conservation of African violet (*Saint paulia ionantha* Wendl.). Advances in Horticultural Science. 4:223-230.

Murashige, T., Skoog, F. 1962. A revised medium for rapid growth and biossays with tobacco tissue cultures. Physiologia Plantarum. 15:473-497.

Shibli, R.A., Shatnawi, M.A., Subaih, W., Ajlouni, M.M. 2006. *In vitro* conservation and cryoconservation of plant genetic resources: a review. World Journal of Agricultural Sciences. 2(4):372-382.

Singh, M., Sherpa, A., Hallan, V., Zaidi, A. 2007. A potyvirus in *Cymbidium* spp. in northern India. Australasian Plant Disease Notes. 2:11-13.

Sukendah, Cedo, M.L. 2005. *In vitro* conservation of coconut (*Cocos nucifera* L.) embryos in culture media. Biotropia. 25:11–21.

Tehrim, S., Sajid, G.M. 2011. *In vitro* establishment, conservation and its implications for grape germplasm biodiversity. Romanian Biotechnological Letters. 16(6):6781-6789.

Tyagi, R.K., Rajni, G., Sanayaima, R., Singh, R., Tandon R., Agrawal, A. 2009. Micropropagation and slow growth conservation of cardamom (*Elettaria cardamomum* Maton). In Vitro cellular & Developmental Biology - Plant. 45:721-729.

Soil Aggregate Fractions and Organic Carbon Pools as Influenced by Tree Diversity in Forest Reserve of Semi Arid Nigeria

H. M Lawal

Department of Soil Science, Ahmadu Bello University, Zaria Nigeria.
Email: lawalhalima2@gmail.com and leem8810@yahoo.com
**Corresponding author*

SUMMARY

The ability of trees to sequester carbon vary from species to species, this study therefore evaluates the influence of seven tree species (*Acacia senegal, Azadirachta indica, Eucalyptus camadulensis, Khaya senegalensis, Prosopis africana, Tamarindus indica* and *Tectona grandis*) on soil aggregate fractions and various pools of carbon in soil at three soil depth (0-5, 5-15 and 15-25cm). Results revealed that most soil quality enhancing properties concentrated more at the soil surface (0-5cm) relative to other depths. Soil planted to *T. indica* stood out in possessing high proportion of large and small macro aggregates, mean weight diameter (MWD) and least proportion of silt plus clay particles. High proportion of fine particulate organic carbon, particulate organic carbon, silt plus clay associated organic carbon and non hydrolysable organic carbons were attributed to soil under *T. indica. Tamarindus indica* might be a better choice of the seven tree species in mitigating green house effects by sequestering atmospheric carbon and improving soil aggregate stability.

Key words: Leaf litter; soil aggregate stability; organic carbon pools; carbon sequestration; savanna.

INTRODUCTION

Human activities such as fossil fuel burning, bush burning, deforestation and land use change in the past few decades have increased the amount of carbon dioxide, (CO_2) by about 30% and other greenhouse gases (GHGs) in the atmosphere; leading to extreme climate change that may have devastating effects on global economy and life forms.

The Kyoto protocol on climate change of 1997,

established the principle that carbon sequestration can be used by participating nations to help meet their respective net emission reduction targets for CO_2 and other GHGs by the year 2012 (Kyoto protocol, 1997). Growing trees to sequester CO_2 could provide relatively net emission reduction because trees act as sinks for atmospheric CO_2 through photosynthesis. Litters on forest floor upon undergoing the process of decomposition to add carbon to the soil and consequently improve soil aggregate stability, since the soil has a greater potential to sequester carbon compared to vegetation and atmosphere (Swift, 2001). Hence non disturbance of forest soils increases the potential for C to remain relatively longer in such soils thus, mitigating green house effect (Six *et al.*, 2002).

Soil aggregation is important for soil to stay productive, it maintains the surface integrity of the soil and facilitates infiltration rather runoff (Franzluebbers et al., 2000) in addition to enhancing root growth and movement of soil water and gases. Carbon in plant residues and tree litter is sequestered in soil through decomposition and conversion into soil organic carbon in aggregates. It has been documented that aggregation and soil organic carbon (SOC) sequestration are suitable indicators for evaluating changes in soil quality (Unger, 1997). Forestation improves SOC storage (Six et al., 2000; Solomon et al., 2002). Since litter on forest floor promote the formation and stabilization of soil macro aggregates and facilitate SOC sequestration.

Organic matter (OM) stored in macro aggregates can become physically protected from decomposition by isolating OM resources from the activity of soil fauna (Beare *et al.*, 1994; Elliot 1986). Furthermore, litter returned to the soil are exposed to variety of decomposition step processes contributing to a gradient in soil biochemical potential from active to passive (Paul *et al.*, 2003). Mineralizable C and N are active components that can be use to characterized biochemical availability of organic resources.

Newly planted or regenerated forests, in the absence of major disturbance, will continue to take up carbon for 20 to 50 years or more after establishment, depending on the species and site conditions (IPCC, 2000). The complex relationship between soil organic carbon (SOC) sequestration and aggregate stability is crucial to understand SOC dynamics and turnover. Understanding how plant specie influences soil organic matter dynamics is important to predicting how soil carbon sequestration and fertility will respond to management decisions and global environment changes/mitigation of GHGs. There is a dearth of information on the effect of tree species on soil aggregate characteristics in the tropical environment. This study therefore assessed the influence of tree diversity on soil aggregate fractions and various concentrations / pools of carbon in varying aggregates sizes.

MATERIALS AND METHODS

Site description

The study was carried out at Afaka Forest Reserve in Kaduna State, Northern Guinea Savanna ecological zone (Keay, 1959) of Nigeria. It has a mean annual rainfall of 1270 mm and lies between coordinates 10° 33' and 10° 40' North and 07° 15' East, altitude 593 m above sea level. The natural vegetation cover consist of *Isoberlina doka* as the dominant tree species on the deeper soils, *Monoles kerstingii* on the eroded areas bordering gully formations. While *Uapaca togoensis* and *parinari curatellifora* were common (Tomlison, 1957). The topography has a <2% slope and the soil type is Alfisol; plinthustalfs, sandy loam soil.

Soil Sampling

Soil samples were collected from seven selected 19years old tree species, the tree species and the land area they occupied in parenthesis are as listed below:

Eucalyptus camadulensis (2.20ha); *Acacia senegal* (1.95ha); *Prosopis Africana* (2.00ha); *Tamarindus indica* (1.48ha); *Khaya senegalensis* (1.72ha); *Azadirachta indica* (1.75ha); *Tectona grandis* (2.00ha).

A stratified random sampling method was adopted to collect composite soil samples from three (3) subplots which represented three replications, soil were collected from 0-5 cm, 5-15 cm and 15-25 cm depth using a traditional hoe and bucket auger.

Water stable aggregates

Aggregate stability was determined by wet sieving (Elliot, 1986). A 200 g of air dried bulk soil (which had been sieved through a 5 mm sieve) was wetted by rapid immersion then passed through 2 mm, 0.25 mm and 0.053 mm sieves, with an average stroke per minute of 36, 14 and 3 for each of the sieve sizes respectively.

The <0.053mm aggregate fraction was allowed to settle down and the water was decanted gently. The fractionated aggregates were oven dried at 60°C to a constant weight. Aggregate fractions >0.053mm were corrected for sand and mean weight diameter determined (MWD). However, large macro aggregate

>2mm were not used in computing or calculating MWD, being that the proportion of aggregates >2mm recovered after the wet sieving were too small in weight to be corrected for sand (sand free fraction). The proportional weight of sand free aggregates is given as:

$$\frac{\text{Weight of fractioned aggregate - \% sand content}}{\text{Weight of bulk soil - \% sand content}}$$

Equation 1 (Masri and Ryan, 2006)

Mean weight diameter (MWD) was determined thus:

$$MWD = \sum_{1=1}^{n} xi\ wi$$

Equation 2

Where:
xi = mean diameter of sieve proceeding and preceding
Wi = proportional weight of sand free aggregates

Soil organic matter fractionation

Soil organic carbon content was determined in the aggregate fractions (2-0.25 mm, 0.25-0.053 mm, <0.053 mm) and the bulk soil by the dichromate (wet) oxidation method of Nelson and Sommers (1982). The organic carbon in the aggregate fractions in parenthesis, were referred to as Fine particulate organic carbon, FPOC, (2 – 0.25mm), Intra-aggregate pParticulate organic carbon, IPOC (0.25 – 0.053mm) and silt plus clay associated organic carbon (< 0.053mm). Sand free carbon concentration was thereafter calculated in aggregate fractions greater than 0.053 mm as:

$$\text{Sand free C fraction} = \frac{\text{C fraction}}{\text{1-[sand proportion] fraction}}$$

Equation 3 (Denef *et.al;* 2001)

Acid hydrolysis was employed to determine the Non hydrolysable carbon (NHC) (Tan *et al*., 2004). One gram of the bulk soil sample was passed through a 250 μm sieve and digested with 25 ml of 6N HCl (Paul *et al*., 2001) Using a digestion block at a temperature of 100°C until only the soil residue was left. The residue was centrifuged twice with 40 ml of distilled water for 30 minutes, to remove HCl. The residue was dried at 60°C over night and weighed, after which carbon was determined in the acid hydrolyzed samples by dichromate (wet) oxidation method (Nelson and Sommers, 1982).

Data analysis

Data collected in this study were subjected to statistical analysis of variance as described by Snedecor and Cochran (1967). Using the SAS computer package (SAS, 1998) the differences between the treatment means were evaluated using Duncan's Multiple Range Test (DMRT) (Duncan, 1955).

RESULTS AND DISCUSSION

Water stable aggregate

The effects of sampling depth and tree diversity on water stable aggregates (WSA) is presented in Table 1, statistical analysis of data revealed no significant difference among the evaluated depths with respect to large macro aggregates (>2mm) and mean weight diameter (MWD). Small macro aggregates (>0.25mm) decrease by 10.3% with increase in sampling depth from 5 – 15cm to 15 – 25cm. The high proportion of smaller macro aggregates at 0-5 and 5 – 15cm depth may be due to accumulation and non removal of residues on soil surfaces, which serve as mulch and seemed to improve soil aggregation. Beare *et al.* (1994) also observed that accumulation of residues at soil surfaces improved soil aggregation and reduced losses of soil organic matter. Litter on the forest floor may provides uniform temperature and moisture conditions which could encourages earthworm activities and lead to the production of burrows or biopores, thereby increasing infiltration rate and decreasing runoff which would consequently enhance soil aggregation. Furthermore the presence of organic matter at these depths in the form of leaf litter and decaying roots exert profound influence on the physical chemical and biological properties of soil, due to the non humic substances produced during decomposition of organic matter which plays a vital role in soil aggregation. It was also found that the proportion of micro aggregate (>0.053mm) and silt and clay aggregate fractions (<0.053mm) increased significantly ($P \geq 0.01$) by 12.8% and 20.3% respectively, with increasing sampling depth from 5 – 15cm to 15 - 25cm. This could be ascribed to reduced organic matter contents at these depths since the translocation of deposited leaf litter into the subsurface soil is slow, thus the influence of leaf residue in binding soil aggregate at this sub depth is limited. Furthermore, at lower depths there could also be the possibility that fragments of plant materials becomes encrusted with clay particles and microbial products to form micro aggregates, as clay was found to increase with depth in these soils (Six *et. al.* 2004).

The effects of tree diversity on water stable aggregates (Table 2) revealed that soils planted to

Prosopis africana and *Tectona grandis*, significantly increased large macro aggregate fraction (>2mm) which was statistically similar with soils planted to *Tamarindus indica* and *Eucalyptus camadulensis*. However, *Acacia senegal Azadirachta indica* and *Khaya senegalensis* had the lowest level of macro aggregation. Soils under *Azadirachta indica* and *T. indica* had significantly higher small range macro aggregates (>0.25mm). Different tree species had no significant influence on the content of micro aggregates (>0.053mm). However, the least proportion of silt and clay aggregates (<0.053mm) was recorded in soil planted to *T. indica*. The MWD was significantly higher in *Azadirachta indica* and was found to be statistically similar with that of *T. indica* plantations. The consistency of *T. indica* to be among the tree species, recording high proportions of large and small macro aggregates, MWD and least proportion of silt and clay aggregates could be attributed to the numerous small dense foliage of *T. indica* which may decompose rapidly. Furthermore, the evergreen nature of the tree would sustain the soil organic matter content through higher level of deposition and decomposition of deposited litter, thereby increasing the rate of organic matter addition to soil as well as organic matter derived binding agent, and consequently, an improvement in soil aggregation. This supports the findings of Paul *et al.* (2003) and Six *et al.* (2002). They reported that soil aggregation was enhanced as soil organic matter increased, due to increased production of organic matter derived binding agent resulting from the activity of microbes on deposited residues in soils. More so, the strong correlation ($P \leq 0.01$) between these various aggregate fractions and their aggregate associated carbon may have been responsible for their better soil aggregate stability.

The proportions of micro aggregate (>0.053mm) was not affected by tree diversity. Six *et al.* (2004) asserted that micro aggregate stability is less dependent on agricultural management because of their binding agent and hierarchical order. The high proportion of silt and clay aggregate as well as the least MWD in soil planted to *Tectona grandis* (Table 2) despite its deciduous nature, could be attributed to the production and deposition of large leathery/fibrous leaves of the tree species (Keay, 1989), which are properties of non-easily decomposable litter. Hence, the production of organic matter derived binding agent in its underlying soils will be very slow and consequently lower soil aggregation relative to the other tree species evaluated in · this study even after 19 years of establishment

Table 1. Effect of sampling depth on water stable soil aggregates of a forested ecosystem.

	Proportional Weight of sand free aggregate fractions				
Depth (cm)	>2mm	>0.25mm	>0.053mm	<0.053mm	MWD
0 – 5	0.196	0.396a	0.258b	0.140b	0.488
5 – 15	0.175	0.400a	0.264b	0.135b	0.479
15 – 25	0.174	0.359b	0.298a	0.162a	0.452
SE±	0.0127	0.00785	0.0111	0.00489	0.0129
Significant level	NS	**	*	**	NS

Means followed by the same letter(s) within the same column are not significant at 5% level

** = Significant at 1% level of probability

NS = Not significant

MWD= Mean weight diameter (mm)

Table 2. Effect of tree diversity on water stable aggregates of a forested ecosystem.

	Proportional Weight of sand free aggregate fractions				
Tree specie	>2mm	>0.25mm	>0.053mm	<0.053mm	MWD
Acacia senegal	0.127b	0.375b	0.323	0.170a	0.472bc
Azadirachta indica	0.135b	0.456a	0.252	0.131c	0.552a
Eucalyptus camadulensis	0.189ab	0.377b	0.303	0.138bc	0.471bc
Khaya senegalensis	0.159b	0.362b	0.279	0.151abc	0.457cd
Prosopis africana	0.241a	0.355b	0.245	0.159ab	0.441cd
Tamarindus indica	0.183ab	0.426a	0.257	0.108d	0.521ab
Tectona grandis	0.237a	0.344b	0.254	0.162a	0.400d
SE±	0.0195	0.0120	0.0169	0.00747	0.0198
Significant level	**	**	NS	**	*

Means followed by the same letter(s) within the same column are not significant at 5% level of probability.

** = Significant at 1% level of probability

NS = Not significant

MWD = Mean weight diameter (mm)

Soil organic carbon pools

The soil organic carbon in the top soil (0-5cm layer) decreased significantly with increasing depth for all aggregate associated carbon (Fine particulate, intra aggregate particulate and silt plus clay), except for Non hydrolysable carbon (NHC) fraction where further increase in sampling depth had no significant influence on SOC (Table 3). The high SOC content of the top soil (0 – 5cm) could be ascribed to deposition of leaf litter on the forest floor, in addition to non cultivation of the soil, which allowed accumulation of OM on soil surface. This agreed with Hubbert *et al.* (2001) who reported that most significant chemical changes associated with plantation forestry occurred at or near the surface of the soil, and is related to the supply of organic matter.

Tamarindus indica plantation had highest total organic carbon (TOC) content followed by *Eucalyptus camadulensis* and *Prosopis Africana* (Table 4). Soil under *T. indica* was among soils with significantly high fine particulate organic carbon,intra aggregate particulate organic carbon and silt plus clay associated organic carbon. It was better than soils planted to *Khaya senegalensis* in all the carbon pools (fine particulate, intra aggregate particulate and silt plus clay organic carbon) except for NHC pool, where *T. indica* and *K. senegalensis* were statistically similar. The several months of flowering (i.e November – December and May to August) of *T. Indica* producing flowers with fleshy slender stalk and hairy ovaries, in addition to its numerous non-leathery leaflets (Keay, 1989; ICRAF, 1992), increased the propensity of *T. indica* biomass to decomposition, thus releasing nutrients into the underlying soil as well as increasing organic carbon content. Furthermore the evergreen dense foliage of *T. indica* may be responsible for its high carbon content, through increased potential of photosynthesis by enhancing atmospheric carbon (CO_2) utilization thus stimulating carbon allocation to soil (Silvola and Ahlholm, 1995). In addition, this may encourage high carbon sequestration through the biomass of *T. Indica,* which when returned to soil as leaf litter and tree branches on forest floor, would improve carbon pools and consequently, the CO_2 effect in the environment is mitigated.

Table 3. Effect of sampling depth on organic carbon content in soil of a forested ecosystem.

	Soil organic carbon (gkg^{-1})				
Depth (cm)	Total organic carbon	Fine particulate	Intra aggregate particulate	Silt plus clay	Non hydrolysable
0 – 5	9. 22a	20.91a	16.93a	10.78a	4.02a
5 – 15	7.57b	13.55b	12.67b	8.90a	3.37ab
15 – 25	6.08c	7.22c	8.38c	6.47b	2.78b
SE±	0.271	1.458	0.991	0.068	0.076
Significant level	**	**	**	**	**

Means followed by the same letter(s) within the same column are not significant at 5% level of probability.

* = Significant at 5% level of probability

** = Significant at 1% level of probability

Table 4. Effect of Tree diversity on organic carbon content in soil of a forested ecosystem.

	Soil organic carbon (gkg^{-1})				
Tree species	Total organic carbon	Fine particulate	Intra aggregate particulate	Silt plus clay	Non hydrolysable
Acacia senegal	6.41d	15.85a	12.86ab	8.96a	3.62ab
Azadirachta indica	6.75d	13.14ab	12.84ab	9.51a	3.16b
Eucalyptus camadulensis	8.89b	19.40a	15.31ab	8.90a	3.11b
Khaya senegalensis	6.01d	7.91b	5.41c	6.11b	3.62ab
Prosopis Africana	8.04bc	12.46ab	13.65ab	8.60a	4.43ab
Tamarindus indica	10.31a	14.68ab	16.97a	9.51a	4.81a
Tectona grandis	6.92cd	13.81ab	11.57b	8.71a	1.83c
SE±	0.414	2.227	1.514	0.104	0.116
Significant level	**	*	**	*	**

Means followed by the same letter(s) within the same column are not significant at 5% level of probability.

* = Significant at 5% level of probability

** = Significant at 1% level of probability

The least carbon content was obtained from soils under *Khaya senegalensis.* Though the crown of this tree is evergreen, but it possesses a non-easily decomposable biomass, such as its larger, dark shiny leathery foliage, and its production of hard woody fruits compared to *T. indica*. These features of *K. senegalensis* would not allow its biomass deposited on forest floor to be easily accessible to microbes and therefore will culminate to its less organic matter turnover in soil, and consequently less SOC sequestrations, as indicated in its total and all other aggregate associated carbon pools (Table 4). Hobbie *et al.* (2007) observed that tree species have differing effects on soil C and N cycling through litter due to inter specific variation in litter chemistry.

The least non hydrolysable SOC was observed in *Tectona grandis* plantation. Though the tree specie is deciduous in nature (not maintaining an ever green crown all year round); therefore, it may be possible that there is a reduction in its photosynthetic ability and hence less carbon may be sequester from the atmosphere for subsequent storage in soil. In addition, physical examination of litters indicates that, the very broad fibrous and leathery leaves (Keay,1989) shed on forest floor, is non-easily decomposable and consequently this will hinder rapid release of nutrient into the soil. This is further confirmed by the findings of Okoro and Osagie (1999). They reported lower organic carbon under *Tectona grandis* plantation than natural vegetation, which was attributed to slow mineralization of *T. grandis* litter.

Interaction effect on soil Properties

Generally, TOC decreased as the depth of soil sampling increases from 0 – 5 to 15 – 25 cm depth (Table 5). Expectedly TOC was higher at surface soil as leaves and other debris decompose on soil surface, though it may be leached down soil profile with time, but it will still remain higher on top soil, since debris and detritus are continually deposited and mineralized. Further more, there may be possibility of higher microbial activity at soil surface (Six *et al.*, 2002) due to the presence of abundant debris and detritus. At 0 – 5 cm depth, significantly higher TOC was found on plots grown to *Eucalyptus camadulensis, Prosopis africana* and *Tamarindus indica,* relative to *Acacia senegal, Azadirachta indica* and *Khaya senegalensis.* The lower TOC content in the latter tree species (*A. senegal and K. senegalensis*) could be ascribed to the poor decomposing properties (leathery leaves, drier debris, hard woody fruits, and xerophytic nature) of these trees which were observed via physical assessment of deposited litter. However at greater depths (5 – 15 and 15 – 25cm), TOC on field grown to *Tamarindus indica* was significantly higher than other tree species evaluated in this study. It may be possible that *T. indica* sloughs off more roots at greater soil depths relative to the other tree species studied. Balesdent and Balabane (1996) observed that roots sloughed off each year continually add C and N to the soil. More so rhizosphere deposition such as root exudates may be a significant source of SOC and may have great influence on SOC concentration.

Table 5. Interaction between sampling depth and tree species (D x S) on Total Organic Carbon (gkg^{-1}).

	Depth (cm)		
Tree species	0 – 5	5 – 15	15 – 25
Acacia Senegal	7.66d-f	6.25e-g	5.33fg
Azadirachta indica	7.73d-f	6.45e-g	6.07e-g
Eucalyptus camadulensis	12.50a	8.12c	6.05e-g
Khaya senegalensis	6.27e-g	6.83d-g	4.93g
Prosopis Africana	10.44a-c	7.65d-f	6.02e-g
Tamarindus indica	10.67ab	11.07ab	9.20bc
Tectona grandis	9.24b-d	6.58e-g	4.93g
SE $\pm$ 0.71788			

Means followed by different letter(s) are significant at 5% level of probability.

CONCLUSION

Soil organic carbon pools and aggregate stability indices were significantly enhanced at the top soil (0–5cm), These enhanced soil properties were attributed to the frequent and continual deposition of litters and debris on forest floor and the non incorporation of same into sub soil.

This study suggests that tree diversity have great influence on soil organic carbon and pools and aggregate stability and therefore their restorative ability. Tree species that increase carbon sequestration in soil will in turn improve soil quality and may provide a good opportunity to combat soil degradation and loss of biodiversity and improve food security. It was observed that the tissue chemistry of tree species particularly, that of the leaves may be an important mechanism contributing to ecosystem functions. Further research on this area with more trees will need to be pursued.

Tamarindus indica could be a better choice of tree, of seven tree species evaluated in this study. In regenerating a degraded soil by afforestation and consequently sequestering atmospheric carbon to reduce global warming, deciding on a good choice of tree species and planting the tree in time, will go a long way in addressing the rate at which soil degradation and green house effect will be mitigated.

REFERENCES

Beare, M.H., Hendrix, P.F. and Coleman, D.C. 1994. Water Stable Aggregates and Organic Matter Fractions in Conventional and no Tillage Soils. Soils Science Society American Journal. 58: 777-786.

Denef, K., Six, J., Bossuyt, H., Frey, S.D., Elliot, E.T., Merckx, R. and Paustian K. 2001. Influence of dry-wet cycles on the interrelationship between. aggregate, particulate organic matter and microbial community dynamics. Soil Biology and Biochemistry. 33: 1599 – 1666.

Duncan, D.B. 1955. Multiple Range and Multiple “F”-test Biometrics II: - 42.

Elliott, E.T. 1986. Aggregate structure and carbon, nitrogen and phosphorus in native and cultivated soils. Soil Science Soc. American journal. 50: 627–633.

Franzluebbers, A.J., Wright, S.F. and Stuedeman, J.A. 2000. Soil aggregation and Glomalin under pastures in the southern piedmont USA Soil Science Society American Journal 64: 1018-1026.

Hobbie, S.E., Ogdahl, M., Chorover, J., Chadwick, O.A., Oleksyn, J., Zytkowiak, R. and Reich, P.B. 2007. Tree species effects on soil organic matter dynamics: the role of soil cation composition. Ecosystems. 10: 999-1018.

Hubbert, K.R., Graham, R.C. and Anderson, M.A. 2001. Soil and weathered bedrock. Components of Jeffrey pine plantation substrate. Soil Science Society American Journal. 65: 1255 – 1262.

Intergovernmental Panel on Climate Change (IPCC) 1990. Scientific Assessment of climate change: Policy makes summary (Geneva and Nairobi). World Meteorological Organization and United Nations Environment Programme, 1990.

Intergovernmental Panel on Climate Change (IPCC) 2000. Land use, Land use change and forestry. Intergovernmental panel on climate change special report. Cambridge Univ. Press Cambridge. U.K. PP 20.

International Center for Research in Agro forestry (ICRAF) 1992. A selection of useful trees and shrubs for Kenya: Notes on their identification, Propagation and Management for use by Farming and Pastoral Communities. PP 226.

Keay, R.W.J. 1959. An outline of Nigerian vegetation 3rd edition London PP 45.

Keay, R.W.J 1989. Tree of Nigeria Oxford University Press New York PP 476.

Kyoto Protocol 1997. Kyoto Protocol to the United Nations Framework, Convention on climate change Kyoto 11/12/1997. http://wwwunfcccorg/.

Masri, Z. and Ryan, J. 2006. Soil organic matter and related physical properties in a Mediterranean wheat based rotation trial. Soil and Tillage Research 87: 146 – 154.

Nelson, D.W., and Sommers, L.E 1982: Total Carbon, Organic Carbon and Organic Matter. In page, A., Miller, R.H. and Keeney, D.R. (eds) Methods of Soil Analysis. Part 2, Chemical and microbiological methods. Agronomy monograph no. 9 second edition. Madison, USA. PP 539-580.

Okoro, S.P., Osagie, C.O. 1999. Effects of selected monoculture plantation species on humid tropical soils of southern Nigeria. Nigerian Journal of Forestry 29(23): 73-79.

Paul, E.A., Collins, H.P., and Leavitt S.W. 2001. Dynamics of resistant soil carbon of Midwestern agricultural soils measured by naturally occurring C abundance. Geoderma 104: 239 -256.

Paul, E.A., Morris, S.J., Six J., Paustian, K., and Gregorich, E.G. 2003. Interpretation of soil carbon and Nitrogen dynamics in Agricultural and Afforested soils *Soil Science Society of America Journal* 67: 1620 – 1628.

SAS Institute (1998). SAS/STAT user's guide Version 8 Vol.2. SAS incorporation Cary. NC.

Silvola J., and Ahlholm, U. 1995. Combined effects of CO_2 concentration and nutrient uptake of birch seedlings (*Betula pendula*) Plant and Soil. 168-169: 547-553.

Six, J., Elliott, E.T and Paulstain K. 2000. Soil micro aggregate turnover and micro aggregate formation: A mechanism for C sequestration under no tillage agriculture. Soil Biology and Biochemistry. 32: 2099 – 2103.

Six, J. Callewqert, P. Lenders, S., DeGryze, S., and Paustian, K. 2002. Measuring and understanding carbon storage in Afforested soils by physical fractionation. Soil Science Society American Journal 66: 1981 – 1987.

Six, J., Bossuyt, H., Degryze, S., and Denef, K. 2004. A history of research on the link between (micro) aggregate. Soil biota and soil organic matter dynamics. Soil Tillage Research. 79: 7 – 39

Snedecor, G.W. and Cochran, W.G. 1967. Statistical methods 6th ed. Iowa State University Press, U.S.A 585 pp.

Solomon, D., Fritzsche F., Tekalign M., Lehman J. and Zech W. 2002. Soil Organic Matter composition in the Sub humid Ethiopian Highlands as influenced by Deforestation and Agricultural Management. Soil Science Society American Journal.66: 68 – 82.

Swift, R.S. 2001. Sequestration of carbon by soil. Soil Science. 166 (11): 858-871.

Tan, Z.X., Lal, R., Izaurralde, C. and Post, W.M. 2004. Biochemical Protected Soil organic carbon at the North Applalachian Experimental watershed. Soil Science. 169 (6): 1-10.

Tomlison, P.R. 1957. A preliminary soil report on Afaka Forest reserve Experimental Area (Zaria Province, Northern Nigeria). Bull. No.2.

Unger, P.W. 1997. Management induced aggregation and organic carbon concentration in the surface layer of a Torrertic paleustoll. Soil Tillage Research, 42: 185-208.

10

Bush Encroachment Control Demonstrations and Management Implications on Herbaceous Species in Savannas of Southern Ethiopia

A. Angassa[13]*, G. Oba[2] and A. Tolera[1]

[1]*Department of Animal and Range Sciences, Hawassa University, P.O Box 1519, Awassa, Ethiopia.*

[2]*Noragric, Department of International Environment and Development Studies, Norwegian University of Life Sciences, PO Box 5003, N-1432 Ås, Norway,*

[3]*Botanical Institute, University of Cologne*

**E-mail: ayanaangassa@yahoo.com*

**Corresponding Author*

SUMMARY

Proper management of bush control methods need to be effectively demonstrated to resource users. In this paper, we evaluated effects of disturbance control (CO) (i.e., government policy approach in the conservation of natural resources), tree cutting and fire (C+F), tree cutting, fire and grazing (C+F+G), the old traditional method of applying fire and grazing (F+G) by herders, grazing with bush cover (GBC) (i.e., current system of land use system by herders in Borana) and tree cutting alone (C) (i.e., Non-governmental Organizations approach in rangeland improvement) on herbaceous species composition, abundance, biomass, basal cover and species diversity in southern Ethiopia. The disturbance control demonstration showed no advantage in terms of biomass and basal cover over other treatments, although herbaceous species richness was improved. Bush removal methods such as tree cutting and fire, tree cutting followed by fire and grazing, fire and grazing, and tree cutting alone improved herbaceous biomass, basal cover and species diversity, while grazing alone greatly reduced herbaceous biomass. Tree cutting and fire treatments seemed superior in terms of herbaceous biomass, while conservation of herbaceous species diversity was improved more by the traditional method of fire and grazing, and tree cutting. With the exception of species richness, protection from disturbance showed no advantage. Only fire and grazing methods can reasonably be applied for the control of bush encroachment with the overall objective of promoting biomass production and species diversity.

Key words: Disturbance control; tree cutting, fire, grazing; policy implication; public education; Ethiopia.

INTRODUCTION

Savanna grasslands are threatened by bush encroachment worldwide (Scholes and Archer, 1997; Oba et al., 2000; Smit and Rethman, 2000; Sheuyange et al., 2005; Wiegand et al., 2005) through the suppression of herbaceous biomass production and probably reduced biodiversity (Smit, 2003; Hagos and Smit, 2005; Ansley and Castellano, 2006). The threat to the pastoral economy is often the main reason for the control of bush encroachment (Olson and Whitson, 2002). Bush encroachment control is a disturbance that reduces the threat of bush encroachment by disrupting the invasive woody plant community structure through transformations of biotic environments and habitat conditions in which colonization of the disturbed microhabitat takes place (Pickett and White, 1985). Bush encroachment control methods shift the savanna vegetation from dominance by woody vegetation to dominance by herbaceous vegetation thereby creating suitable habitat for grazers (Fulbright, 1996; Hudak, 1999; Angassa and Oba, 2008). The shift towards herbaceous species dominance in turn induces shifts in species that are intolerant of bush cover, while the species that tolerate bush cover might decline (Clarke et al., 2005). The changes could cause partial or total reduction of plant biomass (Grime, 1977) by shifting vegetation structure and composition (Łaska, 2001). Disturbance can also produce changes in the life history strategies of individual species in response to intensities of disturbance forces (Grime, 1974; Grime, 1977) and the created micro-environmental conditions (Vetaas, 1992).

In savanna ecosystems, different bush control methods including hand removal of trees (Berlow et al., 2003; Smit, 2003, 2004), use of fire (Sheuyange et al., 2005), combination of hand removal and burning (Duggin and Gentle, 1998; Olson and Whitson, 2002), tree cutting combined with fire and grazing (Kobayashi et al., 1997; Sawadogo et al., 2002; Sawadogo et al., 2005), fire combined with grazing (Fuhlendorf and Engle, 2004; Sawadogo et al., 2005) and grazing control (Fulbright, 1996; Smit, 2004; Bates, 2005) are commonly used to increase herbaceous production and species diversity (Smit, 2003; Ansley and Castellano, 2006). Tree cutting and prescribed fire followed by rest from grazing enhanced herbaceous plant biodiversity (Bates, 2005). Others (Kobayashi et al., 1997; Sawadogo et al., 2005) also studied effects of tree cutting, fire and grazing on the responses of herbaceous plants. Previous studies (Hutchinson et al., 2005) have observed changes in herbaceous species richness following prescribed fire, while Fuhlendorf and Engle (2004) reported the importance of fire and grazing in the restoration of savanna grasslands. Cutting is effective in increasing herbaceous biomass, cover and diversity (Bates et al., 2000). Changes in relative abundance of herbaceous species over time may be a good indicator of species' responses to disturbances compared to pre-disturbance populations. Elsewhere, the importance of bush encroachment control in improving the relative abundance of herbaceous species has been shown (Hudak, 1999). Relative abundance serves as an indicator of population of individual plants that could vary in response to disturbances. The effects of different bush control methods on herbaceous production and diversity can therefore be evaluated in terms of herbaceous species performances (Fulbright, 1996).

Bush encroachment control methods are management systems (Fulbright, 1996) that might have varied policy implications for bush control (Olson and Whitson, 2002). Understanding the potential role of different bush encroachment control methods for promoting herbaceous species composition requires recognition of the objectives of resource users and policymakers. The resource users are interested in livestock production through increased plant productivity, while the goal of policymakers is environmental preservation. The public is likely to adopt management systems that are more efficient in improving forage for livestock grazing, while official government policy promotes the conservation of biodiversity. The two goals may be in conflict. Although the effects of bush encroachment control methods on herbaceous plants have been well documented (Hadar et al., 1999; Angassa, 2002; Sawadogo et al., 2002; Smit, 2003, 2004; Sawadogo et al., 2005; Ansley and Castellano, 2006), information regarding the potential role of such demonstrations in influencing management policies and public education is limited.

In the Borana pastoral region of southern Ethiopia, which is the focus of the present study, we are not aware of other efforts in which bush encroachment control were purposely used to evaluate responses of individual herbaceous species to recommend management policy. Researchers and the extension departments concerned with grazing lands rarely design bush encroachment controls to promote the management of bush cover or discuss the educational benefits of different methods in terms of long-term changes in the relative abundance of herbaceous species and biomass for livestock forage. Monitoring the effects of various control methods is useful in highlighting single and combined effects of disturbances on herbaceous vegetation, by providing evidence of changes in species composition over time and assessing management policy and public educational relevance.

We evaluated the effects of six demonstration methods including disturbance control (CO) (i.e., exclude grazing and other disturbances), tree cutting and fire (C+F), tree cutting, fire and grazing (C+F+G), fire and grazing (F+G) (which is the old traditional method practiced by pastoralists before the official ban on the use of fire), grazing with bush cover (GBC) (which is representing the current methods of pastoral land use), and tree cutting alone (C) on: (a) changes in species composition and the relative abundance of individuals of each herbaceous species; (b) trends in the relative abundance of herbaceous species; (c) performances of individual herbaceous species; and (d) changes in biomass, basal cover and species diversity between the pre- and post- treatment sampling periods. The findings are then discussed in the light of policy guidelines for bush encroachment control, as well as recommendations for sound public education on the post-treatment management of the savanna ecosystems in southern Ethiopia.

MATERIALS AND METHODS

Study area

We demonstrated six bush encroachment control treatments replicated across the government cattle breeding ranch (in Dida-Tuyura 04°57.383′N, 38°12.403E′) and community ranch (in Dambala-Wachu, 04°29.432′N, 38°16.339′E) in southern Ethiopia. The topography varies from uplands (> 1500 m a.s.l.) to bottomlands (< 1400 m a.s.l.). The major soil types are Chromic and Orthic Luvisols (45%), Calcaric and Eutric Regosols (15.25%), and Lithosols together with Humic, Mollic and Vertic Andosols (FAO, 1986; Coppock, 1994). The climate is variable with the main rainy season from March - May (*ganna*) and the short rains from September - November (*hagaya*). The mean annual rainfall for the period 1980 to 2000 was 513±39 mm for Dida-Tuyura and 359±30 mm for Dambala-Wachu. In both sites, bush encroachment pose problems for livestock grazing. Before the two range sites were converted into ranches and prior to the official ban on fire in the 1970s, the pastoral community periodically conducted range burning late in the dry season (the dry season occurs between December and early March). Under the ranch management system, the sites, similar to the communal rangelands lacked regular burning. At the time of the research, the stocking density of the Dida-Tuyura ranch was 12.0 Tropical Livestock Unit km^{-2} (TLU km^{-2}, 1 TLU = 250 kg bovine) compared to 17.6 TLU km^{-2} in Dambala-Wachu (Angassa and Oba, 2007).

The dominant vegetation-type of the ranches is *Acacia* grassland and consists mostly of perennial grasses and woody cover. The grass cover is dominated by *Cenchrus ciliaris, Chrysopogon aucheri* and *Panicum coloratum*, while the woody cover consists mainly of *Acacia drepanolobium* on black soil sites, and *A. brevispica* and *A. horrida* in the uplands (Coppock, 1994). There are also patches of *Balanites aegyptica*, and several species of *Commiphora* and *Terminalia* at the lower altitudes. The higher parts of the hills used to be covered with *Juniperus procera* and *Olea europaea cuspidata* forest (Coppock, 1994).

We selected the ranch sites considering their use as demonstration plots to the local communities for extension services. We used the opportunity to demonstrate different bush encroachment control methods where this can also serve the extension department for further education and communication to resource users to encourage local participation in future management of bush-control.

Sampling methods

Six demonstrations practices were conducted, including disturbance control (CO) that represented the government policy to exclude fire, grazing and tree cutting, in which grazing was excluded by fencing. The second demonstration practice combined tree cutting and fire (C+F) and the third practice was tree cutting followed by fire and grazing (C+F+G). The fourth demonstration represented the old traditional system that combines fire and grazing management (F+G), for which fencing was not necessary. The fifth demonstration, grazing with bush cover (GBC), represented the current traditional system (i.e., after government banned the use of fire) of rangeland management. On both ranches, grazing was conducted using free ranging animals. For plots in the sixth demonstration (cutting alone denoted by C), all trees were hand cleared followed by fencing to protect the area from livestock grazing. All tree clearings were unselective, involving removal of all trees. This is the method of bush encroachment control commonly used by Non-Governmental Organizations (NGOs) for rangeland rehabilitation.

In the demonstration where we combined tree cutting followed by fire and grazing (C+F+G), the woody plants were chopped up into finer pieces and then piled on tree stumps to create a fine fuel load for burning. The piles of woody plants were allowed to dry for one month before fire was ignited. The C+F demonstration plots were fenced in June 2003 to exclude livestock to allow the accumulation of fuel loads before burning. The fourth demonstration represented the traditional use of fire (before official fire suppression) and grazing (F+G), in which livestock was allowed to graze the burnt plots. The three burn demonstrations (i.e., C+F, C+F+G and F+G) were conducted late in the dry season. Details of the burning treatments are presented in Part 1 of this experiment (Angassa and Oba, 2009).

For all demonstrations, we randomly distributed twenty-four 50 x 50 m plots in both topographic locations. In each of the larger plots, we randomly located three smaller 1 x 1 m plots for sampling herbaceous vegetation during pre-treatment in 2003. Sampling was repeated on the same plots during the post-demonstration phases in 2004 and 2005. During each sampling process, all the herbaceous species were identified (where species could not be identified the voucher samples were collected for later identification), and numbers of individuals of each species (i.e., relative abundance) per m^2 and the species richness were determined per plot per census (i.e., total count of all species per plot). Unidentified herbs and forbs, hereafter referred to as "other herbaceous species" and "herbaceous legumes", were also included in species richness. The species diversity index, $H' = -(\sum (pilnpi))$ (Shannon, 1948) was calculated using the frequency of each plant species (where pi represents the proportion of individual species in each plot). The basal cover of root crown in relation to the bare ground was estimated for each species or categories of species (i.e., including unidentified herbaceous and leguminous species). All the herbaceous plants in the sample plots were hand harvested. The combined samples of the species were air dried followed by oven drying at 65°C for 48 hours, for the purpose of dry matter estimation. In this study, we focused on the performances of individual herbaceous species between the pre- and post-demonstration treatments. During the sampling periods, the treatments were aimed at demonstrating to the local communities on how different bush encroachment control methods influenced the performances of different species. More importantly, we used the results to inform the general public using our simplified experimental design.

Analyses

We report mean values (i.e., by combining data from the two ranches and the two landscape locations), by sampling dates, by demonstration treatments. We used a general linear model in SAS to analyze the pre- and post-disturbance mean (±SE) of the relative abundance of individual species. We used descriptive statistics to calculate relative changes (i.e., differences in relative abundance – represented by numbers of individuals of each species between the pre- and post-disturbance) of individual species m^{-2}. A chi-square test was used to analyze changes in the relative abundance of individual species, to determine their susceptibility (i.e., species performance) to the effects of the demonstration (Sweet and Tacheba, 1984). Least significant difference (LSD) was used to analyze changes in biomass, basal cover and diversity between the pre- and the post-disturbance for each demonstration.

RESULTS

Disturbance control

Prior to protection, the disturbance control (CO) demonstration had 9 herbaceous species, which after two years of protection increased to 18 (Tables 1 and 2). Species recorded after establishing protection included: *Aristida adoensis, Bothriochloa radicans, Digitaria milanjiana, Enteropogon somalensis, Harpachne schimperi, Heteropogon contortus* and *Panicum turgidum* (Tables 2 and 3). *Cenchrus ciliaris, Chloris roxburghiana, Leptothrium senegalense, Panicum coloratum* and *Sporobolus pyramidalis* slightly increased in relative abundance in response to protection, while *Eragrostis papposa* and *Panicum maximum* were reduced (Tables 1, 2, 3 and 4). We found no changes in the relative abundance of *Chrysopogon aucheri* and *Themeda triandra* in response to protection (Tables 1, 2 and 3). Before protection, the most abundant species was *Eragrostis papposa* followed by *Panicum maximum*, but both of them declined in response to protection. *Bothriochloa radicans* was the most abundant species after two years of protection (Tables 1 and 2). Protection did not increase herbaceous biomass, basal cover and diversity (Figs. 1 a, b, c).

Tree cutting and fire

The tree cutting and fire (C+F) demonstration increased herbaceous species richness from 9 (i.e., pre-disturbance) to 15 (i.e., after disturbance) (Tables 1 and 2). Of the initial herbaceous species, *Cenchrus ciliaris* increased in relative abundance, while about 78% of the pre-treatment species showed a decline (Tables 1, 2 and 3). Only in the case of *Sporobolus pyramidalis* was the decline significant (Tables 3 and 4). Generally, the most abundant species was *Sporobolus pyramidalis* followed by *Eragrostis papposa* preceding disturbance, while *Chrysopogon aucheri* did not vary in relative abundance during post-disturbance (Tables 1 and 2). Herbaceous biomass and species diversity, but not basal cover, were greatly promoted during the post-disturbance period (i.e., in 2005) compared to the pre-disturbance period (Figs 1 a, b, c).

Tree cutting - fire and grazing

In response to tree cutting, fire and grazing (C+F+G), herbaceous species richness increased from 9 (pre-disturbance) to 17 (after disturbance) (Tables 1 and 2). The species *Bothriochloa radicans, Cynodon dactylon, Enteropogon somalensis, Eragrostis sennii, Heteropogon contortus, Panicum turgidum* and *Sporobolus pyramidalis* were not captured during the pre-demonstration sampling, but were recorded after disturbance (Tables 2 and 3). Among the initial herbaceous species, only *Leptothrium senegalense* was greatly promoted by the combined effects of tree cutting, fire and grazing (Table 3). Hence, tree cutting, fire and grazing reduced relative abundance of most herbaceous species at first, followed later by increases (Tables 1, 2 and 3). The decline of *Chloris roxburghiana, Eragrostis papposa* and *Panicum coloratum* was significant (Tables 3 and 4). The species absent from the post-demonstration period was *Aristida adoensis,* while the species that declined (e.g., *Eragrostis papposa* and *Lintonia nutans*). *Eragrostis papposa* appeared to be the most abundant species prior to disturbance, exception being *Sporobolus pyramidalis* with increased relative abundance following disturbance (Tables 1 and 2). During the post-demonstration periods herbaceous biomass and diversity, but not basal cover, showed increases (Figs 1 a, b, c).

Fire and grazing

The fire and grazing (F+G) demonstration promoted herbaceous species richness from 11 (pre-disturbance) to 21 (after disturbance) (Tables 1, 2 and 3). *Enteropogon somalensis, Panicum maximum* and *Pennisetum stramineum* increased in abundance during post-disturbance (Tables 3 and 4). However, about 82% of the individuals declined in relative abundance during the post-disturbance periods (Tables 1, 2 and 3). Among the species that showed significant declining in trends were *Cenchrus ciliaris, Sporobolus pyramidalis* and *Themeda triandra* (Tables 3 and 4). The species most vulnerable to fire and grazing and also absent from the post-demonstration periods were *Aristida adoensis* and *Leptothrium senegalense* (Tables 2 and 3). The two most disturbance tolerant species were *Heteropogon contortus* and *Sporobolus pyramidalis*. Herbaceous biomass and diversity, but not basal cover, increased in response to fire and grazing (Figs 1 a, b, c).

Grazing with bush cover

Grazing with bush cover (GBC) promoted herbaceous species richness from 11 to 20 species (Tables 1, 2 and 3). Among the initial herbaceous populations, the species that positively responded to grazing with bush cover were *Cenchrus ciliaris, Hyparrhenia hirta* and *Panicum coloratum* (Tables 3 and 4). Among the key species that showed significant declines in relative abundance were *Chrysopogon aucheri, Eragrostis papposa, Pennisetum stramineum, Sporobolus pyramidalis and Themeda triandra* (Tables 3 and 4), while *Lintonia nutans* showed stability in response to grazing (Tables 1, 2 and 3). *Chrysopogon aucheri* was the most abundant species during pre-disturbance, while *Sporobolus pyramidalis* was more abundant after disturbance (Tables 1 and 2). Grazing with bush cover reduced herbaceous biomass and basal cover, while species diversity was promoted (Figs 1a, b, c).

Table 1. Pre-disturbance mean (± SE) of herbaceous species in terms of abundance of individuals recorded in 2003 in different demonstration plots (mean of 12 plots of $1m^2$) in Borana, southern Ethiopia.

	Demonstration plots					
Species list	Control[a]	C+F[b]	C+F+G[c]	F+G[d]	GBC[e]	C[f]
Aristida adoensis Hochst.			4±2.6	2±1.1		8±3.4
Cenchrus ciliaris L.	10±4.2	14±11.8	12±8.1	43±8.1	3±1.2	6±2.5
Chrysopogon aucheri (Boiss.) Stapf.	9±4.2	25±9.1	18±4.8	14±2.0	45±5.9	21±7.7
Chloris roxburghiana Schult.	3±1.4	3±1.8	2±1.6	1±0.1		6±1.4
Eragrostis papposa (Roem. & Schult.) Stued.	48±17.4	52±11.2	63±14.8	17±8.1	13±10.2	19±7.7
Heteropogon contortus (L.) Beauv. ex. R. & Sch.				78±38.1		
Hyparrhenia hirta (L.) Stapf.					6±5.2	
Leptothrium senegalense (Kunth) Clayton	4±2.7	12±5.8	1±0.6	2±1.2		2±1.4
Lintonia nutans Stapf.		12±5.8	37±5.6		2±1.2	18±13.4
Panicum maximum Jacq.	31±14.7			1±0.1		
Panicum coloratum L.	8±7.4	8±5.8	16±5.6	9±2.0	2±1.2	
Pennisetum mezianum (Vahl) Lanza & Mattei					4±1.2	
Pennisetum stramineum Peter					14±11.2	
Sporobolus pyramidalis Beauv.	14±4.7	63±15.2		78±27.0	41±7.2	24±13.4
Themeda triandra Forssk.	9±4.7	32±15.8	31±18.1	10±7.0	6±5.2	14±9.5
Other herbaceous species (unidentified)					13±10.2	5±3.4

aControl = exclusion of cutting, grazing and fire, bC+F = combination of cutting and fire, cC+F+G = combination of cutting, fire and grazing, dF+G = combination of fire and grazing, eGBC = Grazing under increased bush cover, fC = Cutting alone

Table 2. Post-disturbance mean (± SE) of herbaceous species in terms of abundance of individuals recorded in 2004 and 2005 in response to the different demonstration methods (mean of 12 plots of $1 m^2$) in Borana, southern Ethiopia.

	Demonstration plots					
Species list	Control	C+F	C+F+G	F+G	GBC	C
Aristida adoensis Hochst.	3±2.6	3±1.2			1±0.6	3±1.2
Bothriochloa radicans (Lehm.) A.Camus	43±7.9		8±5.5		17±7.7	22±9.4
Cenchrus ciliaris L.	17±4.0	18±6.5	11±5.2	15±6.0	13±4.9	15±4.7
Chrysopogon aucheri (Boiss.) Stapf.	9±3.2	26±6.5	15±4.4	9±4.8	12±4.1	11±3.3
Chloris roxburghiana Schult.	5±2.6	1±0.6	1±0.5	2±1.5		8±2.4
Cynodon dactylon (L.) Pers.			12±10.5	1±0.5	5±1.8	4±3.4
Cyperus species Vahl.				2±1.5		4±1.4
Digitaria milanjiana (Rendle) Stapf.	1±0.9					
Eleusine jaegeri Pilg.						4±4.4
Enteropogon somalensis Chiov.	12±7.9	5±3.2	6±1.5	9±8.5	12±7.7	6±4.4
Eragrostis papposa (Roem. & Schult.) Stued.	2±1.7	8±3.2	1±0.5	5±1.5	3±1.8	7±4.4
Eragrostis sennii Chiov.			1±0.5	1±0.5		
Harpachne schimperi Hochst. ex. A. Rich.	1±0.9			1±0.5	3±1.8	3±2.6
Heteropogon contortus (L.) Beauv. ex. R. & Sch.	5±4.7		15±10.5	36±9.5	7±5.8	8±4.4
Hyparrhenia hirta (L.) Stapf.					16±10.8	
Leptothrium senegalense (Kunth) Clayton	5±3.6	1±0.9	11±10.5		1±0.8	
Lintonia nutans Stapf.		1±0.7	1±0.5		2±1.7	
Panicum maximum Jacq.	12±7.9			13±13.5	1±0.8	10±9.4
Panicum coloratum L.	11±5.6	5±3.4	14±7.4	7±3.5	13±6.3	5±2.4
Panicum turgidum Forssk.	9±7.9	17±14.5	4±1.5	2±1.5		2±1.5
Pennisetum mezianum (Vahl) Lanza & Mattei				2±1.5	3±1.8	1±0.5
Pennisetum stramineum Peter		18±14.5		9±3.5	11±10.8	
Sporobolus pyramidalis Beauv.	23±4.6	25±6.5	19±7.4	17±5.5	19±6.2	18±3.6
Themeda triandra Forssk.	9±5.6	6±4.2	18±6.1	9±7.9	4±3.7	9±5.4
Herbaceous legumes (unidentified)	7±4.0	6±3.7	13±6.1	17±7.9	8±4.9	4±3.2
Other herbaceous species (unidentified)	6±3.9	18±14.5	12±10.5	18±7.9	6±4.8	9±6.6

Tree cutting

Compared to the initial populations, herbaceous species richness doubled during the post-demonstration period in response to tree clearing (Tables 1, 2 and 3). Among the newly recorded species, the changes were significant for *Bothriochloa radicans, Heteropogon contortus* and *Panicum maximum* (Tables 3 and 4). Among the initial herbaceous species recorded (n = 10 herbaceous species), about 70% exhibited declining trends (Tables 1, 2 and 3). Overall, *Aristida adoensis, Chrysopogon aucheri, Eragrostis papposa, Lintonia nutans and Themeda triandra* were greatly reduced in response to tree cutting (Tables 3 and 4). The species most sensitive to the tree cutting treatments were Leptothrium senegalense and Lintonia nutans (Tables 2 and 3), while the most stable species were *Sporobolus pyramidalis* and *Bothriochloa radicans*. Herbaceous biomass and diversity increased by nearly 50% compared to the pre-cutting period, while basal cover showed no improvement (Figs 1 a, b, c).

DISCUSSION

Disturbance control

The official policy of banning the use of fire was aimed at conserving forest cover, but was less concerned with forage production. For the local herders, bush encroachment threatens forage production, which is the feed for the grazing livestock. We were therefore interested to know if the post-disturbance effects of bush encroachment control could be reflected by the performances of individual herbaceous species. We showed that species richness under the disturbance control method was comparable to the disturbance treatments (see below). The dynamics of herbaceous species under bush encroachment are likely to have been influenced by environmental conditions associated with bush cover. Bush cover creates microhabitats that favour some species that were less common in the open (disturbed) habitats. Species that were tolerant to bush cover were promoted, while those that were intolerant were reduced in the absence of disturbance (Osem et al., 2002; Clarke et al., 2005). The dynamics of herbaceous species under bush cover probably reflect the role played by the soil seed bank (Solomon et al., 2006), which might account for the new species. We also considered potential competition between herbaceous species and bushy plants. We confirmed herder concerns about the negative role of bush cover on herbaceous biomass production. Our data also showed that herbaceous plant basal cover and herbaceous species diversity are not promoted under bush cover (Jacobs and Schloeder, 2003). The evidence might suggest that protection from human induced disturbances, as promoted by state conservation policies, has failed to promote production of herbaceous biomass, which is contrary to herder objectives.

Table 3. Trends of individual species and average relative change in terms of abundance in response to the different demonstration methods (mean of 12 plots of $1m^2$) in Borana, southern Ethiopia. Average relative change = average no. of individuals recorded during post-demonstration – average no. of individuals observed during pre-demonstration.

	Demonstration plots					
Species list	Control[a]	C+F[b]	C+F+G[c]	F+G[d]	GBC[e]	C[f]
Aristida adoensis Hochst.	3*	3*	–	–	1*	↓5
Bothriochloa radicans (Lehm.) A.Camus	43*		8*		17*	22*
Cenchrus ciliaris L.	↑7	↑4	↓1	↓28	↑10	↑9
Chrysopogon aucheri (Boiss.) Stapf.	nc	↑1	↓3	↓5	↓33	↓10
Chloris roxburghiana Schult.	↑2	↓2	↓1	↑1		↑2
Cynodon dactylon (L.) Pers.			12*	1*	5*	4*
Cyperus species Vahl.				2*		4*
Digitaria milanjiana (Rendle) Stapf.	1*					
Eleusine jaegeri Pilg.						4*
Enteropogon somalensis Chiov.	12*	5*	6*	9*	12*	6*
Eragrostis papposa (Roem. & Schult.) Stued.	↓46	↓44	↓62	↓12	↓10	↓12
Eragrostis sennii Chiov.			1*	1*		
Harpachne schimperi Hochst. ex. A. Rich.	1*			1*	3*	3*
Heteropogon contortus (L.) Beauv. ex. R. & Sch.	5*		15*	↓42	7*	8*
Hyparrhenia hirta (L.) Stapf.					↑10	
Leptothrium senegalense (Kunth) Clayton	↑1	↓11	↑10	–	1*	–
Lintonia nutans Stapf.		↓11	↓36		nc	–
Panicum maximum Jacq.	↓19			↑12	1*	10*
Panicum coloratum L.	↑3	↓3	↓2	↓2	↑11	5*
Panicum turgidum Forssk.	9*	17*	4*	2*		2*
Pennisetum mezianum (Vahl) Lanza & Mattei				2*	↓1	1*
Pennisetum stramineum Peter		18*		9*	↓3	
Sporobolus pyramidalis Beauv.	↑9	↓38	19*	↓61	↓22	↓6
Themeda triandra Forssk.	nc	↓26	↓13	↓1	↓2	↓5
Herbaceous legumes (unidentified)	7*	6*	13*	17*	8*	4 *
Other herbaceous species (unidentified)	6*	18*	12*	18*	↓7	↑4

↑ = Increase, ↓ = Decline, – = lost, nc = No change, * = indicates species regeneration

aControl = exclusion of cutting, grazing and fire, bC+F = combination of cutting and fire, cC+F+G = combination of cutting, fire and grazing, dF+G = combination of fire and grazing, eGBC = Grazing under increased bush cover, fC = Cutting alone

Tree cutting and fire

The control of bush encroachment by means of tree cutting and fire increased the richness and relative abundance of herbaceous species during post-disturbance periods. The new herbaceous species that were recorded in response to the tree cutting and fire demonstrations are probably those that decreased under bush encroachment. These species responded more positively to disturbance, which might suggest that they would be poor competitors under conditions of bush encroachment (Duggin and Gentle, 1998; Berlow et al., 2003). Under these disturbance conditions, the bush cover tolerant species became poor competitors. The initial decline in the relative abundance of most herbaceous species to tree cutting and fire was probably related to shifts in the micro-environment following the removal of ecologically important trees. The bush cover tolerant species were reduced, while the intolerant species were promoted. Furthermore, the species that declined were probably more sensitive to fire than the other (fire tolerant) species. We confirmed that combinations of tree cutting and fire promoted herbaceous biomass more than grazing and fire or cutting alone (see Fig. 1a). Woody plant encroachment control followed by resting would improve the restoration of herbaceous biomass and plant biodiversity over time. In terms of management, the C+F demonstration methods created a dilemma between conserving disturbance-sensitive herbaceous species and promoting the disturbance-tolerant species. For the herders, the demonstration showed more beneficial gains in terms of biomass in the long-term, although in the short-term the benefits were less. This demonstration suggests that the management strategy needs to be planned so that rest follows treatments, at least for a year or two, before grazing by livestock is resumed. This kind of management strategy has been promoted by the community during the pre-fire ban decades (i.e., before the 1970s) and is more appropriate for range enclosures near settlements than the larger communal grazing lands (Angassa and Oba, 2008).

Table 4. Chi-square (χ^2-tests) for herbaceous species in terms of performance of individuals in response to the effects of different demonstration plots in Borana, southern Ethiopia.

	Demonstration plots					
Species list	Control[a]	C+F[b]	C+F+G[c]	F+G[d]	GBC[e]	C[f]
Aristida adoensis Hochst.	0.81 NS	1.07 NS	0.71 NS	0.77 NS	NS	5.40*
Bothriochloa radicans (Lehm.) A.Camus	8.48**		0.13 NS		21.73***	6.60*
Cenchrus ciliaris L.	7.07*	0.03 NS	0.02 NS	15.53***	13.71***	2.06 NS
Chrysopogon aucheri (Boiss.) Stapf.	0.02 NS	0.49 NS	0.21 NS	2.74 NS	13.38***	8.53**
Chloris roxburghiana Schult.	2.60 NS	0.69 NS	6.74*	NS		10.36**
Cynodon dactylon (L.) Pers.			0.32 NS	NS	NS	0.52 NS
Cyperus species Vahl.				NS		0.52 NS
Digitaria milanjiana (Rendle) Stapf.	0.06 NS					
Eleusine jaegeri Pilg.						2.67 NS
Enteropogon somalensis Chiov.	1.62 NS	0.87 NS	0.20 NS	5.99*	0.85 NS	1.98 NS
Eragrostis papposa (Roem. & Schult.) Stued.	22.07***	1.20 NS	8.66 **	4.66*	1.76 NS	5.79*
Eragrostis sennii Chiov.				0.84 NS		
Harpachne schimperi Hochst. ex. A. Rich.	0.48 NS			0.09 NS	7.25*	2.88 NS
Heteropogon contortus (L.) Beauv. ex. R. & Sch.	NS		0.02 NS	1.75 NS	12.04**	12.60***
Hyparrhenia hirta (L.) Stapf.					13.48***	
Leptothrium senegalense (Kunth) Clayton	2.34 NS	2.37 NS	2.93 NS	NS	2.80 NS	3.63 NS
Lintonia nutans Stapf.		2.06 NS	NS		2.24 NS	12.95***
Panicum maximum Jacq.	18.18***			27.22***	2.48 NS	13.98 ***
Panicum coloratum L.	3.48 NS	1.21 NS	3.86*	11.25**	4.93*	0.28 NS
Panicum turgidum Forssk.	3.67 NS	0.76 NS	1.83 NS	NS		
Pennisetum mezianum (Vahl) Lanza & Mattei					2.84 NS	
Pennisetum stramineum Peter		0.78 NS		14.33**	6.11*	
Sporobolus pyramidalis Beauv.	4.84*	5.64*	1.37 NS	12.78 ***	11.65**	1.22 NS
Themeda triandra Forssk.	NS	0.17 NS	1.10 NS	23.84***	43.18***	6.67*
Herbaceous legumes (unidentified)	0.63 NS	0.08 NS	0.37 NS	3.74 NS	6.99*	6.68*
Other herbaceous species (unidentified)	0.43 NS	1.74 NS	0.01 NS	NS	5.61*	4.55*

* = $P < 0.05$, ** = $P < 0.01$, *** = $P < 0.001$, NS = not significant

aControl = exclusion of cutting, grazing and fire, bC+F = combination of cutting and fire, cC+F+G = combination of cutting, fire and grazing,

dF+G = combination of fire and grazing, eGBC = Grazing under increased bush cover, fC = Cutting alone

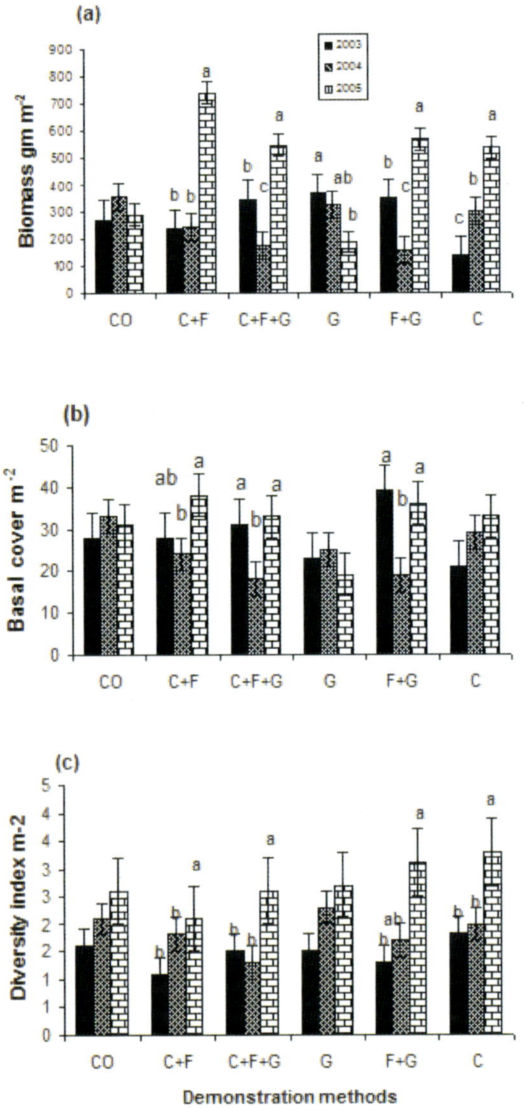

Figure 1. Responses of: (a) herbaceous biomass; (b) basal cover; and (c) species diversity to six bush encroachment control demonstration methods in Borana, southern Ethiopia.

Tree cutting, fire and grazing

The combination of tree cutting, fire and grazing (C+F+G) differed from C+F by introducing grazing into the demonstration, and evaluating plant performances thereafter. This demonstration method represented active management, and by not excluding livestock grazing, was more likely to be acceptable to the herders. Our data showed that herbaceous species richness was highly promoted. The combined treatments might have improved growth conditions of a few species, which increased (11%), while the greater majority of the species that declined in abundance (89%) were probably less tolerant to the treatment combinations. The post-demonstration responses by individual species reflected both the seasonal variability of rainfall (data not shown) and improved environmental conditions for seed germination. For the most sensitive species, tree cutting, fire and grazing disturbances were initially followed by decline in relative abundance (Kobayashi et al., 1997). This decline could be explained either as a result of competition for resources among species (Clarke et al., 2005), seasonal fluctuations in relation to the life history of individual species (Elliott et al., 1997; Deil, 2005; Gőtmark et al., 2005) or variations in the soil seed bank (Solomon et al, 2006). Other species declined due to lack of tolerance to disturbance (Osem et al., 2002). Our results suggest that tree cutting, fire and grazing treatments were soon followed by decline in herbaceous biomass, basal cover and diversity, but in the long-term the effects were positive on herbaceous biomass and diversity.

The decline in herbaceous biomass soon after disturbance may be attributed to the effect of tree removal that reduced bush cover tolerant species, which were dominant prior to the demonstration. It is possible that the treatments reduced the abundance of the bush cover tolerant species and therefore the biomass. After the treatments, the species that tolerated disturbance took a while to build and produce more biomass. For purposes of management, therefore, it appears that grazing might need to be delayed after the hand removal and burning of the woody vegetation. The introduction of grazing after resting might allow a rapid build-up of the biomass. Currently, in southern Ethiopia, this type of management has been initiated through the activities of Non-Governmental Organizations (NGOs) that utilize community labour to clear land in limited areas. However, regulations for grazing systems (e.g. rest followed by grazing) have not been put in place, other than those used by the Borana herders, which involves resting of the treated areas during the wet season and grazing the fallow during the dry season by young and old livestock.

Community adaptation of the system would be constrained by a shortage of labor for land clearing. Thus, the use of these demonstration methods would be more viable in areas near settlements, where the community often grazes calves and old animals or saves fodder reserves for the dry and drought periods. The usual practice is for women to travel long distances to collect hay for calves kept at home. The C+F+G demonstration method could easily be incorporated into the traditional system of pasture management, if it is aimed at community management of fodder for the most vulnerable animal classes (e.g. calves and old animals). However, this method is unlikely to be suited to the wider rangelands that are currently suffering from bush encroachment, due to the cost of land rehabilitation. We therefore emphasize the importance of these demonstration methods for pasture management at settlement levels, which could easily be integrated into the existing systems of land use.

Fire and grazing

The fire and grazing combination represents the old traditional system of rangeland management used by the Borana pastoralists. Traditionally, the society burnt areas of the range periodically where bush cover posed a threat or the grass layer was over-mature and losing its nutritive value. The fire killed young woody plants and opened up the under-story vegetation for better growth of herbaceous vegetation. The system of burning was not random, but planned. It is therefore important to understand the responses of herbaceous species to fire and grazing under conditions comparable to the traditional system of fire use. The method of fire management by the community had to be abandoned because of the official policy of forestland conservation. The response to the F+G treatment showed that this system of bush encroachment management promotes herbaceous species richness (Sawadogo et al., 2005). First, reduced competition from bushy species and increased availability of soil nutrients might have fertilizing effects. Second, fire and grazing could play complementary roles (Fuhlendorf and Engle, 2004). The observed increase in the relative abundance of some species in response to fire and grazing could be interpreted in terms of the removal of old growth and the rejuvenation of disturbance tolerant species. Conversely, the decline of a few species after fire and grazing was probably related to sensitivity to fire damage. The present findings suggest that fire and grazing significantly influenced the relative abundance of most species, probably as a result of increased grazing pressure following fire (Fuhlendorf and Engle, 2004). The findings suggest that fire and grazing could play an important role in maintaining herbaceous species richness and diversity. It was also found that the short-term effects of fire and grazing seem to exert considerable influence on herbaceous biomass and basal cover, which were promoted in the long-term.

The official banning of fire had previously assumed that the use of fire would be damaging to the rangelands. This assumption is not supported when the response of the herbaceous vegetation to fire and grazing is considered in the long-term. Our results suggest that the traditional system is more beneficial from the perspective of biomass production and species diversity, if grazing follows rest after fire. Traditionally, the Borana utilized the burned areas (called *guuba*) after the growth following the early rains and then shifted grazing to older burned areas. Through the rotational use of landscapes with different fire histories (Sheuyange et al., 2005), they maintained the savanna system in productive state.

This demonstration method is widely acceptable to the local communities. It is a system of grazing that can be applied to the wider rangelands without requiring additional labour or other expenses. The difference with our demonstration was that the traditional application of fire was not restricted. Rather, the fire was able to spread beyond the demonstration plots, into those areas where the herders did not intend to burn. Local communities' practices were to protect settlement rangelands from fire in order to conserve fodder for drought periods. In the long-term, the demonstration achieved both conservation of biodiversity and increased biomass production thereby meeting the objectives of the official conservation policy and herder livestock grazing management.

Grazing with bush cover

The current system of range management, in which the use of fire has been banned, promotes bush cover. For the official policymakers, this is the least expensive method of land use, although the exclusion of fire has made the rangelands less productive for the pastoralists in terms of livestock production. The grazing with bush cover system of management showed that herbaceous species richness could be increased. The probable explanation for the addition of new species could be linked to the performance of bush cover tolerant species that were promoted under this demonstration method. An earlier study (Osem et al., 2002) suggested that grazing might favour the recovery of some herbaceous species by suppressing competing species. For example, the observed increase in the relative abundance of a few species under grazing may be attributed to the capacity of individual species to withstand the effect of grazing.

It should be appreciated that bush cover physically constrains livestock grazing, thereby promoting grazing sensitive species that are protected under impenetrable thickets, while in the open areas where the livestock has easier access to grazing, grazing tolerant species were promoted. This means the demonstration might promote two varied populations of grass species; the grazing tolerant and the bush cover tolerant ones. Similarly, previous report (Clarke et al., 2005) indicated a decline in the abundance of herbaceous species under grazing due to selective grazing that promoted disturbance tolerant species over grazing intolerant species. From our results, we inferred that grazing might have negative effects on the relative abundance of some herbaceous species, either in the short- or long-term, depending on the life history of individual species (i.e. annuals and perennials). It was also suggested that grazing can be sustainable on most rangelands, but the long-term effects may lead to the loss of herbaceous species (Fuhlendorf and Engle, 2004). The effect of disturbance such as grazing with bush cover might favour the recovery of some species in the short-term, but in the long-term it may not promote successful establishment. Subsequent failure is probably related to rainfall variability and competition for resources under grazing (Duggin and Gentle, 1998). The significance of the decline in herbaceous biomass may be attributed to the combined effects of grazing pressure and bush cover. Grazing with bush cover reduced the basal cover of herbaceous species, although herbaceous species diversity improved over time.

Tree cutting

The hand clearing of encroached bush should be acceptable to the extension departments, because it treats bush encroachment without introducing fire into the environment. Under the tree cutting method, our results showed an increase in the richness of herbaceous species and the relative abundance of a few of the species among the initial population that were intolerant of bush cover (Dougill and Trodd, 1999). Conversely, a decline in the relative abundance of some herbaceous species in response to tree cutting might reflect shifts in the micro-environment due to the removal of ecologically important trees, thus exposing sensitive species to increased light intensity (particularly for shade loving species). Our results suggest that reduced bush cover could restore herbaceous plant productivity and biodiversity (Bates et al., 2000; Smit, 2004; Ansley and Castellano, 2006). As already indicated, the demonstration method of tree clearing could not be applied to the wider rangelands where bush encroachment is a problem, because of the high costs involved. The local community is more likely to apply this system to local pasture management, such as range enclosures (Angassa and Oba, 2008), as opposed to conducting large-scale land clearing.

CONCLUSIONS

We studied the responses of herbaceous species richness, the relative abundance of species, biomass, basal cover and species diversity in response to different bush encroachment control demonstrations. The results show that protection is comparable to the disturbance demonstrations in terms of herbaceous species richness, while herbaceous biomass accumulation and species diversity were improved by disturbance. Bush encroachment control methods such as tree cutting and fire, tree cutting followed by fire and grazing, and the old traditional method of fire and grazing improved herbaceous biomass, basal cover and diversity, while grazing with bush cover greatly reduced herbaceous biomass. As indicated above, the traditional system of range management is more efficient for improving herbaceous biomass, basal cover and species diversity compared to the current systems of range management imposed on the herders. Thus, future development can use these results to compare benefits in terms of opportunities for the sustainable use of savanna ecosystems and conservation of species diversity in southern Ethiopia. Tree cutting also greatly increased herbaceous biomass and species diversity. Overall, the demonstration methods promoted species richness. The outcomes in terms of herbaceous biomass and the restoration of plant biodiversity in response to the various bush encroachment control treatments have important implications for policy formulation and public education in terms of the future management of savanna ecosystems in southern Ethiopia. By comparing the advantages of various bush encroachment control measures, we conclude that only the fire and grazing method is easy to adopt and reasonable in terms of time and labour required, in order facilitate the control of bush encroachment. The use of fire and grazing seems to be more efficient in promoting forage production, as well as the conservation of biodiversity of herbaceous plants. A bush encroachment control method that promotes forage production and the maintenance of ecosystem diversity, while at the same time is easy to apply is more efficient and has considerable comparative advantages relative to other methods. Defining the comparative advantage of different bush encroachment control methods in terms of forage for livestock and ease of implementation has immediate policy implications. In this regard, priority should be given to the interests of the community rather than focusing on conservation alone. The results of this study clearly show that there is ample scope for policymakers and resource users to improve the management of the savanna ecosystems of southern Ethiopia by focusing on more efficient bush encroachment interventions.

ACKNOWLEDGEMENTS

The fieldwork was supported through grants from the Norwegian State Educational Loan and NUFU-PhD scholarship to Ayana Angassa. The writing was supported by Food Security Centre of Hohenheim University, as part of DAAD (German Academic Exchange Service) Program exceed and the German Federal Ministry for Economic Cooperation and Development (BMZ). Revision of the manuscript was supported by Alexander von Humboldt Foundation Georg Forster Fellowship to A. Angassa. The authors thank Katelo Guyo and the staff of the Southern Rangelands Development Unit (SORDU) for their assistance during fieldwork.

REFERENCES

Angassa, A. 2002. The effect of clearing bushes and shrubs on range condition in Borana, Ethiopia. Tropical Grasslands. 36: 69-76.

Angassa, A. and Oba, G. 2007. Relating long-term rainfall variability to cattle population dynamics in communal rangelands and a government ranch in southern Ethiopia. Agricultural Systems. 94: 715-725.

Angassa, A. and Oba, G. 2008. Effects of management and time on mechanisms of bush encroachment in southern Ethiopia. African Journal of Ecology. 46: 186–196.

Angassa, A. and Oba, G. 2009. Bush encroachment control demonstrations in southern Ethiopia: 1. Woody species survival strategies with implications for herder land management. African Journal of Ecology. 47: 63–76.

Ansley, R. J. and Castellano, M. J. 2006. Strategies for savanna restoration in the southern Great Plains: Effects of fire and herbicides. Restoration Ecology. 14: 420-428.

Bates, J. D., Miller, R. F. and Svejcar, T. J. 2000. Understory dynamics in cut and uncut western juniper woodlands. Journal of Range Management. 53: 119-126.

Bates, J. D. 2005. Herbaceous response to cattle grazing following juniper cutting in Oregon. Range Ecology and Management. 58: 225-233.

Berlow, E. L., D'Antonio, C. M. and Swartz, H. 2003. Response of herbs to shrub removal across natural and experimental variation in soil moisture. Ecological Application. 13: 1375-1387.

Clarke, P. J., Latz, P. K. and Albrecht, D. E. 2005. Long-term changes in semi-arid vegetation: Invasion of an exotic perennial grass has larger effects than rainfall variability. Journal of Vegetation Science. 16: 237-248.

Coppock, D. L. 1994. The Borana plateau of southern Ethiopia: synthesis of pastoral research development and changes, 1980-90. *ILCA (International Livestock Centre for Africa)*: Addis Ababa, Ethiopia.

Deil, U. 2005. A review on habitats, plant traits and vegetation of ephemeral wetlands - a global perspective. Phytocoenologia. 35: 533-705.

Dougill, A. and Trodd, N. 1999. Monitoring and modelling open savannas using multisource information: analyses of Kalahari studies. Global Ecology and Biogeography. 8: 211-221.

Duggin, J. A. and Gentle, C. B. 1998. Experimmental evidence on the importance of disturbance intensity for invasion of *Lantana camara L.* in dry rainforest-open forest ecotones in north-eastern NSW, Australia. Forest Ecology and Management. 109: 279-292.

Elliott, J., Boring, L. R., Swank, W.T. and Haines, B. R. 1997. Successional changes in plant species diversity and composition after clearcutting a southern application watershed. Forest Ecology and Management. 92: 67-85.

FAO/UNESCO/ISRIC 1986. Soil map of the world. Revised legend. Technical Paper 20, ISRIC, Wageningen, the Netherlands.

Fuhlendorf, S. D. and Engle, D. M. 2004. Application of the fire-grazing interaction to restore a shifting mosaic on tallgrass prairie. Journal of Applied Ecology. 41: 604-614.

Fulbright, T. E. 1996. Viewpoint: A theoretical basis for planning woody plant control to maintain species diversity. Journal of Range Management. 49: 554-559.

Grime, J. P. 1974. Vegetation classification by reference to strategies. Nature. 250: 26-31.

Grime, J. P. 1977. Evidence for existence of three primary strategies in plants and its relevance to ecological and evolutionary theory. American Naturalist. 111: 1169-1194.

Gőtmark, F., Paltto, H., Norden, B. and Gőtmark, E. 2005. Evaluating partial cutting in broadleaved temperate forest under strong experimental control: short-term effects on herbaceous plants. Forest Ecology and Management. 214: 124-141.

Hadar, L., Noy-Meir, I. and Perevolotsky, A. 1999. The effect of shrub clearing and grazing on the composition of a Mediterranean plant community: functional groups versus species. Journal of Vegetation Science. 10: 673-682.

Hagos, M.G. and Smit, G.N. 2005. Soil enrichment by *Acacia mellifera subsp detinens* on nutrient poor sandy soil in a semi-arid southern African savanna. Journal of Arid Environments. 61: 47-59.

Hudak, A.T. 1999. Rangeland mismanagement in South Africa: Failure to apply ecological knowledge. Human Ecology. 27: 55-78.

Hutchinson, T. F., Boerner, R.E.J., Sutherland, S., Sutherland, E.K., Ortt L.R. and Inverson, M. 2005. Prescribed fire effects on the herbaceous layer of mixed-oak forests. Canadian Journal of Forest Research-Revue Canadienne De Recherche Forestiere. 35: 877-890.

Jacobs, M. J. and Schloeder, C. A. 2003. Defoliation effects on basal cover and productivity in perennial grasslands of Ethiopia. Plant Ecology. 169: 245-257.

Kobayashi, T., Hori, Y. and Nomoto, N. 1997. Effects of trampling and vegetation removal on species diversity and micro-environment under different shade conditions. Journal of Vegetation Science. 8: 873-880.

Łaska, G. 2001. The disturbance and vegetation dynamics: a review and an alternative framework. Plant Ecology. 157: 77-99.

Oba, G., Post, E.,Syvertsen, P. O. and Stenseth, N. C. 2000. Bush cover and range condition assessments in relation to landscape and grazing in southern Ethiopia. Landscape Ecology. 15: 535-546.

Olson, R. A. and Whitson, T. D. 2002. Restoring structure in late-successional sagebrush communities by thinning with tebuthiuron. Restoration Ecology. 10: 146-155.

Osem, Y., Perevolotsky, A. and Kigel, J. 2002. Grazing effect on diversity of annual plant communities in a semi-arid rangeland: interactions with small-scale spatial and temporal variation in primary productivity. Journal of Ecology. 90: 936-946.

Pickett, S.T.A. and White, P.S. 1985. The Ecology of natural disturbance and patch dynamics. Acadamic Press, New York.

Sawadogo, L., Nygard, R. and Pallo, F. 2002. Effects of livestock and prescribed fire on coppice growth after selective cutting of Sudanian savannah in Burkina Faso. Annual Forage Science. 59: 185-195.

Sawadogo, L., Tiveau, D. and Nygard, R. 2005. Influence of selective tree cutting, livestock and prescribed fire on herbaceous biomass in the savannah woodlands of Burkina Faso, West Africa. Agriculture, Ecosystems and Environment. 105: 335-345.

Scholes, R. J. and Archer, S. R. 1997. Tree-grass interactions in savannas. Annual Review of Ecological Systems. 28: 517 - 544.

Shannon, C. E., A. 1948. Mathematical theory of communication. Bell System Technical Journal. 27: 379 - 423.

Sheuyange, A., Oba, G. and Weladji, B.R. 2005. Effects of anthropogenic fire history on savanna vegetation in northeastern Namibia. Journal of Environmental Management. 75: 189-198.

Smit, G.N. and Rethman, N. F. G. 2000. The influence of tree thinning on the soil water in a semi-arid savanna of southern Africa. Journal of Arid Environments. 44: 41-59.

Smit, G.N. 2003. The importance of *Salvadora australis* in relation to tree thinning in preserving herbaceous plants in a semi-arid Colophospermum mopane savanna. Journal of Arid Environments. 55: 483-501.

Smit, G.N. 2004. An approach to tree thinning to structure southern African savannas for long-term restoration from bush encroachment. Journal of Environmental Managements. 71: 179-191.

Solomon, T.B., Snyman, H.A. and Smit, G. N. 2006. Soil seed bank characteristics in relation to land use systems and distance from water in a semi-arid rangeland of southern Ethiopia. *S.* African journal of Botany. 72: 263-271.

Sweet, R. J. and Tacheba, G. 1984. Bush control with fire in semi-arid savanna in Botswana: In Tothill, J. C., Mott, J. J., (Eds.). International Savanna Symposium, Australia, Canberra, ACT 2601, Commonwealth Agricultural Bureau, 229-237.

Vetaas, O. R. 1992. Micro-site effects of trees and shrubs in dry savannas. Journal of Vegetation Science. 3: 337-344.

Wiegand, K., Ward, D. and Saltz, D. 2005. Multi-scale patterns and bush encroachment in an and savanna with a shallow soil layer. Journal of Vegetation Science. 16: 311-320.

11

Adoption of Agroforestry Systems by Smallholders in Brazilian Amazon

Gisele do Socorro Santos Pompeu[1]*, Leonilde Santos Rosa[2], Mônica Mota Santos[3], Rozi Silva Modesto[4], Thiago Almeida Vieira[5]

[1]*Universidade Federal do Pará , Brazil;*
[2] *Universidade Federal Rural da Amazônia, Brazil;*
[3]*Secretaria Municipal da Gestão do Meio Ambiente e Turismo de Altamira, Brazil;*
[4]*Prefeitura Municipal de Ananindeua, Brazil;*
[5]*Universidade Federal do Oeste do Pará, Brazil*
Email: giflorestal@yahoo.com.br
**Corresponding Author*

SUMMARY

The objective of this study was to identify the factors that have affected the adoption of commercial agroforestry systems (AFS) by smallholders of the municipality of Bragança, state of Pará, Brazil. Tools of Rapid Rural Diagnosis (structured interview and direct observation) and of Participatory Rural Diagnosis (agricultural calendar and ranking) were used.. The majority of smallholders that adopted these systems in Bragança received funding from government banks to purchase seedlings and other inputs; in order of importance, access to financing, utilization and management of agroforestry systems, education of farmers and decision-making are determinant factors for adoption of agroforestry systems implanted in smallholders' areas of the municipality of Bragança, state of Pará.

Key words: Smallholder agriculture; Eastern Amazon; Agroforestry Systems.

INTRODUCTION

Slash and burn agriculture is a widely used practice by smallholders in Amazon region. However, this model of agricultural production has created many doubts about its sustainability, which is hampered by factors such as the use of low level technology, and the negative environmental impacts of agricultural practices, among others.

An alternative to shifting cultivation in Brazilian Amazon are agroforestry systems (AFS). Among the various models of these systems practiced in this region, commercial multistrata systems are highlighted, which are important for income generation for smallholders (Sanguino *et al.*, 2007; Francez, 2007; Bentes-Gama *et al.*, 2005), besides providing various environmental services.

Despite the advantages listed above, the adoption of multistrata AFS for commercial purposes by the farmers in the Brazilian Amazon is low compared to shifting cultivation. This happens due to political, technical, structural, economic and sociocultural barriers (Rosa *et al.*, 2009). These authors highlight the necessity of public policies that support the adoption of AFS based on scientific knowledge and local expertise as well.

Although AFS have been extensively studied in last decades, there is a growing demand for scientific research designed to systematize the experiences of AFS and that enable the identification of constraints and opportunities related to the AFS adoption in the context of family farming in the Brazilian Amazon. In order to understand this problem, the objective of this research was to identify factors that have affected the adoption of commercial agroforestry systems established by smallholders in Bragança, Pará State, in Eastern Amazon, Brazil, aiming to supply subsidy to the programs for smallholder in Amazon region, specially the ones that contemplate the implantation of agroforestry systems.

MATERIAL AND METHODS

The research was carried out in the municipality of Bragança (01° 03' 15" S e 46° 46' 10" W), Pará State, situated in the Bragantina Microregion, Eastern Amazon, Brazil. The climate of Bragança is equatorial super humid, with maximum temperature of 33°C and minim of 18° C, presenting average of 27° C. This municipality presents high rainfall (2,501 mm/year), with rainy period in first six months of the year. The original vegetation of the up land areas was composed for the subtype dense forest that was substituted by secondary forests with different stages, after the deforestations (Governo do Estado do Pará, 2005).

The total population in the urban and rural zone, until the year of 2000, was of 93,779 inhabitants, distributed in 2,333,70km^2, however the percentage of the economically active population was of 38.4% until the year of 2000 (Fundação IBGE, 2000). The economy of the municipality is based in the agricultural and raising cattle activity, and the most important crops economically are: cassava, beans, black pepper, orange and coconut. (Governo do Estado do Pará, 2005).

The research was achieved with 53 smallholders that developed 62 experiences with multistrata agroforestry system on their property. This municipality was selected because it has a significant number of experiences of these kinds of AFS, established by smallholders, compared to others places.

The first field visits occurred in 2004. The following institutions were identified and contacted as key ones: Pará State Technical Support and Rural Extension Company (EMATER-PARÁ), Pará State Secretariat of Agriculture (SAGRI-PA), besides representatives of social movements, such as Rural Workers' Union and also presidents of local associations and cooperatives engaged in agricultural activities in Bragança. The collective participation of these social actors allowed the identification of communities with at least one experience of AFS in this municipality.

The tools used in order to collect the data in rural area were: one structured interview, direct observation, ranking and agricultural calendar with smallholders (Chambers, 1987). The use of different tools and the participation of key people of this municipality favored the localization by triangulation, increasing the accuracy of the information collected.

The collected data were analyzed by frequency distribution analysis and by factor analysis method, using the "Statistical Package for the Social Sciences" (SPSS 13.0®). Factor analysis aims to describe the dependence structure of a set of variables by creating factors that supposedly measure common aspects (Diego, 2003).The application of this statistical method aimed to identify the factors that most affected smallholders AFS adoption. The extraction of factors was performed by the principal components method using the *varimax* criterion.

Adequacy of the method was performed using the Kaiser-Meyer-Olkin (KMO) test and Bartlett's sphericity. The first one calculated the partial correlation coefficient between pairs of variables, eliminating the effect of others, and the second one calculated the overall correlation coefficient, as well as it estimated the overall significance of the correlation matrix.

Among the 33 variables extracted from the field research, only 15 were correlated with each other, such as: AFS composition, source of financial resource, smallholder education level, the reason to deploy the AFS, the purpose of the AFS, purchase of seedlings, land preparation, smallholders sex, geographical origin of the smallholders, cultural practices, age of the smallholder, harvest methods, smallholder's aspirations in relation to AFS, benefits generated and frequency of sale of products. However, only the first eight variables were selected for correlation analysis, the most correlated with each other. After this step, factor analysis was applied using the statistical model in matrix form, according to Dillon and Goldstein (1984).

RESULTS AND DISCUSSION

Financial resources

Approximately 83% of smallholders with commercial AFS in Bragança received funding obtained from financial agents for the purchase of seedlings and other inputs to establish AFS (Figure 1). This figure reveals that more than 11% of smallholders used own resources to obtain these inputs and almost 6% of them established AFS with resources originated of both modalities (financed and own resources).

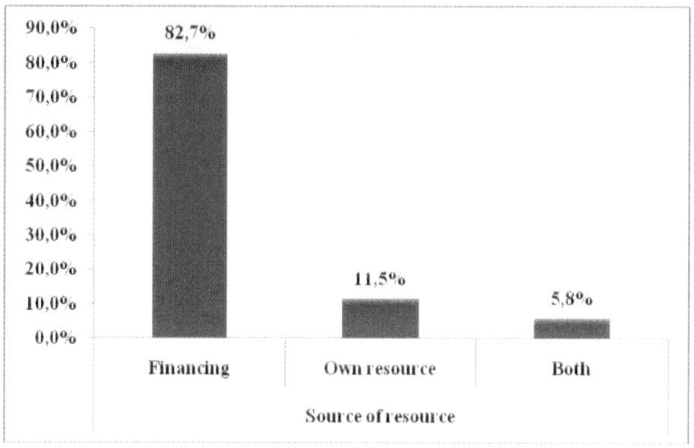

Figure 1. Source of resource used by smallholders to established AFS in Bragança, Pará, Brazil. (n=53).

These results show that smallholders used three modalities of financial resources. In spite of the great importance of the fundings obtained from financial agents for adoption of AFS, it must be recognized that the financial resources coming from smallholders played an important role for the establishment of AFS and diversification of monoculture areas, transforming them in AFS.

Factorial analysis

Analysis of correlation of variables

The results revealed that there was a dependence structure among the eight variables evaluated in this study. (Table 1). The AFS composition showed positive correlation with the source of funding and the purchase of seedlings. However, the most significant positive correlation occurred between the variable source of funding and purchase of seedlings. It was also detected correlation between variable source of funding and purchase of seedlings, once local smallholders have received funding of Constitutional Fund of Financing of the North (FNO-special) to buy the coconut seedlings (*Cocos nucifera* L.) orange seedlings (*Citrus sinensis* (L) Osb.) and black pepper seedlings (*Piper nigrum* Vell.).

According to farmers, seedlings of the black pepper and orange seedlings were purchased on Bragantina microregion. Coconut seedlings were bought in northeastern of Brazil, because the plant nurseries in the microregion Bragantina did not have enough coconut seedlings, required for the project.

It is important to point out that according to the smallholders, they did not took part on the selection of species in the projects supported by the Special FNO, once it was previously determined by financial agents, but they had interest in these species.

The correlation between the purpose of the AFS and the purchase of seedlings was negative and weak. According to almost 86% of local smallholders, the primary purpose of the projects was to sale the products (Figure 2). However, except of black pepper, the remaining fruitful species above mentioned did not produce fruits of the good quality, due to poor quality of the seedlings provided by the financing programs, and, consequently they did not present the expected development, resulting in low or no productivity of the fruits.

The bad quality of seedlings contributed to their high mortality, leading them to introduce commercial species of their own interests which are better adapted to local conditions, leading to diversification of the AFS floristic composition. In this case, the seedlings were produced by the smallholder.

Table 1. Correlation matrix of the variables studied in the municipality of Bragança.

	A	B	C	D	E	F	G	H
A	1.000							
B	0.357	1.000						
C	0.226	-0.095	1.000					
D	0.090	-0.208	0.151	1.000				
E	0.063	-0.180	0.026	0.232	1.000			
F	0.417	0.438	0.282	-0.088	-0.322	1.000		
G	-0.107	-0.173	-0.136	-0.233	0.273	-0.204	1.000	
H	0.081	0.044	0.009	0.065	0.038	0.043	-0.074	1.000

Where: A – AFS Composition; B - Source of financial resources; C - smallholder education level; D - Reason to deploy the AFS, E - Purpose of the AFS; F - Purchase of seedlings; G – Land preparation; H – Sex smallholders.

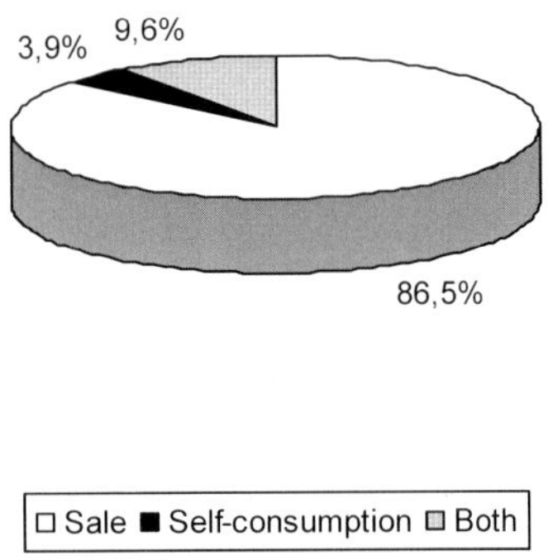

Figure 2. Agroforestry systems purpose in family agriculture areas in Bragança, Pará, Brazil. (n=53).

Almost 4% of farmers have implemented systems only for family consumption, aimed at food security (Figure 2). This happened mainly when the AFS were implemented with own funds. The remaining percentage (almost 10%) aimed to sale and self-consumption.

Extraction of factors

According to Diego (2003), the results obtained by the tests of Kaiser-Meyer-Olkin (KMO) and by Bartlett's sphericity test (Table 2), demonstrated the viability of the sample as well as the adequacy of the use of the Factorial Analysis.

The eigenvalues and the principal components obtained from the spectral decomposition of the correlation matrix are presented in Table 3. The first four components (all with eigenvalues greater than one), together explain approximately 72% of the total variance of the data. The other components explain only around 28% of the total variance.

The matrix of factor loadings, after orthogonal rotation is presented in Table 4. It is observed the creation of four significant factors (Factor 1: Access to finance; Factor 2: AFS use and Management; Factor 3: Education level of smallholders and Factor 4: Decision-making), representing almost 72% of the variance of original data set. A synthesis of the analyisis of factors is presented in Table 4.

Table 2. Results of KMO and Bartlett tests

Measurement of viability of the sample: KMO (*Kaiser-Meyer-Olkin*)		**0.53**
Bartlett's sphericity Test	Chi-square	57.882
	Degree of freedom	28
	p-value	0.001

A summary of the analysis of the factors is presented in Table 5. The AFS composition, the source of financial resources and purchase of seedlings and others materials are closely related and have a direct relation to access to finance (factor 1).Thus, this factor has a significant weight to the adoption decision of the AFS by the smallholders in Bragança, once this factor, itself, explains around 23% of the total variance.

These results correspond to the data presented earlier in Figure 2, which shows that the majority of smallholders in this study received funding from financial agents for the purchase of seedlings and others inputs for the AFS adoption. In spite of that, many smallholders related difficulties in the reception of the financial resource, what, most of the times was received out of the period of planting, what delayed the development of the species in the AFS.

Table 3. Results of the eigenvalues for the extraction of factors, components and total variance explained by the values of the variables under study.

	Eigenvalue e initial variances			**Variance after rotation**		
Components	**Eigenvalue**	**% Variance**	**Cumulative variance**	**Eigenvalue**	**% Variance**	**Cumulative variance**
1	2.106	26.319	26.319	1.876	23.453	23.453
2	1.419	17.739	44.058	1.411	17.641	41.094
3	1.185	14.812	58.870	1.403	17.542	58.637
4	1.029	12.862	71.732	1.048	13.095	**71.732**
5	0.854	10.674	82.406	-	-	-
6	0.576	7.201	89.607	-	-	-
7	0.476	5.952	95.559	-	-	-
8	0.335	4.441	100.000	-	-	-

Table 4. Matrix of factor loadings of the variables after orthogonal rotation by the *Varimax* method

	Variables	Factor 1	Factor 2	Factor 3	Factor 4
1	AFS composition	0.826	-	-	-
2	Source of financial resources	0.610	-	-	-
3	Smallholders' education	-	-	0.656	-
4	Reason to deploy the AFS	-	-	0.767	-
5	Purpose of AFS	-	0.820	-	-
6	Purchase of seedlings	0.753	-	-	-
7	Land preparation	-	0.712	-	-
8	Sex of smallholders	-	-	-	0.858
	% variance	23.453	17.641	17.542	13.095
	Cumulative variance	23.453	41.094	58.637	71.732

Note: The line in the empty spaces represents values less than 0.5.

Table 5. Factors determining the adoption of the AFS deployed in areas of smallholders in Bragança, Pará, Brazil.

Order of factor	Name of factor	Variables
1	Access to financing	1 AFS composition 2 Source of financial resources 6 Purchase of seedlings
2	AFS use and management	5 Purpose of AFS 7 Land preparation
3	Education level of smallholders	3 Smallholders'education 4 Reason to deploy the AFS
4	Decision-making	8 Sex of smallholders

Studies developed by Rosa *et al.* (2006) and Rosa *et al.*, (2009), in areas of smallholders in the microregion of Bragança, showed the same problems, however, these authors point out that government funding programs, such as the FNO-special, were fundamental for the AFS adoption in this microregion.

Studies in Latin America found that financing projects were also decisive in the AFS adoption (Current, 1997). The author emphasizes that credit projects that provide technical assistance combined with minimum incentives are the ones that have obtained the best results in economic terms.

Funding is a major factor in the adoption of commercial agroforestry systems in the tropics. However, the success of these systems need other factors such as social organization, marketing of products and the presence of technical assistance which together constitute a viable option for family agriculture. Bentes-Gama *et al.* (2005) and Rosa *et al.* (2009) point out that technical assistance plays a fundamental role to the AFS adoption.

Almeida (1999) concluded that the main reasons for the AFS adoption in El Salvador were land ownership, presence of technical assistance and the benefits generated by trees. The author reports that farmers who held land ownership were the most interested ones in the AFS adoption.

The factor 2 (AFS use and management) also exerts a strong influence on the AFS adoption in Bragança, for it represents almost 18% of the total variance. This can be explained by the fact that, in many cases, the AFS that were focused to the market, received more attention from the farmer and from the government in terms of mechanization and fertilization, because both aimed to increase the generation of family income.

It is important to emphasize that problems related to marketing like the lack of technical assistance, among others, contributed to the failure of many experiences of AFS in Bragança. Similar results were reported by Rosa *et al.* (2006) and Rosa *et al.* (2009) evaluating the multistrata AFS in the microregion of Bragança, by Vieira (2006) investigating the AFS in Igarapé-Açu-PA, as well as Francez (2007) studying multistrata AFS in Nova Timboteua- PA. In the case of Bragança, smallholders reported that the official enterprise of technical assistance and rural extension of Pará State (EMATER-PA) only provided assistance in the early years of AFS establishment. According to these local smallholders, this short period of the assistance was one of the causes of failure of the AFS.

The education level of smallholders (Factor 3) was responsible for approximately 18% of the total variance. This factor exerts a strong influence on the adoption of the AFS in the study area. These results indicate that the education enabled the credit acquisition for the implantation of AFS, as well as the access to the information.

The results found in Bragança correspond to the ones found by Pereira (2004) in Santo Antônio do Tauá, Pará State. This author observed that the education level of the smallholders had a positive relationship with the implantation of AFS. Similar results were reported by Rosa *et al.* (2009) evaluating the adoption of AFS in the microregion of Bragança, in the state of Pará, observed that the adoption of these systems is directly influenced by education, social organization and cultural aspects. To Franzel *et al.* (2004) the educational level may influence the ability to access information and to obtain funding for the establishment and management of agroforestry systems, thus, it can be decisive in the AFS adoption. Vieira (2006), however, in studying agroforestry systems in Igarapé-Açu, Pará State concluded that the education factor was not decisive for the AFS adoption.

Decision-making (Factor 4), which included only the variable sex of farmers, also had a strong influence on the AFS adoption in this municipality, as it represented almost 13% of the total variance of data, evidencing the force of the patriarchal model in the adoption of technologies, including land use systems as AFS, as well as the importance of organization of men. The total of the number of smallholders who have adopted AFS, 89% were men and only 11% were women.

Vieira *et al.* (2008) and Vieira *et al.* (2009) investigating the role of men and women in agroforestry activities, in the Bragantina microregion, Pará State, found that men take responsibility over commercial AFS decision making, while women are responsible for agroforestry and home garden decisions.

Franzel *et al* (2004) argue that gender and social class are factors directly related to the AFS adoption. Studies conducted in Zambia showed that approximately 23% of women farmers have adopted AFS. The authors attributed to the strength of local women's organization.

Lok (1997), evaluating sustainability in agroforestry systems in Latin America observed that the adoption of these systems is subjected to variables related to family income, the vulnerability of the farmer, as well as to several social and cultural variables that influence directly in the AFS adoption.

The participation of smallholders, including women in AFS adoption needs to be strengthened, as well as local community organizations in order to have access, control, and decision-making on the projects implemented by the governmental official institutions, empowering themselves. Brose (1999) points out that only groups or institutions that work in a participatory way and that want to accept changes in the *status quo*, are capable of introducing projects that involve the smallholders. According to the author, projects that develop this proposal employ participatory tools and establish systems for planning, monitoring and evaluation.

Therefore, the effective participation of smallholders in decision-making of the financing projects can be a way to minimize the failure of many projects in agroforestry systems. Projects outside the local reality of these farmers also contribute to the existence of the problems described above. The proposals tend to be "top down" ones, carrying away the farmers' decision power.

Thus, credit policies besides the ones that foment the implementation of agroforestry systems for family agriculture in the Brazilian Amazon, must take into account smallholders prior knowledge of certain species and/or land use systems, as well as problems related to market so that agricultural and agroforestry become socio-economically and environmentally sustainable.

Similarly, Rosa *et al.* (2009) suggest that political instruments should contemplate the following aspects in relation to AFs: adoption of the appropriate mechanisms for the improvement and the commercialization of the products originated of the AFS; strengthen of the local market; guarantee the agroforestry extension orientated in the scientific and of the common *census* knowledge (traditional knowledge); to provide facilities in the obtaining of official resources of credits and financings for the acquisition of materials and agricultural machines, as well as to stimulate agribusiness opening, and to provide training for the farmers in several knowledge areas related to the smallholders agriculture, mainly in the management of rural properties and agroforestry systems, aiming the empowerment of the same ones.

CONCLUSIONS

The adoption of the commercial multistrata agroforestry systems by smallholders of Bragança was strongly influenced by the financial resources of official agents of government funding, used to purchase seedlings and other inputs.

By order of importance, access to financing, use and management of AFS, education of smallholders and decision-making are the decisive factors in agroforestry systems adoption in areas of family agriculture in Bragança, Pará, Brazil. Men take responsibility over the adoption of commercial multistrata AFS in the Bragança, Pará State. Women have little participation on decision-making and in the control of the resources generated in the family unit, even contributing with the labor force and for family budget.

AKNOWLEDGEMENTS

Thanks the World Agroforestry Centre (ICRAF) for the financial support, in special to Jan Beniest that made possible the development of this research; to the smallholders of Bragança, target of this study, for the care and readiness in receiving the team of the research. To the president, directors of officer of the Unions of Rural Workers of Bragança (*Sindicato de Trabalhadores Rurais de Bragança*), as well as all presidents of associations and cooperatives and the local leaderships, for the information and identification of the families with AFS, as well as for the inestimable logistic support offered to the team of the researchers. To all technicians of EMATER, and technicians of the Secretariat of the Agriculture in Bragança, for information and attention released during the accomplishment of this study.

REFERENCES

Almeida, E. N de.; Galloway, G.; Current, D.; Lok. R.; Prins, C. 1999. Adopción de prácticas agroforestales en el município de San Juan Opico, El salvador. Agroforestería en las Américas, 6: 14-16.

Bentes-Gama, M.M.; Silva, M.L.; Vilcahuamán, L. J. M.; Locatelli, M. 2005. Análise econômica de sistemas agroflorestais na Amazônia Ocidental, Machadinho d'Oeste- RO. Revista Árvore, 29: 401-411.

Brose, M. 1999. Agricultura familiar, desenvolvimento e políticas públicas. EDUNISC, Brazil.

Chambers, R. 1987. Shortcut methods in social information gathering for rural development projects. International Conference on Rapid Rural Appraisal, Khon Kaen. Proceedings... Khon Kaen: Khon Kaen University Press. p.33-46.

Current, D. 1997. Los sistemas agroforestales generam benefícios para las comunidades rurales? Resultados de uma investigación em América Central y Caribe. Agroflorestería em las Américas 4: 8-16.

Diego, M. 2003. Estatística multivariada. 48ª Reunião da RBRAS e 10 SEAGRO. Lavras, Brazil.

Dillon, W. R; Goldstein, M. 1984. Multivariate analisys: methods and applications. 2ªed. v.1. John Wiley & Son, USA.

Francez, D. C. 2007. Sistemas agroflorestais no contexto socioeconômico dos agricultores familiares de Nova Timboteua, Pará. Thesis (Master in Forest Science). Universidade Federal Rural da Amazônia, Brazil.

Franzel, S.; Scherr, S.J.; Coe, R.; Cooper, P. J. M.; Place, F. 2004. Methodos for assessing agroflorestry adoption potential. In: Franzel, S.; Scherr, S. J (eds). Trees on the farms: assessing the potential of agroflorestry practices in África. CAB/ICRAF, Kenya.

Fundação IBGE. Censo Demográfico do Para. Rio de Janeiro: IBGE, 2000.

Governo do Estado do Pará. Estatística Municipal. Disponível em <http//www.pa.gov.br> 30 out. 2005.

Lok, R. M. A. 1997. Base para La diseminación de los sistemas agroflorestales. Agroforestería em las Américas. 4:12-16.

Pereira, C. L. de O. 2004. Viabilidade de sistemas agroflorestais dos agricultores familiares do município de Santo Antonio do Tauá, Pará. In: Agricultura familiar: pesquisa, formação e desenvolvimento. UFPA/CCA/NEAF, 4,:113-118, Brazil.

Rosa, L. S.; Vieira, T. A. ; Santos, A P A ; Menezes, A.A.S. ; Rodriguês, A F ; Perote, J. R. S. ; Lopez, C.V.C. 2009. Limites e oportunidades para a adoção de sistemas agroflorestais pelos agricultores familiares da microrregião Bragantina, PA. In: Roberto Porro. (Org.). Alternativa agroflorestal na Amazônia em transformação. Brasília; EMBRAPA. p. 645-670.

Rosa, L. S.; Vieira, T. A.; Francês, D. C.; Vasconcelos, P. C.; Modesto, R da S.; Santos, M. M dos. 2006. Identification of species and agroforestry systems potentials for Bragantina microregion, in the Amazon region. (Report). UFRA, Brazil. 130 p.

Sanguino, A. C.; Santana, A. C.; Homma, A. K. O.; Barros, P. L. C. ; Kato, O. K.; Amim, M. M. G. H. 2007. Análise econômica de investimentos em sistemas de produção agroflorestal no estado do Pará. Revista Ciências Agrárias, Belém, n. 47, p. 23-47.

Vieira, T. A. 2006. Sistemas agroflorestais em áreas de agricultores familiares no município de Igarapé-Açu, Pará: adoção, composição florística e gênero. Thesis (Master in Forest Science). Universidade Federal Rural da Amazônia, Brazil, 102p.

Vieira, T. A.; Rosa, L. S.; Modesto, R. S.; Santos, M. M. 2008. Gênero e sistemas agroflorestais: o caso de Igarapé-Açu, Pará, Brasil. Revista Ciências Agrárias, Belém, n. 50, p. 143-154.

Vieira, T. A.; Rosa, L. S.; Modesto, R. S.; Santos, M. M.; E. R. A., Ventura, A B, Araújo, S. L. F. 2009. Relações de gênero em sistemas agroflorestais: o caso da microrregião Bragantina, Pará. In: Roberto Porro. (Org.). Alternativa agroflorestal na Amazônia em transformação. 1 ed. EMBRAPA Informação Tecnológica, p. 732-744, Brazil.

12

Effects of Crude Oil on the Growth of *Brachiaria Mutica* and *Leucaena Leucocephala* and on Soil and Plant Macronutrients

María del Carmen Rivera-Cruz[1*], Eduardo Maldonado-Chávez[2] and Antonio Trujillo-Narcía[1,2]

[1]*Colegio de Postgraduados Campus Tabasco. Periférico Carlos A. Molina s/n km. 3.5 CP. 86570. H. Cárdenas, Tabasco, México.*
[2]*Universidad Popular de la Chontalpa. Carretera Cárdenas-Huimanguillo km 2.0 CP. 86500. H. Cárdenas, Tabasco, México.*
**Corresponding Author: mariari@colpos.mx*

SUMMARY

This study assessed the effect of 150, 50000 and 78000 mg kg^{-1} of fresh crude oil (FO) and 78456 mg kg^{-1} of weathered crude oil (WO) on the growth of *Brachiaria mutica* and *Leucaena leucocephala*. The variables measured were plant height, dry biomass, nitrogen (N), phosphorus (P) and potassium (K) contents in soil and plant, density of rhizospheric bacteria and fungi, and FO and WO degradation. Statistical differences were found ($P \leq 0.05$) in all variables. *Leucaena leucocephala* was more sensitive to oil toxicity than *B. mutica*. In *L. leucocephala* FO caused the death of the plant from the second month, while WO led to a 6-fold reduction in height. The FO had no effect on biomass accumulation of *B. mutica* but it reduced it in *L. leucocephala*, whereas WO decreased biomass in both species; N, P and K accumulation decreased in *B. mutica* in soil with WO and in *L. leucocephala* in soils with FO and WO. Density of rhizosphere bacteria was inhibited by WO in both species; there were fewer fungi in soil with FO in *B. mutica* and in soil with WO in *L. leucocephala*. Degradation of FO and WO was higher in *B. mutica* rhizosphere soil, so this grass may be an alternative for decontaminating soils polluted with FO and WO.

Key words: Fresh oil; nitrogen; oil degradation; phosphorus; potassium; weathered oil.

INTRODUCTION

Brachiaria mutica (Para grass) and *Leucaena leucocephala* (White Leadtree) are two native species from tropical regions. Both are important forage-producing species for cattle (Sotomayor-Ríos and Schank, 2001; Wencomo *et al*., 2009), in addition to providing a number of environmental services (SEMARNAT, 2009). *Brachiaria mutica* (Bm) requires waterlogged soils with poor drainage to thrive, and produces up to 20 t ha^{-1} dry matter per year (Farías, 2006). *Leucaena leucocephala* (Ll) is used as a protein source; although it is cultivated in a wide variety of soils, this species prefers deep clayey or loam soils, tolerates draught, adapts to moist sites receiving 2300 mm of rainfall per year, and produces up to 20 t ha^{-1} dry matter per year (González *et al.*, 2003). In Southeast Mexico, Bm is cultivated as pasture for grazing cattle, whereas Ll is used as living fence in livestock pasture lands as well as interspersed with perennial plants in family orchards for human consumption.

Brachiaria mutica has been reported to thrive in soil polluted with weathered crude oil, whereas Ll has been experimentally exposed to crude oil (Rivera-Cruz *et al.*, 2005). Plant exposure to oil-polluted soil results in inhibition of growth and biomass production in grasses (Maldonado-Chávez *et al.*, 2010; Tang *et al.*, 2011) and legumes (Merkl *et al.*, 2005; Vázquez-Luna *et al.*, 2010). The decrease in plant biomass results indirectly from the difficulty of the plant to adapt to the obstruction of soil pores that causes poor gas exchange, and to a lower water flow, which reduces nutrient supply and availability for the plant (Oluwafemi *et al.*, 2008). Nitrogen (N), phosporus (P) and potassium (K) are essential macronutrients for plant growth and development (Bonilla, 2003), and when present in insufficient amounts the plant is subjected to stress. There are plant species which are tolerant to stress derived from the presence of oil in the soil. Some grasses and legumes adapt to these conditions, since the former have a root system with a large surface area for microorganisms, and the latter form roots that host symbiotic atmospheric nitrogen-fixing bacteria (Philippot and Germon, 2005). Oxidation and reduction processes of inorganic and organic chemicals take place in the rizosphere that result in the breakdown of all oil hydrocarbons (Banks *et al.*, 2000) and the mineralization of macronutrients such as N (Walecka-Hutchison and Walworth, 2007).

Grasses and legumes growing in soil polluted with hydrocarbons from fresh (FO) and weathered (WO) crude oil have been identified in Mexico's oil-producing regions (García-López *et al.*, 2006; Trujillo-Narcía *et al.*, 2006; Trujillo-Narcía *et al.*, 2011); these plants are capable to decontaminate these soils. However, the potential of these plants to absorb N, P and K and the accumulation of macronutrients in foliar biomass remain unknown, as well as the relationship of oil with soil and plant parameters in tropical areas. Nevertheless, it is affirmed that Bm produces higher amount of biomass, higher N, P, and K buildup in foliar biomass, a lower density of rizosphere bacteria and higher total petroleum hydrocarbons (TPH) breakdown compared with Ll, and that there is lower N, P and K availability in soil contaminated with WO.

Therefore, the aims of this study were to assess the effect of FO and WO on the growth of Bm and Ll, on the density of the rizosphere microflora, on the N, P and K availability in soil, on the N, P and K buildup in the plant, as well as to determine the degradation rate for each type of crude oil.

MATERIALS AND METHODS

Soil

The study included two experiments. Experiment I corresponded to the control group and assessed a mollic Gleysol soil containing 150 mg kg^{-1} TPH of likely biogenic origin, collected from Arroyo Hondo, municipality of Cárdenas, Tabasco, México (17° 59' 50" N and 93° 24' 59" W). Mollic Gleysol is a clay-loam, moderately acidic soil with moderate organic matter content, high P, moderate K and high cation-exchange capacity (NOM-021-RECNAT-2000; DOF, 2002).

Experiment II assessed an anthraquic Gleysol soil (Rivera-Cruz *et al.*, 2004) collected at ejido José Narciso Rovirosa, municipality of Huimanguillo, Tabasco, México (18° 04' 54" N and 94° 02' 31" W) in the southern area of La Venta gas-processing complex in Tabasco. This soil has been polluted with oil for 28 years and contains 78456 mg kg^{-1} TPH from WO (WO-TPH); this dose was used as reference to set the higher FO dose in the experiment. Anthraquic Gleysol is clayey, with a high cationic-exchange capacity, according to the procedure by NOM-021-RECNAT-2000 (DOF, 2002), strongly acidic and with a very high organic matter content, low P and moderate K levels (DOF, 2002). Samples from both soil types were collected from the surface horizon (0 to 25 cm) and were dried in the dark at room temperature. Samples were separately fragmented and sieved through a 5 mm metallic sieve (OECD, 2006).

Experiment setup and variables assessed

The two experiments were run for 180 days, each including four treatments and six replicates. Experiments I and II were conducted with Ll and Bm plants, respectively; in both cases, soil moisture was controlled at 30%. *Leucaena leucocephala* seeds and Bm sprouts were planted in conic pots filled with oil-free soil, watered daily and transplanted to definitive pots 40 days after planting. Four treatments were used: T1) mollic Gleysol [150 mg kg^{-1} TPH]; T2) T1+ 50000 mg kg^{-1} TPH from FO [FO-TPH]; T3) T1 + 78000 mg kg^{-1} FO-TPH; and T4) anthraquic Gleysol (78456 mg kg^{-1} WO-TPH). The variables assessed were the following: plant growth, accumulation of N, P and K both in soil and in the plant, microflora density and degradation rate for each oil type.

The T4 treatment (anthraquic Gleysol) was included because WO spills occur frequently in Tabasco's oil fields; this was contrasted with T3, having a similar

FO concentration, in order to investigate whether the variables assessed respond differently according to oil type (FO or WO).

The FO was obtained from the Cinco Presidentes oil field, having 29.7° API and a density of 0.867 g cm^{-3} at 40 °C. On the other hand, the WO was oil accumulated during 28 years in soil from oil spills and leaks, which has been exposed to environmental factors.

Chemical analysis in soil and plants

The TPH were extracted from soil at baseline (day 1) and when the experiment was completed (day 180), to determine the percentage of degradation. The extraction was conducted using 99.5% dichloromethane (Merck®) in a soxhlet apparatus following the EPA method 418.1 modified for soil and sediment (EPA, 1986). The TPH were determined gravimetrically according to the procedure by NMX-AA-134-SCFI-2006 (DOF, 2006). Chemical analyses in soil and plants were conducted in samples collected at the end of the study. Total N in soil was determined using the micro-Kjeldhal method (Page *et al.*, 1982); available P was measured through the Olsen method using a $NaHCO_3$ extraction solution and read in a UV-visible spectrophotometer at 882 nm (Olsen and Sommers, 1982); available K was determined in a flame photometer using neutral $C_2H_4O_2{\cdot}H_3N$ (Jackson, 1973). Plants were washed with distilled water and dried at 72 °C for 72 h, ground and sieved. Total N in the plant was extracted by digestion with HNO_3 (Page *et al.*, 1982) and quantified using the micro-Kjeldhal method (Jones *et al.*, 1992); total K in the plant was measured in an atomic absorption spectrophotometer, and total P was extracted with 3 mL H_2SO_4 and 1 mL H_2O_2 at 360 °C and quantified with the Berthelot reaction and molybdenum blue method (Alcántar and Sandoval, 1999).

Microbiological testing

The density of heterotrophic bacteria and fungi (CFU g^{-1} dry soil) was determined using the serial dilution agar-plate technique (Madigan *et al.*, 2009). Ten grams of sample were used, preparing serial decimal dilutions; a 0.1 mL-aliquot of solution was dispersed per Petri dish in culture medium selective for heterotrophic bacteria (nutrient agar) and for heterotrophic fungi (potato dextrose agar). Plates were incubated at 28 °C for 48 h for bacteria and 72 h for fungi.

Plant growth and biomass

Plant height was measured with a ruler graduated in mm at 30-day intervals for six months, from the base of the stem to the leave bud. Plants were harvested six months after planting (day 180), roots were separated from shoots, washed with water, placed into paper bags and oven-dried at 72 °C for 72 h. Root and shoot dry biomass was determined by weighing in a semi-analytical balance with a precision of 0.01 g. Total dry biomass was obtained by adding up shoot and root dry biomass.

Statistical analyses

Analysis of variance, separation of means (Tukey, $P \leq 0.05$) and correlation analysis of soil and plant variables were conducted with the GLM procedure of the software SAS for Windows ver. 8.1 (SAS Institute Inc., 2000). Soil variables were densities of heterotrophic bacteria and fungi, total N, available P, available K and oil degradation. Plant variables assessed were height, root dry biomass, foliar dry biomass, total dry biomass and total foliar N, P and K content.

RESULTS AND DISCUSSION

Soil macronutrients and microflora

No statistical differences ($P \leq 0.05$) were found in total N content in soil between the control soil and the soils polluted with 50000 and 78000 mg kg^{-1} FO, but a lower amount of total N ($P \leq 0.05$) was found as a result of the effect of 78456 mg kg^{-1} WO in both soil types in which Bm and Ll were grown (Table 1). Total N levels in soil where Bm was planted were 0.36, 0.35, 0.32 and 0.30% for control soil, soil with 50000 mg kg^{-1} FO-TPH, soil with 78000 mg kg^{-1} FO-TPH, and soil with 78456 mg kg^{-1} WO-TPH, respectively. Coefficients of variation were up to four times as high as the average in soils with 50000 and 78000 mg kg^{-1} FO (Table 1), probably because of oil redistribution by water runoff during irrigation. Many microorganisms synthesize and release urease, the enzyme responsible for the hydrolysis of urea (Alexander, 1994), and it is likely that because of this biochemical feature total N was similar in the control soil and in those with pollution levels up to 78000 mg FO.

Available P in soil also showed differences between treatments in both experiments as well as in both oil types. In soil with WO, P availability decreased 12 and 3.5 times relative to soil with 78000 mg FO in which Bm and Ll were grown, respectively (Table 1). This response suggests that the soil solution contains a lower amount of available P compared to the soil with FO or to the control soil containing 150 mg TPH. Available K in the soil solution displayed a positive relationship with FO but decreased as an effect of WO in both experiments (Table 1). Increasing FO levels promoted K availability; this response appears to be related to a higher K mineralization rate from organic or mineral substrates. Regarding the inhibition of K availability in soil with WO, it has been reported that the recalcitrant property of weathered crude oil negatively influences mineralization (Porta *et al.*, 2003). Results of Wang *et al.* (2009) reveal decreasing trends in the amount of total N and available P in soil polluted with oil for 5, 10 and 20 years; N content in these years was 1.17, 0.67 and 0.84 g kg^{-1}, and the corresponding values for P were 0.37, 0.32 and 0.26 g kg^{-1}, respectively. This response may be associated with the density of microorganisms in the three soils (Mohn and Stewart, 2000).

Table 1. Total N, available P and available K in soil polluted with fresh and weathered oil, in which *Brachiaria mutica* and *Leucaena leucocephala* were grown. Samples collected at 180 days after planting.

TPH	*Brachiaria mutica*			*Leucaena leucocephala*		
(mg kg^{-1}	Total N	Available P	Available K	Total N	Available P	Available K
dry weight)	(%)	(mg kg)		(%)	(mg kg)	
150 PB[¶]	0.36 ± 0.1^{a}	20.7 ± 1.7^{b}	24 ± 1.0^{b}	0.34 ± 1^{a}	23 ± 1.2^{a}	36 ± 0.05^{ab}
50,000 FO[§]	0.35 ± 1.4^{a}	26 ± 1.4^{b}	27 ± 1.6^{b}	0.31 ± 1.1^{a}	24 ± 2^{a}	36 ± 1.03^{ab}
78,000 FO	0.36 ± 0.8^{a}	32 ± 1.1^{a}	42 ± 1.5^{a}	0.34 ± 1.1^{a}	24 ± 4.3^{a}	49 ± 0.16^{a}
78,456 WO[‡]	0.30 ± 0.6^{b}	2.7 ± 2^{c}	28 ± 0.5^{b}	0.27 ± 0.6^{b}	7 ± 1^{b}	27 ± 0.2^{b}

[¶]Likely biogenic oil, [§]fresh oil and [‡]weathered oil.
[a,b,c]Means with a different superscript are statistically different (Tukey, $P \leq 0.05$, a>b, n=6).

The data obtained in this investigation confirm the positive effect of rizosphere bacteria on total N and available P in the presence of Bm, and only on available P in the association of bacteria with Ll (Table 2).

Table 2. Pearson's correlation between *Brachiaria mutica* and *Leucaena leucocephala* variables, soil microorganisms and macronutrients in samples collected on day 180 after planting.

Species/	Plant	Total	Density		Nutrients in soil			Nutrients in plant		
variable	height	Biomass	Fungi	Bacteria	Total N	Avail. P	Avail. K	N	P	K
Brachiaria										
TPH-degradat	-.654*	-.748**	-.624*	NS***	NS	NS	NS	-.919**	.732**	.831**
Plant height		.964**	.601*	.615*	.764**	NS	NS	NS	NS	NS
Total biomass			.651*	NS	.665*	NS	NS	.622*	NS	NS
Dens. fungi				NS	NS	NS	NS	NS	NS	NS
Dens. bacteria					.907**	.817**	NS	NS	.597*	NS
N-soil						.705*	NS	NS	NS	NS
P-soil							NS	NS	.716**	NS
K-soil								NS	NS	NS
N-plant									-.680*	-.835**
P-plant										.678*
Leucaena										
TPH-degradat	NS	NS	NS	.661*	NS	NS	NS	-.593*	-.594*	NS
Plant height		.990**	NS	NS	NS	NS	NS	.736**	.843**	.841**
Total biomass			NS	NS	NS	NS	NS	.636*	.763**	.803**
Dens. fungi				.578*	NS	NS	NS	-.615*	-.603*	NS
Dens. bacteria					NS	.807**	-.618*	-.868**	-.781**	NS
N-soil						.670*	NS	NS	NS	NS
P-soil							-.866**	NS	NS	NS
K-soil								NS	NS	NS
N-plant									.971**	.831**
P-plant										.881**

*Correlation is significant at $P \leq 0.05$ and **at $P \leq 0.01$, n=12. ***NS: no significance.

The average density of heterotrophic bacteria and fungi in Bm and Ll rizosphere soil revealed statistically significant differences ($P \leq 0.05$) (Table 3). The highest population densities of both fungi and bacteria associated with these two plant species corresponded to oil-free soil, and the lowest figures were recorded in soil containing 78456 mg kg^{-1} WO. This response is likely related to the lower amount of air in soil, and also to the decrease in moisture retention which restrains aerobic microbial reproduction, given that highly compacted aggregates tend to occur because of the resin and asphaltene content associated with WO in soil (Del'Arco and França, 2001).

Table 3. Density of heterotrophic bacteria and fungi in *Brachiaria mutica* and *Leucaena leucocephala* rizospheres. Samples collected on day 180 after planting.

TPH (mg kg^{-1} dry weight)	*Brachiaria mutica*		*Leucaena leucocephala*	
	Bacteria	Fungi	Bacteria	Fungi
	CFU per gram dry rizosphere			
150 PB[¶]	$1100x10^4 \pm 649x10^{4\ a}$	430 ± 685^a	$200x10^4 \pm 169x10^{4\ a}$	$10x10^2 \pm 14x10^{2\ a}$
50,000 FO[§]	$900x10^4 \pm 585x10^{4\ b}$	76 ± 44^c	$160x10^4 \pm 127x10^{4\ b}$	$14x10^2 \pm 10x10^{2\ b}$
78,000 FO	$161x10^4 \pm 34x10^{4\ c}$	70 ± 65^c	$99x10^4 \pm 69x10^{4\ c}$	$19x10^2 \pm 18x10^{2\ c}$
78,456 WO[‡]	$10x10^4 \pm 8x10^{4\ c}$	110 ± 89^b	$31x10^4 \pm 34x10^{4\ d}$	$9x10^2 \pm 7x10^{2\ d}$

[¶]Likely biogenic oil, [§]fresh oil and [‡]weathered oil.
[a,b,c]Means with a different superscript are statistically different (Tukey, $P \leq 0.05$, a>b, n=6).

Plant growth, biomass and nutrient buildup

In Bm and Ll, plant height at 1 month was similar across treatments ($P \leq 0.05$; Table 4). This lack of response was likely due to the fact that the root was still adapting to the stressing conditions in soil, as well as because the root of the transplanted seedling was still surrounded by oil-free soil where it grew in the seedbed. Statistical differences ($P \leq 0.05$) were evident from month 2 (Table 4). Through time, plant growth in Bm was inversely related to FO dose; however, plant growth was similar in soil with WO and in soil with 78000 mg kg^{-1} FO after two, three and four months ($P \leq 0.05$), and differed after five and six months. Two and three months after transplanting Ll in soil with 50000 and 78000 mg kg^{-1} FO, plants dried out completely; however, plants became adapted to 78456 mg WO, although plant height was six times smaller compared to plants grown in the control soil. This toxic effect was likely caused by the light and medium fractions contained in fresh oil, which according to Freedman (1989), when in contact with the plant kill the foliage and some woody tissues, since these form an impermeable coating similar to wax around soil aggregates, restraining water retention and, hence, water availability for the plant. Consequently, according to Li *et al.* (1997), the efficiency of the plant regarding water use decreases, and plant growth may be seriously affected. However, not all perennial tissues are damaged by oil to the extent of being necrotized. Such was the case of Bm, which displayed tolerance to FO, likely as a result of root vigor (Walton *et al.*, 1994). According to Banks *et al.* (2000), grasses have fibrous root systems which provide a broad surface area for rizosphere microorganisms, so that the plant becomes more stress-tolerant and can maintain its productivity.

The production of dry biomass of root and leaf of Bm was similar ($P \leq 0.05$) in control treatments and in 50000 and 78000 mg of FO, but biomass in plants established in soil with 78456 mg of WO was different (Figure 1). As for Ll, the results show that FO caused decreased root and leaf biomass and WO caused less reduction in biomass. Foliar biomass of Bm compared with the control increased 2.9% in soil with 50000 mg FO. These results agree with Lindau and Delaune (2000), who found that petroleum hydrocarbons increased biomass and stem density of *Spartina lancifolia* with respect to the control site. The biomass of Bm decreased 6.1 and 63.5% by effect of exposure to 78000 mg kg^{-1} FO and WO, respectively. Similar results were obtained by Rivera-Cruz *et al.* (2004) in grass (*Echinochloa polystachya*) by the effect of 50000 and 100000 mg kg^{-1} FO, where dry matter was reduced 51 and 53%, respectively. The root and foliar biomass of Ll were not harvested because the plants died from the effect of the second month of the doses of 50000 and 78000 mg kg^{-1} FO. The dose of 78456 mg kg^{-1} WO resulted in the reduction of 94.7 and 98.3% of roots and leaves, respectively (Figure 1).

The accumulation of N, P and K in the foliage of Bm and Ll showed significant differences ($P \leq 0.05$; Table 5) as a result of exposure to FO and WO for 180 days. A highly significant relationship was observed between plant height and biomass and N, P and K buildup in the foliage of both species (Table 2). The highest nutrient accumulation occurred in both species grown in oil-free soil. In Bm, N and P buildup was unaffected by FO, but K accumulation dropped as an effect of 78456 mg kg^{-1} WO. The FO caused toxic effects on Ll, and both plant species displayed a negative effect on the accumulation of the three macronutrients as a result of the presence of WO in soil (Tables 2 and 5).

Table 4. Plant height in *Brachiaria mutica* and *Leucaena leucocephala* in soil polluted with fresh and weathered soil at six different time points.

TPH (mg kg^{-1} dry weight)	Month 1 (cm)	2	3	4	5	6
Brachiaria						
150 PB[¶]	30.3 ± 4.5^a	87 ± 14.1^a	127.7 ± 29.6^a	166.8 ± 33.9^a	194.6 ± 37.8^a	218 ± 34.9^a
50,000 FO[§]	29.7 ± 3.6^a	70.7 ± 8.2^{ab}	91.7 ± 8.1^b	111 ± 15.8^b	136 ± 10.9^b	153 ± 14^b
78,000FO	28.8 ± 4.9^a	53 ± 11.7^b	72.7 ± 14.1^b	98 ± 17.6^b	110.3 ± 13.5^{bc}	130.7 ± 11.9^{bc}
78,456 WO[‡]	30.7 ± 8.6^a	63 ± 11.9^b	86 ± 18.6^b	95.3 ± 22.9^b	101.7 ± 22.8^c	110.3 ± 23.7^c
Leucaena						
150 PB	10.7 ± 1.9^a	20.8 ± 7.1^a	42.8 ± 13.7^a	68 ± 19.6^a	104.3 ± 27.9^a	142.3 ± 25.3^a
50,000 FO	11.8 ± 1.8^a	13.3 ± 1.7^{ab}	17 ± 3.9^b	0^c	0^c	0^c
78,000FO	9.2 ± 1.7^a	0^c	0^c	0^c	0^c	0^c
78,456 WO	9.8 ± 0.6^a	10.6 ± 1.3^b	12.9 ± 1.7^b	15.2 ± 4.6^b	20 ± 8.1^b	23.8 ± 11.2^b

[¶]Likely biogenic oil, [§]fresh oil and [‡]weathered oil.
[a,b,c]Means with a different superscript are statistically different (Tukey, $P \leq 0.05$, a>b, n=6).

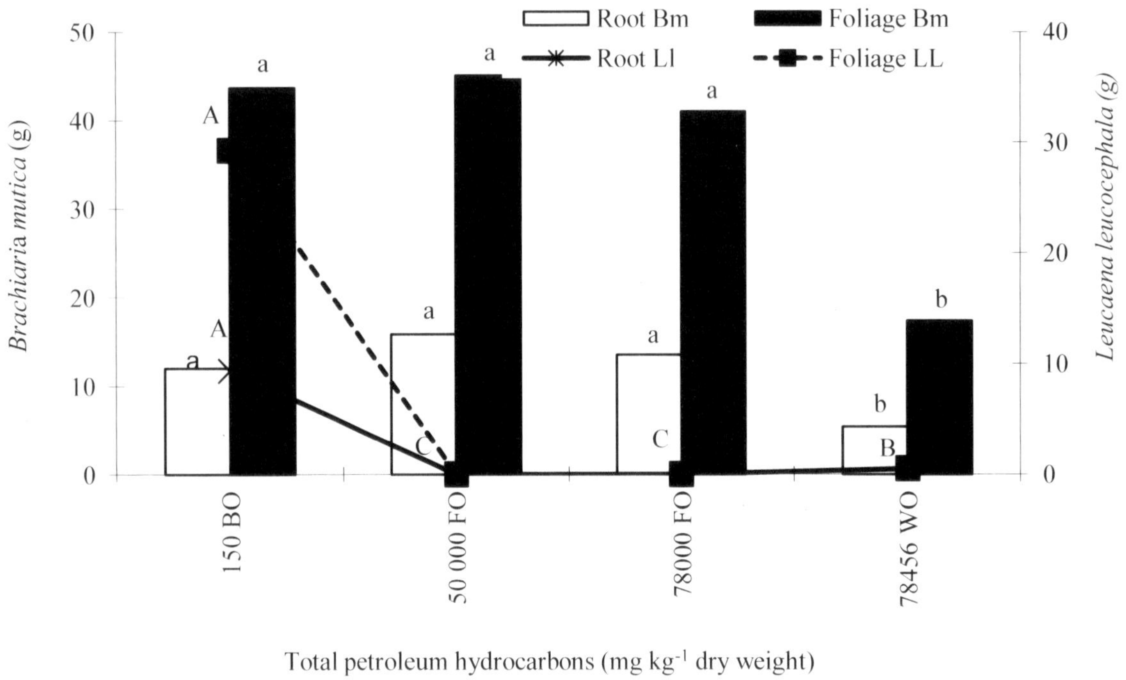

Figure 1. Effect of fresh and weathered oil on root and foliar biomass in *Bracharia mutica* and *Leucaena leucocephala* on Day 180 after transplanting.
Figures with different small letters in each column and different capital letters between points connected by solid lines mean statistically significant differences between treatment means (Tukey's $P \leq 0.05$, a>b, n=6).

Tabla 5. Dose effect of oil present in soil on buildup of three macronutrients in foliar biomass of samples collected on day 180 after planting.

TPH (mg kg^{-1} dry weight)	*Brachiaria mutica*			*Leucaena leucocephala*		
	N	P	K	N	P	K
	%					
150 PB[¶]	0.91 ± 0.3^{ab}	1.57 ± 0.67^{a}	0.12 ± 0.02^{a}	0.5 ± 0.18^{a}	1.1 ± 0.4^{a}	0.96 ± 0.3^{a}
50,000 FO[§]	1.09 ± 0.21^{a}	1.36 ± 0.42^{a}	0.2 ± 0.05^{a}	0^{b}	0^{b}	0^{b}
78,000 FO	1.19 ± 0.16^{a}	1.29 ± 0.41^{a}	0.11 ± 0.05^{b}	0^{b}	0^{b}	0^{b}
78,456 WO	0.55 ± 0.2^{c}	0.37 ± 0.13^{b}	0.04 ± 0.02^{c}	0.02 ± 0.11^{b}	0.01^{b}	0.01^{b}

[¶]Likely biogenic oil, [§]fresh oil and [‡]weathered oil.
[a,b,c]Means with a different superscript are statistically different (Tukey, $P \leq 0.05$, a>b, n=6).

Oil hydrocarbon degradation

The removal of FO-TPH and 78456 mg kg^{-1} of WO was stimulated by both plant species. Significant differences were observed between plant species as well as between oil concentrations in soil ($P \leq 0.05$). *Brachiaria mutica* induced the highest degradation (55%) in soil containing 50000 mg kg^{-1} FO after 120 days; however, when exposed to 78000 mg FO and 78456 mg kg^{-1} WO degradation dropped to 41.3 and 27.6%, respectively (Figure 2). The greater efficiency of grasses in soil restoration was previously reported by Rivera-Cruz *et al.* (2004) and Merkl *et al.* (2005), who pointed out that grasses are effective for this because they produce a dense network of roots reaching as far as 2.7 m deep in soil (Gould and Shaw, 1997), secreting exudates which contribute to the proliferation of microorganisms that are key for the breakdown of soil pollutants, leading to the neutralization of their toxic properties (Davis *et al.*, 2002).

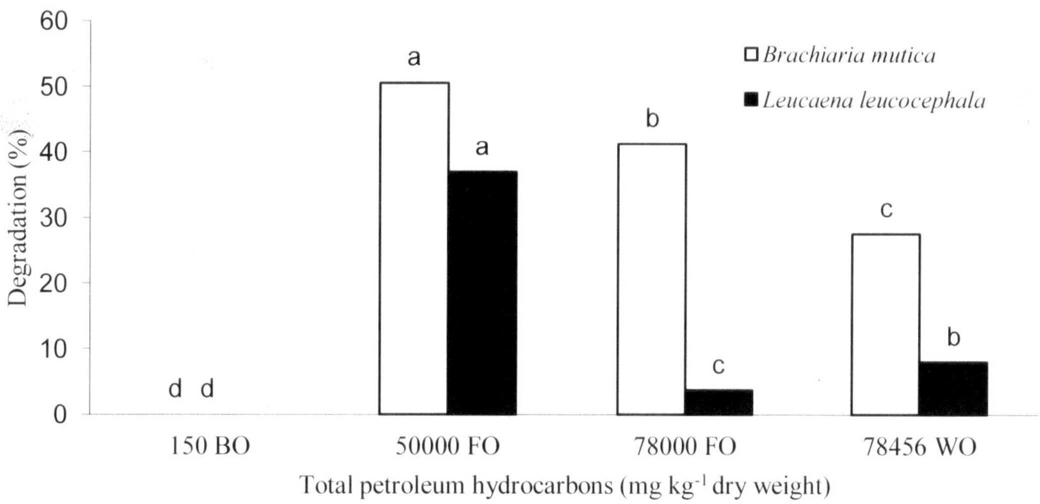

Figure 2. Average degradation of total petroleum hydrocarbons in soil with biogenic oil (BO), fresh oil (FO) and weathered oil (WO) in which *Brachiaria mutica* and *Leucaena leucocephala* were grown, 180 days after planting (Tukey $P \leq 0.05$, a>b, n=6).

In this study, the type of oil proved to be a determining factor in the percentage of TPH removal. In the presence of Bm and at 78000 mg kg^{-1} FO a higher degradation rate and volatilization occurred relative to hydrocarbons derived from 78458 mg kg^{-1} WO. This response is likely due to that WO has a higher proportion of high-molecular weight (MW) hydrocarbons such as C20 to C40 alkanes, polynuclear aromatic hydrocarbons (PAHs), asphaltenes and resins, which are degraded slowly. According to Huesemann (2004), the persistence of these oil fractions is explained by their low solubility, and hence these require long treatments involving sequential processes. On the other hand, FO hydrocarbons contain the hydrocarbons mentioned above plus the fraction of low (C5-C9) and medium (C10-C18) MW n-alkanes and aromatic hydrocarbons that are volatilized and/or degraded relatively easy. Another factor that affects oil degradation rate is soil humidity. Biodegradation is much slower under anaerobic conditions than in the presence of oxygen (Atlas and Bartha, 2002).

Relationship between biological and chemical parameters

For Bm, foliar and total biomass evidenced a high correlation with soil N, P and K levels. The fertility level promoted plant growth, so that it was also correlated with foliar content (Table 2). However, Ll did not display this same relationship with soil macronutrients, but it did in the case of nutrient buildup in foliage (Table 2). The high sensitivity of this leguminous species to FO in the soil limited its growth, contrary to the result when Ll was grown in soil with WO; however, both productivity and N, P and K buildup in foliage dropped. The percentage of TPH degradation in soil displayed a significant negative correlation with fungus density ($r = -0.624$), but not with heterotrophic bacteria density in Bm rizosphere; in the Ll rizosphere, TPH degradation was positively correlated ($r = 0.661$) only with bacterial density (Table 2). Maldonado-Chávez *et al.* (2010) reported similar findings between microbial consortia and oil hydrocarbon degradation.

The grass Bm constitutes an option suitable to be cultivated in soils polluted with FO and WO, because it stimulates decontamination and has the potential to produce forage in oil-producing zones. However, further studies are required to investigate whether this plant species accumulates hydrocarbons, which would affect the quality of livestock production.

CONCLUSION

Doses of 50000 and 78000 mg kg^{-1} FO-TPH caused the death of Ll; however, this legume survived at 78456 mg kg^{-1} WO-TPH. *Brachiaria mutica* survived at all TPH doses, achieving different heights as a result of the effect of dose and type of FO- and WO-TPH. The dose of 50000 mg kg^{-1} FO stimulated plant biomass production.

The Bm rizosphere induced the best response to the effect of the two FO doses and the WO dose on total N and available P and K in soil and accumulated in foliage. This suggests a physiological adaptation ability of this plant species to oil-polluted soils.

The positive relationship of plant height and foliar biomass production in Bm with density of heterotrophic bacteria and fungi in rhizospheric soil, with macronutrient content in soil and with those accumulated in foliage are all positive indicators of adaptation to pollution.

ACKNOWLEDGMENTS

This research work received financial support from the Colegio de Postgraduados Campus Tabasco.

REFERENCES

Alcántar, G.G., Sandoval, V.M. 1999. Manual de análisis químico de tejido vegetal. Publicación Especial 10. Sociedad Mexicana de la Ciencia del Suelo, A. C. Chapingo, Estado de México. México.

Alexander, M. 1994. Introducción a la Microbiología del Suelo. 2a. reimp. AGT Editor, S.A. México.

Atlas R.M., Bartha R. 2002. Ecología microbiana y Microbiología ambiental. 4a. ed. Pearson Educación, S. A. España.

Banks, M.K., Govindaraju, R.S., Schwab, A.P., Kulakow, P. 2000. Field demonstration. In: Fiorenza, C.S., Oubre, L. and Ward, C.H. (eds.). Phytoremediation of Hydrocarbon-contaminated Soil. Lewis Publishers, USA. pp. 1-88.

Bonilla, I. 2003. Introducción a la nutrición mineral de las plantas. Los elementos minerales. In: Azcón-Bieto, J. y Talón, M. (eds.). Fundamentos de fisiología vegetal. McGraw-Hill.Interamericana, España. pp. 83-97.

Davis, L.C., Castro-Diaz, S., Zhang, Q., Erickson, L.E. 2002. Benefits of vegetation for soils with organic contaminants. Critical Review in Plant Sciences. 21:457-491.

Del'Arco, J.P., França, F.P. 2001. Influence of oil contamination levels on hydrocarbon biodegradation in sandy sediment. Environmental Pollution. 110:515-519.

DOF (Diario Oficial de la Federación). 2002. NOM-021-RECNAT-2000 Especificaciones de fertilidad, salinidad y clasificación de suelos. Estudios de muestreo y análisis. Diario Oficial de la Federación. México.

DOF. 2006. NMX-AA-134-SCFI-2006. Suelos-hidrocarburos fracción pesada por extracción y gravimetría – método de prueba. Diario Oficial de la Federación. México.

EPA (Environmental Protection Agency). 1986. Method 418.1 mod. Petroleum Hydrocarbon, Total Recoverable. Spectrophotometric Infrared.

Farías, M.J. 2006. Manejo de pastos y forrajes en la ganadería de doble propósito. In: X Seminario de Pastos y Forrajes. Postgrado de Producción Animal. Facultad de Agronomía, Universidad de Zulia. Venezuela. pp. 1-9.

Freedman, B. 1989. Environmental Ecology: The Impacts of Soil Pollution and Other Stresses on Ecosystem Structure Function. Academic Press, Inc., USA.

García-López, E., Zavala-Cruz, J., Palma-López, D.J. 2006. Caracterización de las comunidades vegetales en un área afectada por derrames de hidrocarburos. Terra Latinoamericana. 24:17-26.

González, I., Faría-Mármol, J., Morillo, D., Mavarez, O., Noguera, N., Fuenmayor, E. 2003. Efecto de frecuencias de riego y corte sobre el rendimiento de materia seca en *Leucaena leucocephala* (Lam.) De Wit. Revista de la Facultad de Agronomía. 20:364-375.

Gould, F.W., Shaw, R.B. 1997. Gramíneas. Clasificación Sistemática. Cuevas, R.A. (trad.). 1a. ed. AGT Editor, S. A., México.

Huesemann, H.M. 2004. Biodegradation and Bioremediation of Petroleum Pollutants in Soil. In: Singh, A. and Ward, P.O. (eds.) Applied Bioremediation and Phytoremediation. Springer, Germany. pp. 13-33.

Jackson, M.L. 1973. Soil Chemical Analysis. 1st ed. Prentice Hall of India, India.

Jones, J.B., Wolf, B., Mills, H.A. 1992. Plant Analysis Handbook. A Practical Sampling, Preparation, Analysis, and Interpretation Guide. Micro-Macro Publishing, Inc., USA.

Li, X., Feng, Y., Sawatsky, N. 1997. Importance of soil-water relations in assessing the endpoint of bioremediated soils. Plant and Soil. 192:251-261.

Lindau, C.W., Delaune, R.D. 2000. Vegetative response of *Sagittaria lancifolia* to burning of applied crude oil. Water Air and Soil Pollution. 121:161-172.

Madigan, M.T., Martinko, J.M., Dunlap, P.V., Clark, D.P. 2009. Brock. Biología de los microorganismos. 12a ed. Pearson Educación, S.A., España.

Maldonado-Chávez, E., Rivera-Cruz, M.C., Izquierdo-Reyes, F., Palma-López, D.J. 2010. Efecto de rizósfera, microorganismos y fertilización en la biorremediación y fitorremediación de suelos con petróleos crudo nuevo e intemperizado. Universidad y Ciencia Trópico Húmedo. 26:121-136.

Merkl, N., Schultze-Kraft, R., Infante, C. 2005. Assessment of tropical grasses and legumes for phytoremediation of petroleum-contaminated soils. Water Air and Soil Pollution. 165:195-209.

Mohn, W.W., Stewart, G.R. 2000. Limiting factors for hydrocarbon biodegradation at low temperature in Arctic soils. Soil Biology and Biochemistry. 32:1161-1172.

OECD (Organisation for Economic Co-operation and Development). 2006. OECD Guidelines for the Testing of Chemicals. Terrestrial Plant Test: Seedling Emergence and Seedling Growth Test. 208. France.

Oluwafemi, D.O., Matthew, O., Sunday, A.A., Olukayode, O.A., Ganiyu, O.O. 2008. Microbial population changes in tropical agricultural soil experimentally contaminated with crude petroleum. African Journal of Biotechnology. 7:4512-4520.

Olsen, S.R., Sommers, L.E. 1982. Phosphorus. In: Page, A.L., Miller, R.H. and Keeny, D.R. (eds.). Methods of Soil Analysis. Part 2. Chemical and Microbiological Properties. 2nd ed. ASA. SSSA, USA. pp. 403-430.

Page, A.L., Miller, R.H., Keeney, D.R. 1982. Nitrogen total. In: Page, A.L., Miller, R.H. and Keeny, D.R. (eds.). Methods of Soil Analysis. Part 2. Chemical and Microbiological Properties. 2nd ed. ASA. SSSA, USA. pp. 595-629.

Philippot, L., Germon, J.C. 2005. Contribution of Bacteria to Initial Input and Cycling of Nitrogen in Soils. In: Buscot, F. and Varma, A. (eds.). Microorganisms in Soils: Roles in

Genesis and Functions. Springer-Verlag Berlin Heidelberg, Germany. pp. 159-176.

Porta, C.J., López-Acevedo, R.M., Roquero, L.C. 2003. Edafología para la Agricultura y el Medio Ambiente. 3ª ed. Editorial Mundi-Prensa, España.

Rivera-Cruz, M.C., Ferrera-Cerrato, R., Sánchez-García, P., Volke-Haller, V., Fernández-Linares, L., Rodríguez-Vázquez, R. 2004. Descontaminación de suelos con petróleo crudo mediante microorganismos autóctonos y pasto alemán [*Echinochloa polystachya* (H.B.K.) Hitchc.]. Agrociencia. 38:1-12.

Rivera-Cruz, M.C., Trujillo-Narcía, A., Miranda de la C., M.A., Maldonado, C.E. 2005. Evaluación toxicológica de suelos contaminados con petróleos nuevo e intemperizado mediante ensayos con leguminosas. Interciencia. 30:326-331.

SAS Institute Inc. 2000. The SAS System for Windows, Release 8.01. SAS Institute Inc, Cary, NC, USA.

SEMARNAT (Secretaría del Medio Ambiente y Recursos Naturales). 2009. Informe de la situación del medio ambiente en México. Edición 2008. Compendio de Estadísticas Ambientales. Secretaría del Medio Ambiente y Recursos Naturales. México.

Sotomayor-Ríos, A., Schank, C.S. 2001. Constraints and Developments in the Enhancement of Tropical Forage Grasses of Economic Importance. In: Sotomayor-Ríos, A. and Pitman, W.D. (eds).Tropical Forange Plants: Development and use. CRC Press, USA. pp. 107-117.

Tang, J., Wang, M., Wang, F., Sun, Q., Zhou, Q. 2011. Eco-toxicity of petroleum hydrocarbon contaminated soil. Journal of Environmental Sciences. 23:845-851.

Trujillo-Narcía A., Rivera-Cruz M.C., Maldonado, C.E. 2006. Efecto de la restauración de suelo contaminado con petróleo en el suelo y en la vegetación en Tabasco, México. In: Gallardo, L.J.J. (ed. y coord.). Medioambiente en Iberoamérica. Visión desde la Física y la Química en los albores del Siglo XXI. Tomo III. Sociedad Iberoamericana de Física y Química Ambiental. España. pp. 353-361.

Trujillo-Narcía, A., Jiménez-López, G., Rivera-Cruz, M.C., Dorantes-Avelino, R. 2011. Inventario florístico y tolerancia vegetal a petróleo crudo derramado en suelos en Tabasco, México. In: Memoria VI Congreso Iberoamericano de Física y Química Ambiental. Cancún, México. pp. 52-58.

Vázquez-Luna, D., Castelán-Estrada, M., Rivera-Cruz, M.C., Ortiz-Ceballos, A.I., Izquierdo, R. F. 2010. *Crotalaria incana* L. y *Leucaena leucocephala* Lam. (Leguminosae): especies indicadoras de toxicidad por hidrocarburos de petróleo en el suelo. Revista Internacional de Contaminación Ambiental. 26:183-191.

Walecka-Hutchison, C.M., Walworth, J.L. 2007. Evaluating the effects of gross nitrogen mineralization, immobilization, and nitrification on nitrogen fertilizer availability in soil experimentally contaminated with diesel. Biodegradation. 18:133-144.

Walton, T.B., Guthrie, E.A., Hoylman, A.M. 1994. Toxicant degradation in rhizosphere In: ACS Symposium Series 563: Biorremediation through Rhizosphere Technology. Anderson, T.A. and Coats, J.R. (eds). American Chemical Society. Chicago Illinois. USA. pp. 11-26.

Wang, Z., Yang, C., Kelly-Hooper, F., Hollebone, B.P., Peng, X., Brown, C.E., Landriault, M., Sun, J., Yang, Z. 2009. Forensic differentiation of biogenic organic compounds from petroleum hydrocarbons in biogenic and petrogenic compounds cross-contaminated soils and sediments. Journal of Chromatography. 1216:1174-1191.

Wencomo, B.H., Cepeda, B., Ramírez, J. 2009. Producción de semillas de *Leucaena* spp en suelo ácido. Pastos y Forrajes. 32:1-7.

13

Botanical and Chemical Composition of Common Grass Species around Dip–Tank Areas in Semi–Arid Communal Rangelands of Swaziland

Solomon T. Beyene[1], Victor Mlambo[2*]

[1]*University of Fort Hare, Faculty of Science and Agriculture, Department of Livestock and Pasture, P/Bag x1314, Alice, 5700 South Africa*

[2]*Department of Food Production, Faculty of Science and Agriculture, University of the West Indies, St. Augustine, Trinidad and Tobago.*

[*]*Corresponding author: Victor.Mlambo@sta.uwi.edu or victormlambo@yahoo.co.uk*

SUMMARY

Communal rangelands in arid and semi–arid areas of southern Africa are the source of virtually all feeds for wild and domestic ruminants. However, the quality and quantity of feed fluctuates in response to rainfall patterns. Little is known about the capacity of these rangelands to support ruminant productivity during the 4-month dry season. This study assesse the botanical composition of grasses and nutritive values of selected grass species growing in the communal rangelands of Swaziland at the start of the dry season. Vegetation survey and sampling were carried out in two semi–arid ecological zones (Lower Middleveld (LMV) and Lowveld (LV) savannas). Dry matter (DM), organic matter (OM), crude protein (CP), minerals, neutral detergent fibre (NDF) and acid detergent fibre (ADF) contents were determined. Of the 20 grass species identified in the two zones, 9 were perennials and 11 were annuals. Only 3 the grass species were classified as highly palatable. Grass species showed significant ($P<0.05$) (LMV) and non–significant (LV) variations in the DM yield. In both zones the highest NDF level was measured in *Panicum maximum* and *Urochloa mosambicensis* and lowest in *Digitaria longiflora* and *Eleusine coracana*. In the LV region, *P. maximum* ranked first in CP content, while *E. coracana* ranked last. In the LMV region, the level of CP was highest in *P. maximum*, and lowest in *D. argyrograpta*. In most instances, macro and trace elements showed variations ($P<0.05$) between grass species. Grass species collected from the two ecological zones showed great variations in NDF, ADF, CP, macro and microminerals. $P<0.05$– It was concluded that most grasses were deficient in protein, phosphorus and potassium. It is recommended that protein and mineral supplements should be offered to animals to optimize the utilization of grasses and improve animal productivity during the dry season.

Key words: botanical composition; nutrient content; semi-arid savannas; biomass yield; nutrient supplementation.

INTRODUCTION

Livestock production plays a significant role in the livelihood of farmers in the arid and semi arid communal rangelands of Swaziland. The livestock production system is predominantly extensive, with virtually all cattle and small ruminants relying on the natural rangelands and crop residues for feed. In the savannas of southern Africa, grasses make up a significant proportion of the diet of domestic ruminants in both the wet and dry seasons. Grasses in the savannas of Swaziland are generally dominated by perennials although a significant proportion of annuals occur in few areas (Solomon *et al*., 2007). Holecheck *et al.* (2001) reported that annual grasses are seldom considered as having good forage value as perennial grasses. Yet, some annuals grow in the dry periods when perennials are dormant and contribute significantly to the nutrition of livestock, while others provide good quality nutrition in the growing season.

The productivity of ruminants on communal grazing lands largely depends on the quality and the quantity of the available forage. Pasture quality is related to the amount of nutrient in the pasture that can be available for grazing (Walton, 1983). Optimum nutrition is dependent upon a delicate balance of four basic factors i.e. the animal's nutrient requirements, nutrient content of the feedstuffs consumed, digestibility of the feed stuffs consumed, and the amount consumed (Vallentine, 1990). These factors are in turn affected by botanical and chemical composition of the range forage, both of which vary with season. Annual and perennial grass species grow rapidly during the growing rainy season and decline in growth rate, production, and nutritive value as they mature towards the end of the growing season (Tefera *et al*., 2009).

Information on the distribution and botanical composition of the vegetation in the key communal grazing lands such as those surrounding the dip–tank areas are vital to identify the key grass species that contribute to the diets of grazing animals. In the communal areas of Swaziland, ticks are controlled chemically through a compulsory programme of dipping cattle and livestock movement controls (Musisi and Lawrence, 1995) . The country has approximately 800 registered dip–tanks (plunge dips and spray races) at which all cattle are mandatorily dipped at weekly intervals in summer and bi–weekly in winter. These dip–tanks are located in the middle of the grazing areas where ruminant livestock derive their nutrition for long periods during the year. Understanding the nutritional dynamics of foragesaround the dip–tank areas is vital for proper management and sustainable use of the grazing resources. In the savannas of Swaziland, however, there is little data that describes the botanical composition and nutritional dynamics of grasses in general and the status of nutrients at the beginning of the dry season. Knowledge of the distribution of grass species and their nutritive value will assist in their timely utilization, help predict nutrient deficiencies, and suggest supplementation needs. The objective of this study was to determine the botanical and chemical composition of grasses around dip–tank areas in the Lowveld and Lower Middleveld savanna ecosystems of Swaziland. Veld (or veldt) is an Afrikaans word used to describe open, flat fields covered with grass and/or shrubs in southern Africa.

MATERIAL AND METHODS

Study area

The study was conducted on Swazi Nation Land (SNL) located in the northern part of Swaziland. The study areas fall in the semi—arid Lower Middleveld (LMV) and Lowveld (LV) savanna ecological zones. Altitude in the LMV ranges between 400–600 m above sea level, and annual rainfall ranges between 625 – 725 mm. The LV lies in the altitude range from 250 – 400 m and has relatively low mean annual rainfall (400–600 mm). In both study areas, mean annual temperature varies from an average of 18 — 26 ^{o}C in summer and can drop to as low as 10 ^{o}C in winter (Environmental Centre for Swaziland, 1993). About 12% of the study area in the LV has good or fair soils that are mainly dominated by sandstone or claystone, while the LMV is dominated by gneiss (Environmental Centre for Swaziland, 1993). The vegetation of the study areas is mainly characterized by broad–leafed and mixed savanna vegetation (Kunene, 2005).

Site selection and determination grass species composition

Vegetation survey was conducted around three and two dip–tank areas in the LV and LMV ecological zones, respectively.. The sites were identified with the aid of a geographical information system of Swaziland. Five sites were selected around each dip–tank area, and a 20 x 50 m (1 000 m^2) plot was marked in each site, making a total of 25 plots . Grass species composition was estimated using a step point method (Hardy and Walker, 1991). The nearest plant and basal strikes were recorded from 200 point observations per plot. Point observations were spaced by 2 m interval and records were made over the length of each plot in straight parallel lines about 1 m distance apart. The palatability of grasses was classified based on the succession theory described by Dyksterhuis (1949) and on ecological information for the arid to semi–arid regions of Southern Africa (Tainton *et al*.,1980; Vorster,1982). Accordingly, the species were grouped into (i) highly palatable species - those which occur in rangeland in good condition and decrease with overgrazing (decreasers), (ii) moderately palatable species - those which occur in rangeland in good condition and increase with moderate overgrazing (increaser IIa), and (iii) less and poorly palatable species - those which occur in rangeland in good condition and increase with severe/extreme overgrazing (increasers IIb and IIc, respectively).Vegetation survey was carried out during the late growing season (February–March, 2008) when most grasses were in bloom to aid in their identification. Based on the survey result, the following grass species dominated the study areas: *Bothriochloa radicans, Cynodon dactylon, Digitaria argyrograpta, Digitaria longiflora, Eleusine coracana, Panicum maximum,* and *Urochloa mosambicensis* (Table 1). These species were targeted for chemical composition analysis.

Grass sampling for dry matter and chemical analysis

Grass samples for each species were collected once from four 0.25m^2 quadrates distributed randomly within each plot. Samples were cut to a stubble height of 20 mm. The samples from each plot were bulked and oven dried at 72°C for 48 hours. Grass samples were then ground to pass through a 1 mm sieve and kept in paper bags for one week at room temperature pending analysis. Grasses were harvested towards the end of the growing season (April, 2008), thus the grasses had been growing for 3 months, the length of the growing (rainfall) season.

Chemical analysis

Grass samples were analyzed for dry matter (DM), organic matter (OM), crude protein (CP), neutral detergent fibre (NDF), acid detergent fibre (ADF), macro–minerals (phosphorus (P), magnesium (Mg), calcium (Ca), potassium (K)), and micro–minerals (iron (Fe), copper (Cu), manganese (Mn) and zinc (Zn)). The DM content was measured by placing 1 g of sample in a pre–weighed porcelain crucible in an oven set at 100°C overnight. The loss in weight was used as a measure of moisture content. The OM content was determined by igniting the sample in a muffle furnace set at 500°C overnight. The loss in weight was used as a measure of OM content. Neutral detergent fibre (NDF) and ADF were determined by refluxing 2 g grass samples with neutral detergent and acid detergent solutions, respectively, for 1 hour according to Van Soest *et al*. (1991). Neutral detergent fibre was assayed without sodium sulfite and α–amylase since mature grasses contain very little neutral detergent insoluble protein and starch that can interfere with fibre determination. Both NDF and ADF were expressed exclusive of residual ash. Nitrogen content was determined by a Kjeldahl method (AOAC 1999, method no. 976.06) and converted to CP content by multiplying % N content by 6.25. Phosphorous was analyzed by using the Ultra—violet spectrophotometer (Chemicallab instruments, method no–075–01, Bavaria, Germany), K by flame photometer and Ca, Mg, Fe, Zn and Cu by atomic absorption spectrophotometer (Perkin Emer 1982).

Statistical Analysis

Statistical analysis of all data was done using the general linear models (GLM) procedure of SAS (1996) in a randomised complete block design to test differences between species within an ecological zone, and same species between ecological zones. The main effects were ecological zones and grass species. Plots are nested within ecological zones and served as replicates. Means were separated using the PDIFF option of the least squares means statement of the GLM procedure of SAS.

RESULTS

Botanical composition and dry matter production

In both study areas, 20 grass species were identified, 9 of which were perennials. In terms of palatability grouping, 3 species were classified as highly or less palatable, 7 species as moderately or poorly palatable. (Table 1). In the LV area, the mean frequency of *B. radicans, C. dactylon, D. argyrograpta, D. longiflora, E. coracana, P. maximum,* and *U. mosambicensis* ranged between 5 – 15% and were regarded as commonly occurring species. In the LMV area, *C. dactylon* and *D. longiflora* were recorded as dominant with a mean frequency value of >15% and the remaining species were regarded as common (mean frequency ≥ 5 – 15%) (Table 1). All the above grass species together contributed to about 68% and 78% of the total grass frequency in the LV and LMV areas, respectively (Table 1). The DM yield of the grass species is presented in Table 2. P<0.05In the LV area, *E. coracana* produced the highest (P<0.05) DM followed by *U. mosambicensis*.

Dry matter production did not differ (P>0.05) between all the other grass species (Table 2). In the LMV area, *E. coracana, P. maximum* and *U. mosambicensis* yielded higher (P<0.05) DM than the other species. *Bothriochloa radicans, E. Coracana* and *U. mosambicensis* growing in the LV area produced higher (P<0.05) DM than the LMV area (Table 2).

Table 1. Palatability, life forms and distribution of grass species around dip–tank areas in the northern communal rangelands of Swaziland

Grass species	Life form[a]	Palatability[b]	Species composition (%)	
			Lowveld	Low middleveld
Bothriochloa radicans	P	PP	9.4	5.0
Brachiaria deflexa	A	MP	10.8	1.3
Cynodon dactylon	P	MP	12.7	17.2
Dactyloctenium aegyptium	A	MP	0.5	1.3
Digitaria argyrograpta	P	HP	9.3	8.9
Digitaria longiflora	A	MP	8.3	17.9
Digitaria velutina	A	LP	0.01	2.3
Eleusine coracana	A	PP	5.0	5.0
Eragrostis ciliaris	A	LP	0.2	1.4
Eragrostis cylindriflora	A	PP	1.5	–
Eragrostis racemosa	A	LP	1.7	3.6
Eragrostis superba	P	MP	3.2	4.5
Fingerhuthia africana	A A (MP)	MP	0.3	0.6
Panicum maximum	P	HP	10.8	10.3
Perotis patens	A	PP	0.9	1.9
Sporobolus africanus	P	PP	3.6	2.7
Sporobolus nitens	P	PP	2.9	–
Themeda triandra	P	HP	3.5	1.4
Tragus berteronianus	A	PP	3.4	1.5
Urochloa mosambicensis	P	MP	12.3	13.5

[a] Life forms: A = annual; P = perennial
[b]Palatability: HP = highly palatable; MP = moderately palatable; LP = less palatable; PP =poorly palatable.

Table 2. Mean dry matter production (kg/ha) of the selected grass species in the two study areas.

Grass species	Lowveld	Low Middleveld
B. radicans	676.8 ± 89.9^{Ca}	383.8 ± 84.9^{Bb}
C. dactylon	575.4 ± 89.9^{Ca}	609 ± 69.4^{ABa}
D. argyrograpta	607.2 ± 80.0^{Ca}	518.7 ± 69.4^{Ba}
D. longiflora	556.5 ± 89.9^{Ca}	510.1 ± 60.1^{Ba}
E. coracana	1006.8 ± 87.2^{Aa}	739.8 ± 84.9^{Ab}
P. maximum	651.4 ± 89.0^{Ca}	725.1 ± 69.4^{Aa}
U. mosambicensis	809.1 ± 51.9^{Ba}	677.7 ± 42.5^{Ab}

Lowercase superscripts are used to compare site (ecological zone) means within each grass species while uppercase superscripts are used to compare grass species within each site.
Means with different superscripts differ significantly ($P<0.05$).

Chemical composition

Table 3 presents the NDF, ADF and CP contents of grasses harvested from the LV and LMV ecological zones. In both ecological zones, the highest NDF level was measured in *P. maximum* and *U. mosambicensis* and the lowest in *D. longiflora* and *E. coracana*. *Urochloa mozambicensis* had the greatest ($P<0.05$) ADF in both studied areas, while *D. longiflora* and *E. coracana* had the lowest. In the LV region, *P. maximum* ranked first in CP content (81.9 ± 6.7 g/kg DM), while *E. coracana* ranked last (50.5 ± 1.0 g/kg DM). In the LMV region, the level of CP was highest (95.4 ± 6.1 g/kg DM) in *P. maximum*, and lowest in *D. argyrograpta* (47.9 ± 0.9 g/kg DM). When comparing between the ecological sites,the NDF content of *B. radicans* harvested from the LV was significantly ($P<0.05$) higher than the LMV area, while the reverse was true for *C. dactylon* and *E. coracana*. *Cynodon dactylon, D. argyrograpta, D. longiflora,* and *E. coracana* had significantly ($P<0.05$) higher ADF content in the LMV than the LV area. *Digitaria argyrograpta, D. longiflora, E. coracana* and *P. maximum* harvested from the LV had higher ($P<0.05$) CP content than the same grasses harvested from the LMV area.

The macromineral contents of grasses collected from the two ecological zones are presented in Tables 4 and

5. *Eleusine coracana* (2.4 g/kg DM) and *C. dactylon* (3.3 g/kg DM) had the highest ($P<0.05$) Ca level in the LV and LMV, respectively, while *B. radicans* had the lowest (1.3 g/kg DM) in both areas. In the LV, except for *D. argyrograpta* which had no detectable P, and *E. coracana*, grass species did not show significant differences ($P>0.05$) in the level of P. In the LMV area, grass species except *D. argyrograpta* had similar P values (range: 0.001 – 0.49 g/kg DM). Comparing the results of the ecological zones, Ca content in *C. dactylon* differed ($P<0.05$) between the LV and LMV areas (Table 4). All the grass species had similar P levels across the two ecological zones. *Urochloa mosambicensis* and *E. coracana* had significantly ($P<0.05$) the highest K content in both study areas. *Eleusine coracana* and *P. maximum* displayed the highest Mg content in both ecological zones. The lowest Mg level was measured in *D. longiflora* (LV) and *C. dactylon* (LMV). With regard to ecological zones, *B. radicans* and *C. dactylon* had significantly ($P<0.05$) the highest concentration of K and Mg in the LV area.–. $P<0.05$.

Table 3. Fibre and crude protein content (g/kg DM) of grasses harvested from the Lowveld and Low Middleveld ecological zones.

	NDF		ADF		CP	
Grass species	LV[1]	LMV[2]	LV	LMV	LV	LMV
B. radicans	743.4 ± 27.5^{Aa}	435.6 ± 0.5^{Eb}	430.0 ± 61.6^{Aa}	383.0 ± 0.2^{Da}	64.4 ± 2.3^{BCa}	66.3 ± 0.5^{Ca}
C. dactylon	421.9 ± 26.7^{Cb}	509.9 ± 21.9^{CDa}	364.5 ± 24.7^{BCb}	446.1 ± 12.2^{BCa}	65.0 ± 4.3^{BCa}	72.6 ± 3.9^{Ba}
D. argyrograpta	616.2 ± 90.0^{Ba}	586.7 ± 84.0^{BCa}	375.7 ± 10.3^{Bb}	513 ± 14.1^{Aa}	68.9 ± 6.8^{Ba}	47.9 ± 0.9^{Eb}
D. longiflora	333.2 ± 87.0^{Ca}	491.2 ± 77.6^{DEa}	269.1 ± 82.8^{Cb}	413.4 ± 68.7^{CDa}	57.3 ± 5.3^{CDa}	76.9 ± 5.1^{Bb}
E. coracana	383.8 ± 14.2^{Cb}	455.7 ± 24.8^{Ea}	334.4 ± 19.2^{Cb}	415.6 ± 28.9^{CDa}	50.5 ± 10.0^{Da}	86.1 ± 25.1^{ABCb}
P. maximum	731.5 ± 37.6^{ABa}	694.4 ± 16.7^{Aa}	388.1 ± 10.7^{Ba}	433.4 ± 22.2^{BCa}	81.9 ± 6.7^{Aa}	95.4 ± 6.1^{Ab}
U. mosambicensis	752.9 ± 38.1^{Aa}	$668.3 \pm 65.9^{Aa B}$	462.9 ± 39.7^{Aa}	500.0 ± 64.6^{ABa}	67.7 ± 4.4^{Ba}	59.6 ± 4.3^{Da}

[1] LV–Lowveld, [2] LMV–Lower Middleveld
Lowercase superscripts are used to compare site (ecological zone) means within each grass species while uppercase superscripts are used to compare grass species means within each site.
Means with different superscripts differ significantly ($P<0.05$).

Table 4. Calcium and phosphorus contents (g/kg DM) of grasses harvested from the Lowveld and Low Middleveld ecological zones.

	Calcium		Phosphorus	
Grass species	LV	LMV	LV	LMV
B. radicans	0.94 ± 0.4^{Ca}	1.63 ± 0.4^{Ca}	0.50 ± 0.4^{Aa}	0.26 ± 0.3^{Aa}
C. dactylon	1.53 ± 0.1^{Bb}	3.25 ± 0.1^{Aa}	0.50 ± 0.1^{Aa}	0.27 ± 0.3^{Aa}
D. argyrograpta	1.80 ± 0.6^{ABa}	2.27 ± 0.4^{BCa}	0.0001	0.001
D. longiflora	1.58 ± 0.13^{Ba}	1.72 ± 0.15^{Ca}	0.30 ± 0.2^{ABa}	0.34 ± 0.2^{Aa}
E. coracana	2.41 ± 0.6^{Aa}	2.5 ± 0.1^{Ba}	0.09 ± 0.09^{Ba}	0.30 ± 0.3^{Aa}
P. maximum	1.90 ± 0.3^{ABa}	1.96 ± 0.2^{Ca}	0.39 ± 0.3^{ABa}	0.49 ± 0.4^{Aa}
U. mosambicensis	1.90 ± 0.2^{ABa}	2.22 ± 0.2^{Ba}	0.47 ± 0.2^{Aa}	0.30 ± 0.2^{Aa}

Lowercase superscripts are used to compare sites (ecological zone) within each grass species while uppercase superscripts are used to compare grass species within each site.
Means with different superscripts differ significantly ($P<0.05$).

Table 6 shows the micromineral content of grass species harvested from the two study areas. The levels of Zn and Fe displayed species variation within the LV region. *Cynodon dactylon* had the highest level of Zn followed by *B. radicans* and *P. maximum*. The lowest level of Fe was measured in *C. dactylon* and *B. radicans*. In the LMV area, Zn and Fe did not show significant differences ($P>0.05$) between the grass species. The amount of Cu did not differ ($P>0.05$) between species in both ecological regions. *Bothriochloa radicans* and *C. dactylon* showed the highest Mn content in both ecological zones followed by *D. longiflora*, *E. coracana*, *P. maximum* and *U. mosambicensis*. Grass species, except *B. radicans*, *C. dactylon* (Zn), and *D. argyrograpta* (Cu), had no significant differences in the level of microminerals between the two ecological zones.

Table 7 presents the number of grass species in the two ecological zones whose mineral content fell below, within or above the normal mineral requirement of ruminants. Four grasses in the LV area had Ca levels within the normal animal requirements while this was 6 in the LMV. In both ecological zones, the P level of all grass species was below the normal animal requirement, while the Fe and Zn levels were above the normal requirements.

DISCUSSION

Data on the species composition indicated that although many grass species have been identified in the study areas, only a few species made up the bulk of the dry matter yield. The findings of this study generally does not concur with the concept of Abule et al. (2007) who suggested that the natural community of a rangeland is dominated by one of a few species, which are best adapted to the specific combination of environmental factors of that site. This is not always true because in the current study, all the dominant grass species, except *P. maximum* could have low frequency under natural community, but increase in abundance mainly in response to disturbance such as moderate to heavy grazing.

Table 5. Potassium and magnesium contents (g/kg DM) of grasses harvested from the Lowveld and Low Middleveld ecological zones.

Grass species	Potassium		Magnesium	
	LV	LMV	LV	LMV
B. radicans	3.11 ± 0.6^{ABa}	1.1 ± 0.1^{Db}	3.51 ± 0.8^{Aa}	1.31 ± 0.6^{DEb}
C. dactylon	3.11 ± 0.9^{ABa}	1.06 ± 0.8^{CDb}	2.57 ± 1.5^{ABCa}	0.70 ± 0.2^{Eb}
D. argyrograpta	2.96 ± 0.58^{Bb}	5.37 ± 0.8^{Aa}	2.57 ± 0.37^{Ba}	1.59 ± 0.6^{CDb}
D. longiflora	1.17 ± 0.56^{Ca}	2.13 ± 0.6^{Ca}	1.87 ± 0.1^{Ca}	2.14 ± 0.3^{CDa}
E. coracana	4.1 ± 1.91^{ABa}	4.2 ± 0.8^{ABa}	3.3 ± 0.2^{Aa}	3.3 ± 0.3^{Aa}
P. maximum	3.79 ± 1.5^{ABa}	3.92 ± 0.6^{Ba}	3.67 ± 0.6^{Aa}	2.31 ± 1.3^{ABCDa}
U. mosambicensis	3.84 ± 0.3^{Aa}	4.39 ± 0.4^{ABa}	2.64 ± 0.3^{Ba}	2.39 ± 0.4^{Ba}

Lowercase superscripts are used to compare sites (ecological zone) within each grass species while uppercase superscripts are used to compare grass species within each site.
Means with different superscripts differ significantly ($P<0.05$).

Table 6. Microminera lconcentrations (g/kg DM) of grasses harvested from the Lowveld and Low Middleveld ecological zones.

	Zn		Cu		Fe		Mn	
Grass species	LV	LMV	LV	LMV	LV	LMV	LV	LMV
B. radicans	0.12^{Ba}	0.04^{Ab}	0.02^{Aa}	0.02^{Aa}	0.47^{Ba}	0.44^{Aa}	0.27^{Aa}	0.23^{Aa}
C. dactylon	0.34^{Aa}	0.05^{Ab}	0.03^{Aa}	0.02^{Aa}	0.47^{Ba}	0.44^{Aa}	0.84^{ABa}	0.23^{Aa}
D. argyrograpta	0.09^{Ca}	0.09^{Aa}	0.01^{Aa}	0.04^{Ab}	0.83^{ABa}	0.46^{Aa}	0.14^{Aa}	0.16^{Da}
D. longiflora	0.06^{Da}	0.06^{Aa}	0.003^{Aa}	0.02^{Aa}	0.54^{Aa}	0.47^{Aa}	0.17^{Ba}	0.21^{ABa}
E. coracana	0.1^{Ba}	0.1^{Aa}	0.01^{Aa}	0.01^{Aa}	0.81^{Aa}	0.67^{Aa}	0.1^{Ba}	0.2^{Ba}
P. maximum	0.12^{Ba}	0.12^{Aa}	0.01^{Aa}	0.02^{Aa}	0.85^{ABa}	0.75^{Aa}	0.18^{Ba}	0.18^{Ba}
U. mosambicensis	0.07^{Da}	0.07^{Aa}	0.01^{Aa}	0.01^{Aa}	0.80^{Aa}	0.67^{Aa}	0.19^{Ba}	0.2^{Ba}

Lowercase superscripts are used to compare site (ecological zone) within each grass species while uppercase superscripts are used to compare grass species within each site.
Means with different superscripts differ significantly ($P<0.05$).

Table 7. Number of grass species below, above and within the normal mineral requirements of ruminants

Mineral	Normal requirement[a] (g/kg DM)	Lowveld			Lower middleveld		
		Below	Within	Above	Below	Within	Above
Ca	1.9 – 8.2	3	4	0	1	6	0
K	5 – 10	7	0	0	5	2	0
Mg	1 – 2.5	0	1	6	1	5	1
P	1.2 – 4.8	7	0	0	7	0	0
Cu	0.007 – 0.011	1	4	2	0	2	5
Fe	0.03 – 0.05	0	0	7	0	0	7
Zn	0.02 – 0.04	0	0	7	0	0	7

[a]Recommended mineral requirements for all classes of ruminants suggested by National Research Council (1996) and summarized by McDowell (1997).

The findings that grasses in this study had high fibre (NDF and ADF) and low protein content is in agreement with Gonzalez Ronquillo *et al.* (1998) and Mlay *et al.* (2006). These results are typical for mature tropical grasses found in the rangelands of Swaziland which are available for about 7 months grazing starting from the end of summer to the beginning of spring . –It is obvious that the utilization of mature grasses without supplementation will cause protein deficiency in ruminants. Protein deficiency reduces the cellulolytic activity of rumen microbes resulting in the slow breakdown of fibre and low passage rate (Nogueira Filho *et al.*, 2000). Ultimately, the intake of these grasses will be inadequate to support the maintenance functions of ruminants leading to loss in weight. When protein deficiency is combined with mineral imbalances the result is low productivity of ruminant animals. It is therefore, essential that communal farmers explore the use of non–conventional protein supplements such as forage from trees that tend to retain green leaves throughout the long dry season. Some significant variations in terms of fibre contents were observed between grass species. Although *D. longiflora* and *E. coracana* had the least concentration of NDF and ADF, which would have made these species easier to digest, they also had low protein content especially in the LV area.

The presence of great variations in the level of macro–elements among the grass species within each ecological zone reflected the studies of Nsinamwa *et al.* (2005), Javed *et al.* (2008) and Tefera *et al.* (2009) in similar ecological zones. Except for P, significant variations were also noticed for the same grass species harvested from the LV and LMV areas. Indeed, the inter–species (within ecological zone) and intra–species (between ecological zones) variations were not consistent.

Table 8. Calcium: Phosphorus ratio of grass species harvested from the Lowveld and Lower middleveld areas.

Species	Lowveld	Lower middleveld
B. radicans	1.9 : 1	3.3 : 1
C. dactylon	3.1 : 1	12.0 : 1
D. argyrograpta	1800 : 1	2270 : 1
D. longiflora	5.3 : 1	5.1 : 1
E. coracana	26.8 : 1	8.3 : 1
P. maximum	4.9 : 1	4.0 : 1
U. mosambicensis	4.0 : 1	7.4 : 1

Optimal Ca: P ratio for ruminant animals ranges between 1 : 2 – 2 : 1.

In the LV area, *E. coracana, P. maximum* and *U. mosambicensis* had Ca level (range: 1.9–2.4 g/kg DM) that fall within the normal requirement for ruminants (Table 7), while the remaining species had Ca level below the normal requirement (range: 0.94 –1.8 g/kg DM). In the LMV area, all grass species, except *B. radicans* and *D. longiflora* had Ca level within the range of ruminant requirement. The Ca content reported in this study falls within the range reported by Moleele (1998) and Nsinamwa *et al.* (2005) in Botswana, Javed *et al.* (2008) in Pakistan, and Tefera *et al.* (2009) in Swaziland rangelands. Except for *D. longiflora*, Mg levels of grasses from the LV area were above the animal diet requirements as suggested by National Research Council (1996) and McDowell (1997). Most grasses harvested from the LMV area had Mg level within the dietary requirement of ruminants. Mean Mg concentrations (range: 0.7–3.7 g/kg DM) found in this study overlap with those reported in Botswana (Nsinamwa *et al.*, 2005) (range 0.4–2.4 g/kg DM), and in Swaziland savannas (Tefera *et al.*, 2009) (range: 0.4–20.0 g/kg DM).

All grass species harvested from the two ecological zones were deficient inK and P. The K levels recorded in this study are within the range reported by Nsinamwa *et al.* (2005) (range: 0.4–2.4 g/kg DM), but far lower than those reported by Tefera *et al.* (2009) (range: 35.8–123 g/kg DM). The consistent low P content across all grasses might be due to the low P content in the soil. According to Murdoch (1970), Sutcliffe (1975), and Tefera (2010), most savanna rangelands in Swaziland are deficient in P. The low P concentrations can also be linked to the advanced stage of maturity of the plants (Minson, 1990). Grass samples were collected towards the end of the rainy, growing season. Lower dietary P has been associated with reduced feed intake by ruminants (Little, 1982). According to Moleele (1998), however, it is the Ca: P ratio which is considered a better index for grass utilization than the absolute concentrations of the two minerals. Calcium is closely related to P metabolism in the formation of bones. In this study, Ca: P ratio was variable between ecological zones, with higher ratios being observed in the LMV than the LV area. The highest ratio was obtained (Table 8) in *D. argyrograpta* at 1 800: 1 in the LV area and 2270: 1 in the LMV area. In general, most grasses had high Ca: P ratio which is outside the range required for optimal absorption of both minerals. It is therefore expected that animals grazing on these communal lands may encounter problems associated with mineral imbalances. In these regions, supplementary feeds during the dry season originate from cereal by–products and crop residues that have low Ca and high P levels, and these feed stuffs are likely to correct the imbalances. It has been suggested that high Ca: P ratio is associated with poor conception and infertility in cattle (Khanal and Subba, 2001), bone diseases or hormonal imbalances reflecting a chronic disturbance.

The significant variations in the concentrations of micro elements among grass species found in this study were also reported in Botswana (Nsinamwa *et al.* 2005), Pakistan (Zafar *et al.* 2007; Javed *et al.* 2008), and Swaziland (Tefera *et al.*, 2009). These variations among and within forage species reflect differences in soil nutrient uptake (Zafar *et al.*, 2007). –In general, for most grass species, micro–nutrient concentrations are either within or above the dietary requirements for ruminants as suggested by National Research Council (1996) and McDowell (1997). In particular, as regards for Zn and Fe, all grass species harvested from both study areas had above critical levels sufficient to meet the requirement of the ruminants. The observed ranges of microelement concentrations found in this study correspond to findings by Tefera *et al.* (2009). In contrast, Nsinamwa *et al.* (2005), reported a far greater values for Zn (range: 1.67–36.3 g/kg DM), Fe (Range: 4.9–58.9 g/kg DM) and Cu (1.2–12.6 g/kg DM).

CONCLUSION

Our study indicates that most forage species in the study areas would meet the macro and micro–nutrient requirements for grazing ruminants managed in the communal grazing lands. Phosphorus and K, however, were deficient in all forage species. Correcting P deficiencies by cost–effective supplementing could result in acceptable Ca: P ratio. Provision of mineral licks could ensure that P and K requirements of animals are met. In order to enhance utilization of the high–fibre, low–protein grasses; it is desirable that farmers feed animals conventional and non–conventional protein supplements. This study also highlights the need for further research into the *in vitro* ruminal fermentation of grass species that occur in the study areas in order to provide information on the ability of the grasses to support microbial activity and assess the potential availability of nutrients. In addition, future studies should also define the correlation between mineral content of the forages (grasses and browses), the soil and the body tissues of grazing animals.

ACKNOWLEDGEMENTS

This publication is an output from a research project funded by the University of Swaziland Research Centre. The views expressed are not necessarily those of UNISWA Research Centre.

REFERENCES

AOAC 1999 Official Methods of Analysis of AOAC International, 16th ed. Method number 976.06. Association of Official Analytical Chemists, Arlington, VA, USA.

Dyksterhuis, E.J. 1949. Condition and management of rangeland based on quantitative ecology. Journal of Range Management 2: 104–115.

Environmental Centre for Swaziland 1993. The Swaziland Environment Action Plan (SEAP). http://www.ecs.co.sz/seap/projects_seap_chapter2.htm. Accessed: 2 June 2011.

Gonzâlez Ronquillo, M., Fondevilla, M., Barrios Urdaneta, A., Newman, Y. 1998. *In vitro* gas production from buffel grass (*Cenchrius ciliarius* L.) fermentation in relation to the cutting interval, the level of nitrogen fertilisation and the season of growth. Animal Feed Science and Technology 72: 19–32

Hardy, M.B., Walker, L.S. 1991. Determining sample size for assessing species composition in grassland. Journal of the Grassland Society of South Africa 8: 70–73.

Holecheck, J.L., Pieper, R., Herbel, C.H. 2001. Range management principles and practices. 4th ed. Prentice–Hall. Upper Saddle River, N.J. USA.

Javed, I.S., Inam–Ur, R., Haq, W., Muhammad, Y., Ijaz, J. 2008. Mineral composition, palatability and digestibility of free rangeland grasses of Northern Grasslands of Pakistan. Pakistan Journal of Botany 40: 2059–2070.

Khanal, R.C., Subba, D.B. 2001. Nutritional evaluation of leaves from some major fodder trees cultivated in the hills of Nepal. Animal Feed Science and Technology 92: 17–32.

Kunene, M.L.B. 2005. Studies on the environmental impact of the cattle dip–tank system in the rangelands of the Swazi Nation Land. M.Sc. Thesis, University of Swaziland, Swaziland.

Little, D.A. 1982. Utilization of minerals. In: Hacker J B (ed), Nutritional limits to animal production from pastures. Commonwealth Agricultural Bureaux, Farnham Royal, UK. pp.259–283.

McDowell, L.R. 1997. Minerals for grazing animals in tropical regions. Extension Bulettin. Department of Animal Science, Centre for Tropical Agriculture, University of Florida, Gainesville, USA.

Minson, D.J. 1990. Forage in ruminant nutrition. Academic Press. San Diego, California, USA.

Mlay, P.S., Pereka, A., Phiri, E.C., Balthazary, S., Igusti, J., Hvelplund, T., Weisbjerg, M.R., Madsen, J. 2006. Feed value of selected tropical grasses, legumes and concentrates. Veterinarski Arhiv 76: 53–63.

Moleele, N.M. 1998. Encroacher woody plant browse as feed for cattle: cattle diet composition for three seasons at Olifant Drift, South–east Botswana. Journal of Arid Environments 40: 255–268.

Murdoch, G. 1970. Soils and land capability in Swaziland. Ministry of Agriculture Mbabane, Swaziland.

Musisi, F.L., Lawrence, J.A., 1995. Prospects for control of tickborne diseases in cattle by immunization in eastern, central, and southern Africa. Agriculture and Human Values 12: 95 – 106.

National Research Council 1996. Nutrient requirements for beef cattle. 7th edn. National Academy Press, Washington, DC, USA.

Nogueira Filho, J.C.M., Fondevila, M., Barrios Urdaneta, A., Gonzâlez Ronquillo, M. 2000. *In vitro* microbial fermentation of tropical grasses at an advanced maturity stage. Animal Feed Science and Technology 83: 145–157.

Nsinamwa, M. , Moleele, N.M., Sebego, R. J. 2005. Vegetation patterns and nutrients in relation to grazing pressure and soils in the sandveld and hardveld communal grazing areas of Botswana. African Journal of Range Forage Science 22: 17–28.

Perkin–Elmer, I. 1982. Perkin–Elmer AAS manual (model 2389) Norwalk, Connecticut, USA.

SAS® 1996. User's guide: Statistics, Version 6.12. SAS Institute, Inc. Cary, NC, USA.

Solomon, T.B., Dlamini, B.J., Dlamini, A.M., Dlamini, W.M. 2007. Evaluation of vegetation diversity, veld condition, soil seed bank characteristics and diet Selection of cattle in Swazi rangelands. Final Research Project Report UNISWA Research Center, University of Swaziland, Swaziland. pp 112.

Sutcliffe, J.P. 1975. A field guide to soils of Swaziland. Ministry of Agriculture, Mbabane. Swaziland.

Tainton, N.M., Edwards, P.J., Mentis, M.T. 1980. A revised method for assessing veld condition. Proceedings of The Grassland Society of Southern Africa 15: 37–42.

Tefera, S., Dlamini, B.J., Dlamini, A.M. 2010. Changes in soil characteristics and grass layer condition in relation to land management systems in the semi–arid savannas of Swaziland. Journal of Arid Environments 74: 675–684.

Tefera, S., Mlambo, V., Dlamini, B.J., Dlamini, A.M., Koralagama, K.D.N., Mould, F.L. 2009. Chemical composition and *in vitro* ruminal fermentation of selected grasses in the semi–arid savannas of Swaziland. African Journal Range Forage Science 26: 9–27.

Ternout, J.H. 1990. Phosphorous and beef production in northern Australia. 3. Phosphorous in cattle – a review. Tropical Grasslands 24: 159–169.

Vallentine, J.F. 1990. Grazing management. Academic Press, Inc. San Diego, CA, USA.

Van Soest, P.J., Robertson, J.B., Lewis, B.A. 1991. Methods for dietary fiber, neutral detergent fiber, and non–starch polysaccharides in relation to animal nutrition. Journal of Dairy Science 74: 3583–3597.

Vorster, M. 1982. The development of the ecological index method for assessing veld condition in the Karoo. Proceedings of The Grassland Society of Southern Africa 17: 84–89.

Walton, P.D. 1983. Production and management of cultivated forages. Reston Publishing Company Inc., Virginia, USA., pp 335.

Zafar, K., Muhammad, A., Kafeel, A., Irfan, M., Muhammad, D. 2007. Evaluation of micro–minerals composition of different grasses in relation to livestock requirements. Pakistan Journal of Botany 39: 719–728.

14

Nutritional Composition and Antinutritional Factors of Little-Known Species of *Vigna*

Chinnamadasamy Kalidass and Veerabahu Ramasamy Mohan*

Ethnopharmacology unit, Research Department of Botany,

V.O.Chidambaram College, Tuticorin-628008, Tamil Nadu.

vrmohan_2005@yahoo.com

**Corresponding author*

SUMMARY

The seeds of *Vigna trilobata, V. radiata* var. *sublobata, V. umbellata, V. unguiculata* subsp. *cylindrica, V. aconitifolia, V. vexillata* and *V. bourneae* were collected from different geographical regions in Western Ghats, Tamil Nadu. They were analyzed for their proximate and mineral composition, vitamins (niacin and ascorbic acids), fatty acid profiles, amino acid profiles of total seed proteins, *in vitro* protein digestibility (IVPD) and certain antinutritional factors. The crude protein ranged from 18.24 to 26.12%, crude lipid 3.18 to 6.48%, total dietary fibre 3.42 to 7.48%, ash 3.10 to 4.12% and carbohydrates 59.44 to 72.06%. Energy values of the seeds were 1584.87 – 1644.34 kJ100g^{-} DM, which were comparable to those of other legumes. The fatty acid profiles of all the species of Vigna revealed that the seed lipids contained higher concentrations of palmitic, oleic, linoleic and linolenic acids. The essential amino acid profiles of total seed proteins compared favourably with FAO/WHO (1991) requirements, except that there were deficiencies of sulphur containing amino acids in all the species of Vigna. The IVPD of the different Vigna species ranged from 70.38 to 79.12%. Antinutritional substances like total free phenolics, tannins, L-DOPA (3, 4-dihydroxyphenylalanine), phytic acid, hydrogen cynide, trypsin inhibitor activity, oligosaccharides and phytohaemagglutinating activity were also investigated. The antinutritional factors that were detected were thought to have little nutritional significance if the beans are properly processed.

Key words: *Vigna* spp,; amino acid profiles; fatty acid profiles; IVPD; antinutrient.

INTRODUCTION

Inadequate availability and consumption of protein foods in developing countries like India are a major concern as large segments of population of these countries suffer from protein malnutrition. The little known legume crops could make a useful contribution to world's food population because some of them are well adapted to adverse environmental conditions, exhibit resistance to pests and pathogens and have good nutritional qualities. Hence the identification and introduction of such, hitherto, little known grain legumes, including wild legumes consumed by different tribal sects throughout the tropical regions as well as their genetic improvement of the quality and quantity of protein would be a great contribution (Maikhuri *et al.*, 1991; Arinathan *et al.*, 2009; Kala *et al.*, 2010). Information on the chemical composition of these potential protein sources and their possible utilization as human food is inadequate. However, some of these species are actually consumed by several human groups (like Mayan people) in the American continent since centuries ago. Hence in the present study the mature seeds of different *Vigna* species, which were eaten by the tribal people of the Western Ghats, Tamil Nadu, were subjected to biochemical investigation with a view to assessing their nutritional potential.

MATERIALS AND METHODS

Collection of seeds

Seven species of *Vigna* were gathered as mature pods in individual species from natural strands of various geographical regions, Western Ghats, Tamil Nadu, viz, *Vigna trilobata* (L.) Verdc; *V. aconitifolia* (Jacq.) Marechal and *V. vexillata* (L.) A.Rich from Siruvani reserve forest, Coimbatore district; *Vigna radiata* (L.) Wilczek var. *sublobata* (Roxb.) Verdc. and *V. bourneae* Gamble from Seithur reserve forest and *Vigna unguiculata* (L.) Walp. subsp. *cylindrica* (L.) Eselt. from Ayyanarkoil reserve forest, Virudhunagar district and *Vigna umbellata* (Thumb.) Ohwi & Ohashi from Petchiparai reserve forest, Kanniyakumari district. The collected plants were identified taxonomically with the help of various floras. After thoroughly drying in the sun, the pods were thrashed to remove seeds. The seeds, after thorough cleaning and removal of broken seeds and foreign materials, the mature seeds were stored in airtight plastic jars at room temperature (25°C).

Proximate composition

The moisture content (%) was determined by drying 50 transversely cut seeds in an oven at 80°C for 24h. The air-dried samples were powdered separately in a Wiley mill (Scientific Equipment, Delhi, India) to 60-mesh size and stored in screw-capped bottles at room temperature for further analysis.

The nitrogen content was estimated by the micro-Kjeldahl method (Humphries, 1956) and the crude protein content was calculated (N x 6.25). Crude lipid content was determined using Soxhlet apparatus. The ash content was determined by heating 2g of the dried sample in a silica dish at 600°C for 6h (AOAC, 2005). Total dietary fibre (TDF) was estimated by the non-enzymatic-gravimetric method (Li and Cardozo, 1994). To determine the TDF, duplicate 500 mg ground samples were taken in separate 250mL beakers. To each beaker 25 mL water was added and gently stirred until the samples were thoroughly wetted, (i.e. no clumps were noticed). The beakers were covered with Al foil and allowed to stand 90 min without stirring in an incubator maintained at 37°C. After that, 100 mL 95% ethanol were added to each beaker and allowed to stand for 1h at room temperature (25±2°C). The residue was collected under vacuum in a pre-weighed crucible containing filter aid. The residue was washed successively with 20 mL of 78% ethanol, 10 mL of 95% ethanol and 10 ml acetone. The crucible containing the residue was dried ≥2 h at 105°C and then cooled ≥ 2h in a desiccator and weighed. One crucible-containing residue was used for ash determination at 525°C for 5h. The ash-containing crucible was cooled for ≥ 2h in a desiccator and weighed. The residue from the remaining duplicate crucible was used for crude protein determination by the micro-Kjeldahl method as already mentioned. The TDF was calculated as follows.

$$TDF\% = 100 \times \frac{Wr - [(P+A)/100]\, Wr}{Ws}$$

Where Wr = mg residue, P = % protein in the residue; A = % ash in the residue, and Ws = mg sample.

The nitrogen free extract (NFE) was obtained by difference (Muller and Tobin, 1980). The energy value of the seed (kJ) was estimated by multiplying the percentages of crude protein, crude lipid and NFE by the factors 16.7, 37.7 and 16.7, respectively (Siddhuraju *et al.*, 1996).

Analysis of minerals and vitamins

Five hundred milligrams of the ground legume seed was digested with a mixture of 10 mL concentrated nitric acid, 4mL of 60% perchloric acid and 1 mL concentrated sulphuric acid. After cooling, the digest was diluted with 50 mL of deionised H_2O, filtered through Whatman No. 42 filter paper and filtrates were made up to 100 mL in a glass volumetric flask with deionised H_2O. All minerals except phosphorus were analyzed from a triple acid-digested sample by atomic absorption spectrophotometry, ECIL (Electronic Corporation of India Ltd., India) (Issac and Johnson, 1975). The phosphorus content in the triple acid digested extract was determined colorimetrically (Dickman and Bray, 1940).

Ascorbic acid and niacin were extracted and estimated as per the method given by Sadasivam and Manickam (1996). For the extraction of ascorbic acid, 3g air-dried powdered sample were ground with 25 mL of 4% oxalic acid and filtered. Bromine water was added drop by drop to 10 mL of the filtrate until it turned orange-yellow to remove the enolic hydrogen atoms. The excess of bromine was expelled by blowing in air. This filtrate was made up to 25 mL with 4% oxalic acid and used for ascorbic acid estimation. Two milliliters of the extract was made up to 3 mL with distilled H_2O in a test tube. One milliliter of 2% 2, 4-dinitrophenyl hydrazine reagent and a few drops of thiourea were added. The contents of the test tube were mixed thoroughly. After 3hr incubation at 37°C, 7 mL of 80% H_2SO_4 were added to dissolve the osazone crystals and the absorbance was measured at 540nm against a reagent blank.

For the extraction of niacin, 5g air-dried powdered sample was steamed with 30 mL concentrated H_2SO_4 for 30min. After cooling, this suspension was made up to 50 mL with distilled H_2O and filtered. Five milliliters of 60% basic lead acetate was added to 25 mL of the filtrate. The pH was adjusted to 9.5 and centrifuged (10,000g for 10 min at room temperature) to collect the supernatant. Two milliliters of concentrated H_2SO_4 was added to supernatant. The mixture was allowed to stand for 1hr and centrifuged (10,000g for 10 min at room temperature). The 5 mL of 40% $ZnSO_4$ was added to the supernatant. The pH was adjusted to 8.4 and centrifuged (10,000g for 10 min at room temperature) again. Then the pH of the collected supernatant was adjusted to pH 7 and used as the niacin extract. For estimation, 1 mL extract was made up to 6 mL with distilled water in a test tube, 3 mL cyanogen bromides was added and shaken well, followed by addition of 1 mL of 4% aniline. The yellow colour that developed after 5min was measured at 420nm against a reagent blank. The ascorbic acid and niacin contents present in the sample were calculated by referring to a standard graph and expressed as milligrams per 100 grams of powdered samples.

Lipid extraction and fatty acid analysis

The total lipids was extracted from the seeds according to the method by Folch *et al.*, (1957) using chloroform and methanol mixture in ratio of 2: 1 (v/v). Methyl esters were prepared from the total lipids by the method of Metcalfe *et al.*, (1966). Fatty acid analysis was performed by gas chromatography (ASHMACO, Japan; Model No: ABD20A) using an instrument equipped with a flame ionization detector and a glass column (2m X 3mm) packed with 1% diethylene glycol succinate on chromosorb W. The temperature conditions for GC were injector 200°C and detector 210°C. The temperature of the oven was programmed from 180°C and the carrier gas was nitrogen at a flow rate of 30 mL/min. Peaks were identified by comparison with authentic standards, quantified by peak area integration and expressed as weight percentage of total methyl esters; the relative weight percentage of each fatty acid was determined from integrated peak areas.

Amino acid analysis

The total seed protein was extracted by a modified method of Basha *et al.*, (1976). The extracted proteins were purified by precipitation with cold 20% trichloroacetic acid (TCA). A protein sample of 30mg was hydrolysed by 6N HCL (5mL) in an evacuated sealed tube, which was kept in an air oven maintained at 110°C for 24 hr. The sealed tube was broken and the acid removed completely by repeated flash evaporation after the addition of deionized H_2O. Dilution was effected by means of citrate buffer pH 2.2 to such an extent that the solution contained 0.5mg protein mL^{-1}. The solution was passed through a millipore filter (0.45μM) and derivatized with O-phthaldialdehyde by using an automated pre-column (OPA). Amino acids were analysed by a reverse – phase HPLC (Method L 7400, HITACHI, Japan) fitted with a denali C_{18} 5 micron column (4.6X 150mm). The flow rate was 1 mL min^{-1} with fluorescence detector. The cystine content of protein sample was obtained separately by the Liddell and Saville (1959) method. For the determination of tryptophan content of proteins, aliquots containing known amounts of proteins were dispersed into glass ampoules together with 1 mL 5M NaOH. The ampoules were flame sealed and incubated at 110°C for 18 hr. The tryptophans content of the alkaline hydrolysates were determined colorimetrically using the method of Spies and Chambers (1949) as modified by Rama Rao *et al.*, (1974). The contents of the different amino acids were expressed as $g100g^{-1}$ proteins and were compared with FAO/WHO (1991) reference pattern. The essential amino acid score (EAAS) was calculated as follows:

$$\text{EAAS} = \frac{\text{EAAS (g/100g of total protein)}}{\text{EAAS (g/100g reference pattern)}} \times 100$$

Analysis of antinutritional compounds

The antinutritional compounds, total free phenolics (Bray and Thorne, 1954), tannins (Burns, 1971), the non-protein amino acid, L-DOPA (3, 4-dihydroxyphenylalanine) (Brain, 1976), phytic acid (Wheeler and Ferrel, 1971), hydrogen cyanide (Jackson, 1967) were quantified. Trypsin inhibitor activity was determined by the enzyme assay of Kakade *et al.*, (1974) by using benzoil-DL-arginin-*p*-nitroanilide (BAPNA) as a substrate. One trypsin inhibitor unit (TIU) has been expressed as an increase of 0.01 absorbance units per 10ml of reaction mixture at 410nm. Trypsin inhibitor activity has been defined in terms of trypsin units inhibited per mg protein.

Extraction, TLC separation and estimation of Oligosaccharides

Extraction of oligosaccharides was done following the method of Somiari and Balogh (1993). Five grams each raw seed flours of all the samples were extracted separately with 50 mL of 70% (v/v) aqueous ethanol and kept on an orbital shaker at 130 rpm for 13 hr and then filtered through Whatman No. 1 filter paper. Residue was further washed with 25mL of 70% (v/v) ethanol. The filtrates obtained were pooled and vacuum-dried at 45°C. The concentrated sugar syrup was dissolved in five ml of distilled water.

Separation of oligosaccharides was done by TLC. Thirty grams of cellulose-G powder were dissolved in 45 mL of double distilled water and shaken well until the slurry was homogeneous. TLC plates were coated with the slurry and air-dried. Spotting of the sugar samples was done by using micropipettes. 5 μl aliquots of each sample were spotted thrice separately. The plates were developed by using a solvent system of n-propanol, ethyl acetate and distilled H_2O (6:1:3), and dried (Tanaka *et al.*, 1975). The plates were sprayed with α –naphthol (1%, w/v). Plates were dried in a hot-air oven. The separated spots were compared with standard sugar spots. A standard sugar mixture containing raffinose, stachyose and verbacose (Sigma Chemical). Separated sugars that appeared were verbascose, stachyose and raffinose. The sugar spots were scrapped eluted in 2 mL of distilled H_2O kept overnight at room temperature and filtered through Whatman No. 1 filter paper. The filtrates were subjected to quantitative estimation.

The eluted individual oligosaccharides were estimated by the method of Tanaka *et al.*, (1975). One mL of the eluted and filtered sugar solution was treated with one ml of 0.2 M thiobarbituric acid and one ml of concentrated HCL. The tubes were boiled in a water bath for exactly 6 min. After cooling, the oligosaccharide contents were quantified in a Elico UV-Spectrophotometer model SL 150 at 432 nm. Average values of triplicate estimations were calculated and the content of oligosaccharides was expressed on dry weight basis.

Quantitative determination of phytoheamagglutinating (Lectin) activity.

Lectin activity was determined by the method of Almedia *et al.* (1991). One g of air-dried seed flour was stirred with 10 mL of 0.15N sodium chloride solution for 2hr and the pH was adjusted to 4.0. The contents were centrifuged at 10,000 X g for 20 min. and the supernatants were collected separately. The protein content was estimated by the Lowry *et al.* (1951) method.

Human blood (blood groups A, B and O) was procured from the blood bank of Jothi Clinical Laboratory, Tuticorin. Blood erythrocyte suspensions were prepared by washing the blood samples separately with phosphate-buffered saline and centrifuged for 3min at low speed (3,000 g for 10 min at room temperature). Supernatants were removed with Pasteur pipettes. The washing procedure was repeated three times. The washed cells were diluted by one drop of cells with 24drops of phosphate – buffered saline.

The determination of lectin was done by the method of Tan *et al.,* (1983). Clear supernatant (50μl) was poured into the depression (pit) on a microtitration plate and serially diluted 1:2 with normal saline. The human blood erythrocyte (A, B and O blood groups) suspensions (25μl) were added to each of the twenty depressions. The plates were incubated for 3hours at room temperature. After the incubation period, the titer values were recorded. One Haemagglutinating unit (HU) is defined as the least amount of heamagglutinin that will produce positive evidence of agglutination of 25μl of a blood group erythrocyte after 3hr incubation at room temperature. The phytoheamagglutinating activity was expressed as heamagglutinating units (HU)/mg protein.

Determination of *in vitro* protein digestibility (IVPD).

This was determined using the multi-enzyme technique (Hsu *et al.*, 1977). The enzymes used for IVPD were purchased from Sigma Chemical Co., St. Louis, MO, USA. Calculated amounts of the control (casein) and sample were weighed out, hydrated in 10ml of distilled water and refrigerated at 5°C for 1hr. The samples containing protein and enzymes were all adjusted to pH 8.0 at 37°C. The IVPD was determined by the sequential digestion of the samples containing protein with a multi-enzyme mixture [trypsin (porcine pancreatic trypsin–type IX with 14190 BAEE unites per mg protein), α-chymotrypsin (bovine pancreatic chymotrypsin–type II, 60 units per mg powder) and peptidase (porcine intestinal peptidase–grade III, 40 units per g powder)] at 37°C followed by protease (type IV from *Streptomyces griseus*) at 55°C. The pH drop of the samples from pH 8.0 was recorded after 20min of incubation. The IVPD was calculated according to the regression equation $Y = 234.84 - 22.56\ X$, where Y is the % digestibility and X the pH drop.

Statistical analysis

The data were subjected to a one-way analysis of variance (ANOVA) and the significance if difference between means at 5% was determined by Duncan Multiple Range Test (DMRT) using SPSS software (version 11.5; SPSS Inc., Chicago IL, USA). Results were expressed as means values ± standard deviations of three separate determinations.

RESULTS AND DISCUSSION

The crude protein content of *Vigna* species (Table 1) investigated in the present study, *Vigna radiata* var. *sublobata*, *V. aconitifolia* and *V. vexillata* are found to be higher than pulse crops commonly consumed in India such as *Vigna subterranean* and *Glycine max* (Oluwole and Taiwo, 2009; Martin *et al.*, 2010). The crude lipid content of *Vigna radiata var. sublobata, V. aconitifolia, V. umbellata, V. unguiculata* subsp. *cylindrica*, *V. vexillata* and *V. bourneae* is found to be more or less equal to the values of other species like *Vigna radiata* (Noel and Rosario, 1989); *V. trilobata* (Siddhuraju *et al.*, 1992); *V. sesquipedalis, V. calacaratus, V. glabrescens* and *V. sublobata* (Rajaram and Janardhanan, 1990, 1992); *V. capensis* and *V. sinensis* (Mohan and Janadhanan, 1993) and *V. umbellata* (Mohan and Janardhanan, 1994). The total dietary fibre content of *V. trilobata, V. radiata* subsp. *sublobata* and *V. aconitifolia* is found to be higher when compared with the other commonly cultivated pulses such as *Cicer arietinum*, *Macrotyloma uniflorum*, *Pisum sativum*, *Cajanus cajan* and *Vigna mungo* (Premakumari *et al.*, 1984). The ash content of the investigated tribal pulses would be important to the extent that it contains the nutritionally important mineral elements. It appears that all the currently investigated tribal pulses have a high range of carbohydrates (59.44 – 72.06%) because of their low fat content when compared with *Arachis hypogaea* and *Glycine max*, which have less carbohydrate content at 26.1% and 20.9% respectively (Narasinga Rao *et al.*, 1989). All the investigated *Vigna* species have higher energy value than those of *Phaseolus vulgaris, P. limensis, Vigna unguiculata, Cicer arietinum, Pisum sativum* and *Lens culinaris* (Meiners *et al.*, 1976a).

Food legumes have been recognized as important sources of several minerals in Indian diets (Gopalan *et al.*, 1978). Table 2 shows the mineral composition of the samples. In the present investigation, all the *Vigna* species exhibit higher levels of potassium content than that of other legumes *Phaseolus vulgaris*, *P. lunatus*, *Vigna unguiculata*, *Cicer arietinum*, *Pisum sativum* and *Lens culinaris* (Meiners *et al.*, 1976b); *Cicer arietinum* (Alajaji and El-Adawy, 2006; Zia Ul-haq *et al.*, 2007) and also register a higher level of potassium when compared with recommended dietary allowance values NRC/NAS (1980). The high content of potassium can be utilized beneficially in the diets of people who take diuretics to control hypertension and suffer from excessive excretion of potassium through the body fluid (Siddhuraju *et al.*, 2001). The calcium content of *Vigna trilobata* and *Vigna bourneae* is found to be higher than that of the recommended dietary allowances of calcium (>400mg) for children by the Indian Council of Medical Research (ICMR, 1992), All the investigated *Vigna* species contain higher levels of calcium, magnesium and iron when compared with *Cicer arietinum* ((Meiners *et al.*, 1976b) and *Cajanus cajan* (Nwokolo 1987). The phosphorus content is found to be higher than the *Vigna umbellata* var. RBL-40 (Rajaram and Janardhanan, 1990). The ratios of sodium to potassium (Na/K) and calcium to phosphorus (Ca/P) are shown in Tables 2. The Na/K ratio in the body is of great concern for prevention of high blood pressure, Na/K ratio less than one is recommended. Hence, most of the investigated samples would probably reduce high blood pressure disease because they had Na/K less than one. Modern diets which are rich in animal proteins and phosphorus may promote the loss of calcium in the urine (Shills and Young, 1988). This has led to the concept of the Ca/P ratio. If the Ca/P ratio is low (low calcium, high phosphorus intake) more than the normal amount of calcium may be lost in the urine, decreasing the calcium level in bones. Food is considered "good" if the ratio is above one and "poor" if the ratio is less than 0.5 (Nieman *et al.*, 1992). The Ca/P ratio in the present study ranged between 1.23 to 2.35 indicating they would serve as good sources of minerals for bone formation.

The presently investigated *Vigna* species exhibits the highest level of niacin content (Table 3). Which is found to be higher than that of an earlier report in *Cajanus cajan, Dolichos lablab, D. biflorus, Mucuna pruriens*, *Phaseolus mungo, Vigna catjang* and *Vigna* species (Rajyalakshmi and Geervani, 1994) and *Cicer arietinum* (Alajaji and El-Adawy, 2006). The presently investigated tribal pulses also register higher level of ascorbic acid content than *Vigna radiata* and *V. mungo* (Kakati *et al.*, 2010).

The data on fatty acid composition of the total lipids of different species of *Vigna* (Table 4) indicate that palmitic acid, oleic acid, linoleic acid and linolenic acid are the major fatty acids, as in *Glycine max*, *Vigna mungo* and *V. radiata* (Salunkhe *et al.*, 1982).

Table 1: Proximate composition of seven species of *Vigna*. *$

Components	*Vigna trilobata*	*V. radiata* var. *sublobata*	*V. aconitifolia*	*V. umbellata*	*V. unguiculata* subsp. *cylindrica*	*V. vexillata*	*V. bourneae*
Moisture	5.24 ± 0.05^{d}	4.80 ± 0.08^{c}	3.84 ± 0.06^{a}	5.56 ± 0.02^{e}	5.44 ± 0.05^{e}	6.94 ± 0.01^{f}	4.24 ± 0.05^{b}
Crude protein (Kjeldahl N × 6.25)	22.10 ± 0.46^{b}	25.64 ± 0.17^{d}	24.40 ± 0.10^{c}	26.12 ± 0.56^{d}	18.24 ± 0.57^{a}	25.84 ± 0.56^{d}	21.40 ± 0.01^{b}
Crude lipid	6.48 ± 0.59^{d}	3.58 ± 0.56^{a}	5.26 ± 0.53^{b}	4.18 ± 0.50^{a}	3.18 ± 0.01^{a}	5.80 ± 0.17^{d}	4.20 ± 0.02^{a}
TDF (Total Dietary Fibre)	7.48 ± 0.34^{d}	6.12 ± 0.58^{c}	6.78 ± 0.71^{d}	4.74 ± 0.01^{b}	3.42 ± 0.55^{a}	4.89 ± 0.05^{b}	5.25 ± 0.05^{b}
Ash	3.12 ± 0.56^{a}	3.48 ± 0.56^{a}	4.12 ± 0.58^{a}	4.04 ± 0.04^{a}	3.10 ± 0.58^{a}	3.94 ± 0.56^{a}	3.38 ± 0.05^{a}
Nitrogen Free Extractives (NFE)	60.82	61.18	59.44	60.92	72.06	59.53	65.77
Calorific value (kJ100g^{-1} DM)	1629.06	1584.87	1598.43	1611.15	1627.90	1644.34	1614.08

*All values are means of triplicate determinations expressed on a dry weight basis ± denotes standard error.

$ Mean values in the following row sharing a common letter are not statistically significant according to DMRT.

Table 2. Mineral composition of seven species of *Vigna* (mg 100 g^{-1} seed flour) *$

Components	*Vigna trilobata*	*V. radiata* var. *sublobata*	*V. aconitifolia*	*V. umbellata*	*V. unguiculata* subsp. *cylindrica*	*V. vexillata*	*V. bourneae*
Sodium	34.12 ± 0.01^{a}	32.10 ± 0.03^{a}	24.78 ± 0.58^{c}	22.18 ± 0.02^{d}	18.78 ± 0.06^{e}	26.48 ± 0.57^{b}	27.34 ± 0.58^{b}
Potassium	1696.10 ± 5.2^{e}	1948.40 ± 0.01^{c}	2452.78 ± 0.60^{a}	2618.16 ± 2.39^{a}	1848.10 ± 0.57^{d}	2012.62 ± 1.08^{b}	2142.30 ± 0.57^{b}
Calcium	418.46 ± 0.01^{a}	284.20 ± 1.72^{b}	264.12 ± 1.53^{c}	274.14 ± 0.38^{b}	258.54 ± 0.56^{c}	201.18 ± 0.01^{d}	438.14 ± 0.01^{a}
Magnesium	242.00 ± 0.57^{a}	158.24 ± 0.56^{c}	144.10 ± 0.58^{d}	156.10 ± 0.58^{d}	161.20 ± 0.33^{c}	178.10 ± 0.01^{a}	168.20 ± 1.20^{b}
Phosphorus	178.18 ± 0.02^{c}	212.12 ± 0.57^{b}	141.28 ± 0.02^{d}	126.20 ± 0.58^{e}	210.10 ± 0.56^{b}	248.08 ± 0.58^{a}	231.16 ± 0.31^{a}
Iron	9.48 ± 0.11^{c}	8.40 ± 0.01^{d}	6.48 ± 0.01^{f}	6.12 ± 0.57^{f}	11.40 ± 0.01^{a}	10.40 ± 0.58^{b}	7.54 ± 0.01^{e}
Zinc	2.56 ± 0.01^{a}	1.74 ± 0.01^{b}	1.21 ± 0.01^{d}	0.78 ± 0.01^{e}	2.40 ± 0.01^{a}	1.58 ± 0.57^{c}	2.10 ± 0.01^{a}
Copper	1.22 ± 0.05^{a}	0.84 ± 0.01^{c}	0.48 ± 0.02^{f}	0.54 ± 0.01^{f}	0.58 ± 0.01^{e}	1.10 ± 0.01^{b}	0.72 ± 0.01^{d}
Manganese	2.34 ± 0.03^{b}	1.24 ± 0.01^{d}	0.94 ± 0.01^{f}	1.10 ± 0.01^{e}	1.68 ± 0.02^{c}	2.50 ± 0.01^{b}	2.66 ± 0.01^{a}
Na/K	0.02	0.02	0.01	0.01	0.01	0.01	0.01
Ca/P	2.35	1.34	1.87	2.17	1.23	0.81	1.90

*All values are of means of triplicate determination expressed on dry weight basis ± denotes Standard error.

$ Mean values in the following row sharing a common letter are not statistically significant according to DMRT.

The linolenic acid content of *Vigna aconitifolia, V. umbellata, V. vexillata* and *V. bourneae* is more than that of *Vigna unguiculata* (Thangadurai, 2005), where as the palmitic acid and oleic acid contents of *Vigna aconitifolia, V. unguiculata* subsp. *cylindrica, V. vexillata* and *V. bourneae* seems to be equal to that of *Cajanus cajan* (Salunkhe *et al*., 1982). Similarly the linoleic acid content of all the investigated *Vigna* species is relatively high when compared with *Vigna unguiculata* and *Phaseolus vulgaris* (Omogbai, 1990).

Table 3. Vitamins (niacin and ascorbic acid) content of seven species *Vigna* (mg $100g^{-1}$ seed flour) *$

Species	Niacin	Ascorbic acid
Vigna trilobata	$26.20{\pm}0.02^{e}$	$52.04{\pm}0.58^{f}$
V. radiata var. *sublobata*	$32.10{\pm}0.56^{b}$	$62.14{\pm}0.58^{d}$
V. aconitifolia	$38.48{\pm}0.58^{a}$	$79.40{\pm}0.56^{b}$
V. umbellata	$22.30{\pm}0.57^{f}$	$58.10{\pm}0.57^{e}$
V. unguiculata subsp. *cylindrica*	$18.36{\pm}0.01^{g}$	$98.14{\pm}0.57^{a}$
V. vexillata	$30.40{\pm}0.57^{c}$	$66.42{\pm}0.57^{c}$
V. bourneae	$28.34{\pm}0.59^{d}$	$48.36{\pm}0.63^{g}$

*All values are of means of triplicate determination expressed on dry weight basis ± denotes standard error.
$ Mean values in the following row sharing a common letter are not statistically significant according to DMRT.

The essential amino acid profiles of total seed proteins are compared favourably with the FAO/WHO (1991) requirement pattern, except that there were deficiencies of sulphur containing amino acids in all the *Vigna* species and also in threonine content of *Vigna trilobata,* leucine content of *V. aconitifolia*; lysine content of *V. trilobata,* and *V. umbellata* (Table 5). Among the seven species of *Vigna*, the *Vigna radiata* var. *sublobata*; *V. unguiculata* subsp. *cylindrica* and *V. vexillata* register higher amounts of IVPD than the other four species of *Vigna* presently investigated (Table 6) and their protein digestibility is found to be higher than that of *Cajanus cajan* (Singh and Eggum, 1984) and *Cicer arietinum* (Attia *et al*., 1994).

The problem of plant protein digestibility has been suggested to be because of the interplay of several factors such as protease inhibitors, phytates, oxalates, lectins, goitrogens and other antinutritional factors. In societies where legumes are consumed rather than much more expensive animal foods, there is bound to be great concern over the level of antinutrients in the diets. For this reason, in the present investigation an attempt has been made to detect the presence of certain antinutritional factors in different species of *Vigna* (Table 6). The content of total free phenolics in the samples investigated appears to be lower when compared with *Vigna sesquipedalis* (Rajaram and Janardhanan, 1990) and *Phaseolus lunatus* (Egbe and Akinyele, 1990) and comparable to that of *Vigna umbellata* and *V. sinensis* (Rajaram and Janardhanan, 1990). The tannin content of different species of *Vigna* is relatively lower than the domesticated legumes like *Vigna mungo, Cicer arietinum, Vigna unguiculata* and *V. radiata* (Khan *et al*., 1979; Rao and Deosthale, 1982). The content of the non-protein amino acid, L-DOPA, is relatively lower when compared with other tribal pulse reported earlier *M. pruriens* var. *utilis* (Mohan and Janardhanan, 1995; Vadivel and Janardhanan, 2000; Vadivel and Pugalenthi, 2008). It has been demonstrated that in *Mucuna pruriens*, the level of L-DOPA is significantly eliminated by dry heat treatment (Siddhuraju *et al*., 1996), cooking and autoclaving (Vijayakumari *et al*., 1996). Phytic acid content of different species of *Vigna* is found to be low when compared with that of some commonly consumed grain legumes like *Vigna mungo* (Kataria *et al*., 1988), *V. radiata,* (Kataria *et al*., 1989) tribal pulses, *Dolichos lablab* var. *vulgaris* (Vijayakumari *et al*., 1995) and *Mucuna pruriens* var. *utilis* (Janardhanan *et al*., 2003). Hydrogen cyanide is known to cause acute or chronic toxicity. The content of HCN level in the presently investigated tribal pulses is far below the lethal level i.e. 36mg/100g (Oke, 1969) and comparable with those of *Vigna sinensis* and *Pisum sativum* (Montgomery, 1980); *Dolichos lablab* var. *vulgaris, Bauhinia purpurea* (Vijayakumari *et al*., 1995, 1997); *Entada phaseoloides* (Siddhuraju *et al*., 2001). Though the trypsin inhibitor activity has been studied in a number of pulses, the results obtained in the present investigation cannot be compared because the expression of trypsin inhibitor activity, nature and concentration of the substrate, etc. are different. However, based on investigations that have been reported and where the same experimental conditions were used, the trypsin inhibitor activities in cultivated legumes like *Cajanus cajan* (67.1 – 71.3 TIUmg^{-1} protein) (Singh and Eggum, 1984) is higher than that of different *Vigna* species. Trypsin inhibitor activity has the greatest impact on the IVPD of the legumes when the former is known to be heat labile. Stachyose seems to be the principle oligosaccharide of all the *Vigna* species. It is in conformity with the earlier reports in cow pea (Onigbinde and Akinyele, 1983); *Canavalia ensiformis, Phaseolus lunatus* and *Canavalia gladiata* (Ravilleza *et al*., 1990) and *Dolichos lablab* var. *vulgaris* (Vijayakumari *et al*., 1995).

Table 4: Fatty acid profiles of seed lipids of seven species of *Vigna* [a].

Fatty acid (%)	*Vigna trilobata*	*V. radiata* var. *sublobata*	*V. aconitifolia*	*V. umbellata*	*V. unguiculata* subsp. *cylindrica*	*V. vexillata*	*V. bourneae*
Palmitic acid [C16:0]	24.78	16.10	19.28	26.48	20.70	20.40	21.28
Stearic acid [C18:0]	10.50	5.42	9.40	7.10	12.44	7.32	9.68
Oleic acid [C18:1]	18.48	21.40	18.76	11.42	19.24	10.48	13.14
Linoleic acid [C18:2]	28.40	20.12	28.32	29.10	25.10	34.60	29.02
Linolenic acid [C18:3]	16.20	32.00	23.10	24.78	21.04	24.10	25.14
Others [unidentified]	1.64	4.96	1.14	1.12	1.48	3.10	1.74

[a] All values are of two determination.

Table 5. Amino acid profiles of acid-hydrolysed, purified total seed proteins of seven species of *Vigna* ($g100g^{-1}$) [a].

Amino acid	*Vigna trilobata*	EAAS	*V. radiata* var. *sublobata*	EAAS	*V. aconitifolia*	EAAS	*V. umbellata*	EAAS	FAO/WHO (1991) requirement pattern
Glutamic acid	12.46		14.78		12.78		13.20		
Aspartic acid	9.78		13.46		10.37		12.46		
Serine	4.13		4.76		3.01		4.21		
Threonine	3.10	91.17	3.68	108.23	3.60	105.88	4.08	120.00	3.4
Proline	3.46		3.50		4.31		5.42		
Alanine	5.60		4.24		4.72		3.86		
Glycine	4.10		4.10		3.46		3.26		
Valine	5.09	145.42	5.25	150.00	4.94	141.14	5.12	146.28	3.5
Cystine	0.53	}55.6	0.43	}65.60	0.23	}68.40	0.12	}60.00	}2.5
Methionine	0.86		1.21		1.48		1.38		
Isoleucine	3.13	111.76	5.78	206.42	4.78	170.71	4.31	153.92	2.8
Leucine	6.61	100.15	6.93	105.00	5.96	90.30	6.80	103.03	6.6
Tyrosine	3.11	}114.12	3.46	}137.14	3.38	}122.06	2.76	}119.68	}6.3
Phenylalanine	4.08		5.18		4.31		4.78		
Lysine	5.18	89.31	6.31	108.79	5.96	102.75	5.30	91.37	5.8
Histidine	3.16	166.31	3.20	168.42	2.96	155.78	2.36	124.21	1.9
Tryptophan	0.56	50.90	0.66	60.00	0.73	66.36	0.78	70.90	1.1
Arginine	5.10		5.34		4.36		4.46		

Table 5. Continues

Amino acid	*V. unguiculata* subsp. *cylindrica*	EAAS	*V. vexillata*	EAAS	*V. bourneae*	EAAS	FAO/WHO (1991) requirement pattern
Glutamic acid	14.16		15.70		11.30		
Aspartic acid	9.48		11.24		12.48		
Serine	3.78		4.12		4.36		
Threonine	3.54	104.11	4.56	134.11	4.30	126.47	3.4
Proline	5.68		3.46		3.74		
Alanine	5.10		4.17		5.24		
Glycine	3.28		3.33		3.48		
Valine	5.48	156.57	4.98	142.28	5.30	151.42	3.5
Cystine	0.63	}80.40	0.33	}59.20	0.78	}64.40	}2.5
Methionine	1.38		1.15		0.83		
Isoleucine	5.18	185.00	3.36	120.00	4.76	170.00	2.8
Leucine	6.43	97.42	7.04	106.66	6.74	102.12	6.6
Tyrosine	3.17	}142.06	2.86	}136.03	3.14	}132.53	}6.3
Phenylalanine	5.78		5.71		5.12		
Lysine	6.30	108.62	6.06	104.48	5.90	101.72	5.8
Histidine	5.14	270.52	1.86	97.89	4.18	220.00	1.9
Tryptophan	0.94	85.45	0.74	67.27	0.86	78.18	1.1
Arginine	5.30		6.16		5.36		

EAAS: Essential amino acid score, [a] Values are single determination.

Table 6: Data on *in vitro* protein digestibility (IVPD) and antinutritional factors of seven species of *Vigna*. $

Components	*Vigna trilobata*	*V. radiata* var. *sublobata*	*V. aconitifolia*	*V. umbellata*	*V. unguiculata* subsp. *cylindrica*	*V. vexillata*	*V. bourneae*
IVPD (%)[g]	74.30	79.12	70.38	72.40	77.30	76.21	72.68
Total free phenolics[h] g100g^{-1}	0.78±0.01^{b}	1.07±0.01^{f}	1.02±8.82^{e}	0.58±0.01^{a}	0.83±0.02^{c}	0.98±0.01^{d}	1.21±0.01^{f}
Tannins[h] g 100g^{-1}	0.23±0.01^{a}	0.31±0.01^{c}	0.48±0.01^{e}	0.24±0.01^{b}	0.27±0.03^{b}	0.19±0.01^{a}	0.36±0.01^{d}
L-DOPA[h] g 100g^{-1}	0.78±0.02^{c}	0.44±0.06^{a}	0.56±0.01^{b}	0.36±0.01^{a}	1.12±0.03^{d}	0.74±0.01^{c}	1.04±0.01^{d}
Phytic acid[h] mg100g^{-1}	312±1.76^{a}	394±3.48^{d}	376±0.33^{c}	336±0.57^{b}	406±1.45^{e}	414±0.01^{f}	436±0.88^{f}
Hydrogen cyanide[h] mg100g^{-1}	0.18±0.01^{c}	0.24±0.17^{d}	0.26±0.01^{d}	0.09±0.01^{a}	0.27±0.01^{c}	0.12±0.01^{b}	0.14±0.02^{b}
Trypsin inhibitor activity[g] (TIU mg^{-1} protein)	23.19	24.48	31.36	34.30	26.40	33.20	24.60
Oligosaccharide[h] g 100g^{-1}							
Raffinose	0.41±0.01^{b}	0.38±0.01^{c}	0.58±0.01^{a}	0.78±0.01^{d}	0.56±0.01^{d}	0.66±0.01^{b}	1.01±0.01^{c}
Stachyose	1.78±0.01^{b}	1.58±0.02^{d}	1.98±0.01^{c}	2.01±0.01^{c}	1.76±0.01^{a}	2.06±0.01^{b}	2.12±0.01^{d}
Verbascose	1.17±0.01^{b}	0.98±0.01^{c}	1.36±0.01^{e}	1.76±0.01^{d}	1.01±0.01^{a}	1.74±0.01^{b}	1.84±0.02^{c}
Phytohaemagglutinating activity Hu mg^{-1} protein[g]							
A group	34	64	45	58	46	42	54
B group	166	150	148	138	112	124	134
O group	14	24	29	34	13	28	26

[g]All values of two independent experiments, [h]All values are of means of triplicate determination expressed on dry weight basis ± standard error,
$ Mean values in the following row sharing a common letter are not statistically significant according to DMRT.

The lectins of different species of *Vigna* exhibit a high level of agglutination activity specifically in 'B' group compared to other two blood groups 'A' and 'O'. This is in good agreement with earlier reports in the tribal pulse *Dolichos lablab* var. *vulgaris* (Vijayakumari *et al*., 1995). However, dry heat and autoclaved are known to inactivate completely that trypsin inhibitors and phytohaemagglutins in *Mucuna* beans (Siddhuraju *et al*., 1996).

CONCLUSIONS

On the basis of the above findings, it is concluded that the tribal pulses investigated seem to be good source of protein, essential amino acids, essential fatty acids and minerals. The adverse effect of most of the antinutritional factors detected in the present study can be eliminated by moist heat treatment or cooking process, since they are heat labile.

ACKNOWLEDGEMENT

We would like to acknowledge Mrs. Geetha, ATOZ Pharmaceuticals Pvt. Ltd., Ambattur, Chennai, India, for her help.

REFERENCES

Alajaji, S.A., El-Adawy, T.A. 2006. Nutritional composition of chickpea (*Cicer arietinum* L.) as affected by microwave cooking and other traditional cooking methods. Journal of Food Composition and Analysis.19: 806 – 812.

Almedia, N.G., Calderon de la Barca, A.M., Valencia, M.E. 1991. Effect of different heat treatments on the anti-nutritional activity of *Phaseolus vulgaris* (variety ojode Carbra) lution. Journal of Agricultural and Food Chemistry. 39: 1627 – 1630.

AOAC. 2005. Official Methods of Analysis (18th edn.). Association of Official Analytical Chemists. Washington. DC.

Arinathan, V., Mohan, V.R., Maruthupandian, A., Athiperumalsami, T. 2009. Chemical evaluation of raw seeds of certain tribal pulses in Tamil Nadu, India. Tropical and Subtropical Agroecosystem. 10: 287 – 294.

Attia, R.S., El-Tabey Shehata, A.M., Aman, M.E., Hamza, M.A. 1994. Effect of cooking and decortication on the physical proerties, the chemical composition and the nutritive value of chick pea (*Cicer arietinum* L.) Food Chemistry. 50: 125 – 131.

Basha, S.M.M., Cherry, J.P., Young, C.T. 1976. Changes in free amino acids, Carbohydrates and proteins of maturity seeds from various peas (*Arachis hypogaea*) cultivars. Cereal Chemistry. 53: 583 – 597.

Brain, K.R. 1976. Accumulation of L-DOPA in cultures from *Mucuna pruriens.* Plant Science Letters. 7: 157-161.

Bray, H.G., Thorne, W.V. 1954. Analysis of phenolic compounds methods. Biochemical. Analyst, 1: 27-52.

Burns, R.R. 1971. Methods of estimation of tannin in grain Sorghum. Agronomy Journal. 63: 511-512.

Dickman, S.R., Bray, R.H. 1940. Colorimetric determination of phosphate. Industrial Engineering Chemistry Analytical Education. 12: 665-668.

Egbe, I.A., Akinyele, I.O. 1990. Effect of cooking on the antinutritional factors of lima beans (*Phaseolus lunatus*). Food Chemistry. 35: 81 -87.

FAO/WHO. 1991. Protein quality evaluation, (p 66). Rome, Italy: Food and Agricultural Organization of the United Nations.

Folch, J., Lees, M., Solane-Stanly, G.M. 1957. A simple method for the isolation and purification of total lipids from animal tissues. The Journal of Biological Chemistry. 226: 497 – 506.

Gopalan, C., Ramasastri, B.V., Balasubramanian, S.C. 1978. Nutritive value of Indian Foods. Indian Council of Medical Research, Hyderabad India.

Hsu, H.W., Vavak, D.L., Satterlee, L.D., Miller, G.A. 1977. A multi-enzyme technique for estimating protein. Digestibility. Journal of Food Science. 42: 1269 -1271.

Humphries, E.C. 1956. Mineral composition and ash analysis. In: Modern Methods of Plant Analysis (Vol. 1), Pp. 468-502, Peach K, Tracey MV, (Eds) Springer-Verlag, Berlin.

ICMR. 1992. Nutrient requirement and recommended dietary allowances for Indians: a report of the expert group of the Indian Council of

Medicinal Research (ICMR). Hyderabad, India. National Institute of Nutrition.

Issac, R.A., Johnson, W.C. 1975. Collaborative study of wet and dry techniques for the elemental analysis of plant tissue by Atomic Absorption Spectrophotometer. Journal of the Association of Official Analytical Chemists. 58: 436-440.

Jackson, M.L. 1967. Cyanide in Plant tissue. In: Soil Chemical Analysis. Asia Publishing House New Delhi India. pp. 337.

Janardhanan, K., Gurumoorthi, P., Pugalenthi, M. 2003. Nutritional potential of five accessions of a south Indian tribal pulse, *Mucuna pruriens* var. *utilis* I. The effect of processing methods on the content of L-DOPA, phytic acid and oligosaccharides. Tropical and Subtropical Agroecosystems. 1: 141 – 152.

Kakade, M.L., Rackis, J.J., McGhce, J.E., Puski, G. 1974. Determination of trypsin inhibitor activity of soy products: a collaborative analysis of an improved procedure. Cereal Chemistry. 51: 376 -382.

Kakati, P., Deka, S.C., Kotoki, D., Saikia, S. 2010. Effect of traditional methods of processing on the nutrient contents and some antinutritional factors in newly developed cultivars of green gram [*Vigna radiata* (L) Wilezek] and black gram [*Vigna mungo* (L.) Hepper] of Assam, India. International Food Research Journal. 17: 377 – 384.

Kala, B. K., Kalidass, C., Mohan, V.R. 2010. Nutritional and antinutritional potential of five accessions of a South Indian tribal pulse *Mucuna atropurpurea* DC. Tropical and Subtropical Agroecosystem. 12: 339 – 352.

Kataria A, Chauhan, B.M., Gandhi, S. 1988. Effect of domestic processing and cooking on the antinutrients of black gram. Food Chemistry. 30: 149 – 156.

Kataria, A., Chauhan, B.M., Punia, D. 1989. Antinutrients and protein digestibility (*in vitro*) of mungo bean as affected by domestic processing and cooking. Food Chemistry. 32: 9 – 17.

Khan, M.A., Jacobsen, I., Eggum, B.D. 1979. Nutritive value of some improved varieties of legumes. Journal of Science Food and Agriculture. 30: 395-400.

Li, B.W., Cardozo, M.S. 1994. Determination of total dietary fiber in foods and products with little or no starch, nonenzymatic-gravimetric method: collaborative study. Journal of Association of Official Analytical Chemists International. 77: 687-689.

Liddell, H.F., Saville, B. 1959. Colorimetric determination of cysteine. Analysts. 84: 133 -137.

Lowry, O.H., Rorebrough, N.J., Farr, A.L., Randall, R.J. 1951. Protein measurement with folin phenol reagent. The Journal of Biological Chemistry. 193: 265 – 275.

Maikhuri, R.K., Nautiyal, M.C., Khai, M.P. 1991. Lesser known crops of food value in Garhwal Himalaya and a strategy to conserve them. FAO/IBPGER Plant Genetic Resources Newsletter. 86: 33 – 36.

Martin, H., Laswai, H., Kulwa, K. 2010. Nutrient content and acceptability of soyabean based complementary food. African Journal of Food Agriculture Nutrition and Development. 10: 2040 – 2049.

Meiners, C.R., Derise, N.L., Lau, H.C., Crews, M.G., Ritchey, S.J., Murphy, E.W. 1976a. Proximate composition and yield of raw and cooked dry legumes, Journal of Agricultural and Food Chemistry. 24: 1122-1126.

Meiners, C.R., Derise, N.L., Lau, H.C., Crews, M.G., Ritchey, S.J., Murphy, E.W. 1976b. The content of nine mineral elements raw and cooked mature dry legumes. Journal of Agricultural and Food Chemistry. 24: 1126-1130.

Metcalfe, L.D., Schemitz, A.A., Pelka, J.R. 1966. Rapid preparation of fatty acid esters from lipids for gas chromatographic analysis. Analytical Chemistry. 38: 514 – 515.

Mohan, V.R., Janardhanan, K. 1993. Chemical composition and nutritional evaluation of two little-known species of *Vigna*. Food Chemistry. 48: 367 – 371.

Mohan, V.R., Janardhanan, K. 1995. Chemical analysis and nutritional assessment of lesser known pulses of the genus *Mucuna.* Food Chemistry. 52: 275-280.

Montgomery, R.D. 1980. Cyanogens. In: Toxic constitutes of Plant Food Stuffs, 2nd edn, ed. IE Liener, pp. 158 – 160. New York: Academic Press.

Muller, H.G., Tobin, G. 1980. Nutrition and food processing: Croom Helm Ltd., London.

Narasinga Rao, B.S., Deosthale, Y.G., Pant, K.C. 1989. Nutritive value of Indian foods, pp. 47 – 91. Hyderabad, India: National Institute of Nutrition, Indian Council Medical Research.

Nieman, D.C., Butterworth, D.E., Nieman, C.N. 1992. Nutritions. Little Brown and Co., USA., pp: 276-282.

Noel, M.G., Rosario, R.R.D. 1989. Changes in protein composition during germination of mung bean (*Vigna radiata* (L.) Wilczek). The Philippine Agriculturist. 72: 271 – 177.

NRC/NAS. 1980. National Research Council Committee on Dietary Allowances. Recommended Dietary Allowances 9th edn. National Academy of Science Press. Washington, DC. USA.

Nwokolo, E. 1987. Nutritional evaluation of pigeon peas meal. Plant Foods for Human Nutrition. 37: 283-290.

Oke, O.L. 1969. The role of hydrocyanic acid in nutrition. World Review of Nutrient and Dieters. 11: pp: 118-174.

Oluwole, S.I., Taiwo, R.E. 2009. Comparison of nutritional composition and antinutrient status of fermented, germinated and roasted bambara groundnut seeds (*Vigna subterranean*), British Food Journal. 111: 376 – 386.

Omogbai, F.E. 1990. Lipid composition of tropical seeds used in the Nigerian diet. Journal of Science Food and Agriculture. 50: 253 – 255.

Onigbinde, A.O., Akinyele, I.O. 1983. Oligosaccharide content of 20 varieties of cow pea in Nigeria. Journal of Food Science. 48: 1250 – 1251, 1254.

Premakumari, M.N., Fathima, A., Saraswathi, G. 1984. Dietary fibre content of some food materials. Journal of Food Science and Technology. 21: 95 - 96.

Rajaram, N., Janardhanan, K. 1990. Chemical composition and nutritional evaluation of certain under exploited *Vigna* spp. Food Science and Nutrition. 42: 213 – 221.

Rajaram, N., Janardhanan, K. 1992. Agrobotanical and biochemical studies of some less known *Vigna* spp. Acta Botanica Indica. 20: 143 – 149.

Rajyalakshmi, P., Geervani, P. 1994. Nutritive value of the foods cultivated and consumed by the tribals of South India. Plant Foods for Human Nutrition. 46: 53-61.

Rama Rao, M.V., Tara, M.R., Krishnan, C.K. 1974. Colorimetric estimation of tryptophan content of pulses. Journal of Food Science and Technology. 11: 213 – 216.

Rao, P.U., Deosthale, Y.G. 1982. Tannin content of pulses: Varietal differences and effects of germination and cooking. Journal of Science Food and Agriculture. 33: 1013-1016.

Ravilleza, M.J.R., Mendoza, E.M.T., Raymundo, L.C. 1990. Oligosaccharides in several Phillippine indigenous food legumes: determination, localization and removal. Plant Foods for Human Nutrition. 40: Pp. 83 – 93.

Sadasivam, S., Manickam, A. 1996. Biochemical methods, New Age International (P) Limited Publishers New Delhi India. pp. 1-250.

Salunkhe, D.K., Sathe, S.K., Reddy, N.R. 1982. Legume lipids. In: Chemistry and Biochemistry of Legumes, SK Arora. (Eds.), Oxford and IBH Publishing Co., New Delhi, India, pp.51-107.

Shills, M.E.G., Young, V.R. 1988. Moodern nutrition in health and diseases. In: *Nutrition*. Nieman, D.C., Buthepodorth, D.E. and Nieman, C.N. (eds). WMc. Brown publishers, Dubugue, USA. pp. 276- 282.

Siddhuraju, P., Becker, K., Makker, H.S. 2001. Chemical composition, protein fractionation, essential amino acid potential and antimetabolic constituents of an unconventional legume, Gila bean (*Entada phaseoloides* Merrill.) seed kernel. Journal of the Science of Food and Agriculture. 82: 192 -202.

Siddhuraju, P., Vijayakumari, K., Janardhanan, K. 1992. Nutritional and chemical evaluation of raw seeds of the tribal pulse *Vigna trilobata* (L.) Verdc. International Journal of Food Science and Nutrition. 43: 97-103.

Siddhuraju, P., Vijayakumari, K., Janardhanan, K. 1996. Chemical composition and protein quality of the little-known legume, velvet bean (*Mucuna pruriens* (L.) DC). Journal of Agricultural and Food Chemistry. 44: 2636 – 2641.

Singh, V., Eggum, B.O. 1984. Factors affecting the protein quality of pigeon pea (*Cajanus cajan* L.), Plant Foods for Human Nutrition. 34: 273 – 283.

Somiari, R.T., Balogh, E. 1993. Effect of soaking, cooking and alpha- galactoside treatment on the oligosaccharide content of cowpea flours. Journal of Science Food and Agriculture. 61: 339 – 343.

Spies, J.R., Chamber, D.C. 1949. Chemical determination of tryptophan in proteins. Analytical Chemistry. 21: 1249 – 1266.

Tan, N.H., Rahim, Z.H.A., Khor, H.T., Wong, K.C. 1983. Winged bean (*Psophocarpus tetragonolobus*). Tannin level, phytate content and Haemagglutinating activity. Journal of Agricultural and Food Chemistry. 31: 916 – 917.

Tanaka, M., Thanankul, D., Lee, T.C., Chichester, L.O. 1975. A simplified method for the quantitative determination of sucrose, raffinose and stachyose in legume seeds. Journal of Food Science. 40: 1087 – 1088.

Thangadurai, D. 2005. Chemical composition and nutritional potential of *Vigna unguiculata* ssp. *cylindrica* (Fabaceae). Journal of Food Biochemistry. 26: 88 – 89.

Vadivel, V., Janardhanan, K. 2000. Nutritional and anti-nutritional composition of velvet bean: A lesser - known food legume in South India. International Journal of Food Sciences and Nutrition. 51: 279 – 287.

Vadivel, V., Janardhanan, K. 2001. Diversity in nutritional composition of wild jack bean (*Canavalia ensiformis* (L.) DC) seeds collected from South India. Food Chemistry. 74: 507 – 511.

Vadivel, V. and Pugalenthi, M. 2008. Effect of various processing methods on the levels of antinutritional constituents and protein digestibility of *Mucuna pruriens* (L.) dc. var. *utilis* (Wall. ex Wight) Baker ex Burck (velvet bean) seeds. Journal of Food Biochemistry. 32: 795–812.

Vadivel, V., Janardhanan, K., Vijayakumari, K. 1998. Diversity in sword bean (*Canavalia gladiata* (Jacq.) DC.) collected from Tamil Nadu, India, Genetic Resources and Crop Evolution. 45: 63 – 68.

Vijayakumari, K., Siddhuraju, P., Janardhanan, K. 1995. Effects of various water or hydrothermal treatments on certain antinutritional compounds in the seeds of the tribal pulse, *Dolichos lablab* var. *vulgaris* L. Plant Foods for Human Nutrition. 48: 17 – 29.

Vijayakumari, K., Siddhuraju, P., Janardhanan, K. 1996. Effects of different post-harvest treatments on anti-nutritional factors in seeds of the tribal pulse, *Mucuna pruriens* (L.) DC. International Journal of Food Sciences and Nutrition. 47: 263 – 272.

Vijayakumari, K., Siddhuraju, P., Janardhanan, K. 1997. Chemical composition, amino acid content and protein quality of the little-known legume *Bauhinia purpurea* L. Journal of Science Food and Agriculture. 73: 279 – 286.

Wheeler, E.L., Ferrel, R.E. 1971. A method for phytic acid determination in wheat and wheat fractions. Cereal Chemistry. 48: 312 – 320.

Zia-Ul-Haq, M., Iqbal, S., Ahmad, S., Imran, M., Niaz, A., Bangher, M.I. 2007. Nutritional and composition study of desi chickpea *Cicer arietinum* (L.) cultivars grown in Punjab, Pakistan. Food Chemistry. 105: 1357 – 1363.

15

Chemical Composition, *In Vitro* Rumen Fermentation, and Digestibility of Selected Browse Species by Buffaloes (*Bubalus Bubalis* L.) at a Semi-Deciduous Forest in Venezuela

Álvaro Ojeda[1*], Carlos Canelones[1], Dorian Muñoz[1] and José L. Gil[2]

[1]*Facultad de Agronomía, Universidad Central de Venezuela.*

P.O. Box 4579. CP 2101, Maracay, Venezuela. E-mail: ajojeda99@yahoo.com

[2]*Instituto Nacional de Investigaciones Agrícolas, Centro Nacional de Investigaciones Agropecuarias. P.O. Box 4653. Maracay, Venezuela*

**Corresponding author*

SUMMARY

To determine the chemical composition and nutritive value of selected browse species by buffaloes (*Bubalus bubalis* L.) in a silvopastoral system at a semi-deciduous tropical forest (SDTF), epidermal fragments in fecal samples from 14 males of Murrah breed were evaluated. The animals grazed for 91 days in 85 ha located in Portuguesa State, Venezuela, of which 35 ha corresponded to natural grass pastures (969.3 ± 28.6 kg DM ha^{-1}), and the rest to SDTF with 463 plant ha^{-1}. Eleven botanical families and 22 species were identified, 40.9% belonged to the Fabaceae family. The specie *Guazuma ulmifolia* (Sterculiaceae) had the highest relative dominancy (60.4%) and an Importance Value Index of 162.6. 60.4% of epidermal fragments corresponded to herbaceous Poaceae, and the rest was distributed among five different species, with the highest occurrence (94.6%) for *Samanea saman* and *Sida acuta*. There were differences ($P<0.05$) among selected species in the proximal composition (40.9 ± 7.1% dry matter, 16.0 ± 4.1% crude protein and 2.9 ± 1.2% ether extract), neutral detergent fiber (41.6 ± 2.2%), acid detergent fiber (25.9 ± 3.8%), minerals (1.6 ± 0.59% Ca and 0.1 ± 0.02% P), condensed tannins (0.9 ± 0.7%), potential gas production (67.8 ± 37.3 ml g^{-1} DM), and *in vitro* organic matter digestibility (54.2 ± 8.0%). In general, the edible fraction of selected browse species evaluated here was similar to other tropical browse forages and confirms the strategic value of this biomass in buffalo nutrition at silvopastoral system. Furthermore, these browse species could be employed in agroforestry systems to achieve a better integration of livestock production in tropical areas.

Key words: Selectivity; digestibility; tannins; agroforestry.

INTRODUCTION

Buffalo (*Bubalus bubalis* L.) herds are usually managed as a triple purpose systems (milk, meat and draft-power), taking advantage of their rusticity, high fertility rates and ability to adapt to different climatic conditions (Borghese and Mazzi, 2005). Additionally, this specie show higher efficiency in the use of fibrous resources, superior dry matter intake and digestibility due to an adjusted selectivity of the forage quantity and quality, as well as to a higher nitrogen recycling and retention time of the feed (Ichinohe *et al.*, 2004; Corrêa *et al.*, 2008).

In tropical America, buffalo herds are distributed over large extensions of savannas with presence of sparse forest formations, under a rainy regime with marked seasonality which determines a wide annual variability in the quantity, quality and structure of forage biomass produced. During the dry season, to increase the offer and nutritional value of forage, animals are allowed to have access to the forest when annual availability of grasses and other herbaceous plants are reduced. These seasonal silvopastoral systems present higher biological, economic, social and ecological sustainability than other traditional systems (Corrêa *et al.*, 2008).

Due to their protein crude content (16.5 to 23.2%), neutral detergent fiber (39.5 to 49.8%), gross energy (15.0 to 20.3 kJ g^{-1} DM), degradability of dry matter (41.3 to 73.2%), biomass of trees and shrubs present in tropical forests can contribute to improve the quality of the animal diet (García and Medina, 2006). Even though there are no studies that evaluate the selectivity by buffaloes in silvopastoral conditions in tropical dry forests, evaluations with sheep (García and Medina, 2006) and cows (Baldizán *et al.*, 2006) point out that it is related with the participation of secondary metabolites in the foliage, where stand out polyphenols, alkaloids, saponins, terpenes, and phytic phosphorus. In humid forest, it has been found that, in spite of a marked preference for grasses, up to 32.7% of the buffalo diet can be formed by shrubby dicots (Bekhuis *et al.*, 2008).

To improve buffalo efficiency in the seasonal grazing of these tropical forests, it is required to generate information about its feeding habits, as well as the chemical and nutritional characterization of woody species with forage importance. This information would allow building programs for sustainable management that would add value to the forests and promote an active role of the farmers on the biodiversity conservation of these ecosystems. The study therefore, aimed to investigate the chemical composition and nutritional value of woody plants selected by buffaloes grazing a tropical semi-deciduous forest in Venezuela.

MATERIALS AND METHODS

Study site

The study was carried out during the dry season (January to May 2010), at the San Nicolás Experimental Farm (8° 49′ 58′′ N and 69° 48′ 00′′ W) of the Universidad Central de Venezuela, located in Portuguesa State, Venezuela. During the experimental period, the cumulative rainfall was 246.8 mm, mean temperature 27.0 ± 0.72°C, potential evapotranspiration 892.5 mm, and relative humidity of 63%. The soils were classified as Fluventic Haplustepts and Vertic Endoaquepts (Abarca, 2005), with clay fine mixed, non acid and isohypertermic. The total experimental area (85 ha) was delimited by a perimetrical fence and consisted of a semi-deciduous tropical forest with low level of human intervention (50 ha), composed by herbaceous and graminaceous vegetation (35 ha) dominated by *Cynodon nlemfuensis, Sporobolus indicus*, *Axonopus sp.*, *Mimosa pudica*, and *Hyptis suaveolens*. Paddocks had a mean yield of 969.3 ± 28.6 kg DM ha^{-1} and a leaf:stem ratio of 0.53 ± 0.15, evaluated according to Greig-Smith (1983).

Botanical composition and browse species sampling

Botanical composition of the woody species in the forest was determined in 12 transects of 100 to 200 m length and 2 to 4 m wide, by the quadrant point technique (Muller-Dumbois and Ellemberg, 1974) with 10 m distance between points. Considering a height equal or lower than 2 m as the susceptible browsing strata by buffaloes, monthly samples (900 g) of green foliage (leaves, leaflets, reproductive structures, and woody material with diameter less than 6 mm) were collected for chemical analysis. Ten woody plants per species selected by the animals were chosen randomly, observing that all plants had similar age and phenological stage. Leaves and flowers of the browsed species were pressed, labeled, and dried for latter identification at the Botanical Herbarium Victor Manuel Badillo (MY) of the Universidad Central de Venezuela.

Animal management

Fourteen male adult buffaloes of Murrah breed (446.9 ± 19.3 kg LW) were used in continuous grazing with free open access to the forest. Water availability and a commercial blend of minerals and salt (20% Ca, 10% P, 9% Na, 0.6% S, 0.5% Mg, 0.5% Zn, 0.25% Mg, 0.25% Cu, 0.08% F, 80 ppm I, 20 ppm Co, and 20 ppm Se) were offered to the animals on an *ad libitum* basis. To allow selectivity with usual stocking rates for these savannas, additional animals were used to obtain a grazing pressure of 6 kg DM for 100 kg LW.

Fifteen days before evaluation, all animals were dewormed with levamisol chlorhydrate (1 ml for 50 kg LW) and dosed with 1 ml for 100 kg LW of a multivitamins complex (500 000 UI ml^{-1} vit. A, 75 000 UI ml^{-1} vit. D3, and 50 UI ml^{-1} vit. E).

Fecal collection and preparation

Fecal samples were individually taken every 15 d (8:00 to 10:00 h) directly from the rectum. Once collected, samples were placed in plastic bags and stored at -18°C. Later, at the laboratory, material was dehydrated until constant weight in a forced convection oven at 65°C, grounded with a hammer mill to pass a 1 mm screen and then, 2 g were rehydrated with ethylic alcohol of 50% v/v. Samples so treated were screened (180 μ) and plant pigments removed with sodium hypochlorite (Holechek, 1982). After that, 20 ml of glycerin (10%) and four drops of tolouidine blue (1%) were added to prepare 3 semi-permanent plates (24 x 24 mm) for each sample. Evaluation of epidermic fragments was performed with a binocular microscope at 400X, using transversal views of 2 mm wide, 60 mm length and 3 mm between transversals. As reference, at the beginning of the experimental phase, digital photographic patterns of the woody species were development using as identification criteria, the presence, frequency, shape, size and distribution of trichomes, stomata, crystals, and typical epidermal cells.

Botanical and selectivity indexes

In the forest, plant density, dominance and frequency were estimated according to Muller-Dumbois and Ellemberg (1974). The Importance Value Index (IVI), a non-dimensional parameter that estimates the relative ecological success of a species on a plant community, was determined by the following equation (Barbour *et al.*, 1999):

$$IVI = \Sigma\ [\text{Density}\ (\%) + \text{Frequency}\ (\%) + \text{Dominance}\ (\%)]$$

Selectivity of woody species was evaluated through the Ivlev Selectivity Index (ISI), where values varies between +1 (maximum preference) and -1 (total rejection), while records close to zero indicates that the species are selected in direct relationship with its density in the grazing area (Lechowicz, 1982). The formula is:

$$ISI = \left[\frac{\text{Plant 1 epidermic fragments in feces (\%) - IVI Plant 1 (\%)}}{\text{Plant 1 epidermic fragments in feces (\%) + IVI Plant 1 (\%)}}\right]$$

Chemical analyses

For the determination of phenolic compounds and *in vitro* studies, foliage samples from selected browse species were dehydrated at room temperature until constant weight in a dry and dark environment. For the other analysis, samples were dehydrated in a forced convection oven at 65°C. All samples were later ground with a hammer mill to pass a 1 mm screen and stored in amber containers. The dry matter (DM), nitrogen, fat, ash, calcium (Ca), and phosphorous (P) contents were determined using the methods of the AOAC (1999). The organic matter (OM) content was calculated as 100-ash content. Crude protein (CP) was calculated by multiplying nitrogen content with a factor of 6.25. The neutral detergent fiber (NDF) and acid detergent fiber (ADF) were determined by the method of Van Soest *et al.* (1991), adding sodium sulfite to remove the tannins-protein complex (Terril *et al.*, 1994). Extracted total phenols (TP, tannin acid equivalent) were determined by a modified Folin-Ciocalteu assay (Makkar, 2001) and the extracted condensed tannins (CT, leucocyanidin equivalent) were determined using the method of Porter *et al.* (1986).

***In vitro* gas production and digestibility**

Browsed foliage samples were tested for *in vitro* ruminal fermentation with a gas pressure transducer technique as described by Mauricio *et al.* (1999). Foliage DM (1 g), with or without extra 1 g polyethylene glycol with relative molecular mass 8000 (PEG 8000, Merck®), were weighed into serum bottles with 168 ml capacity. Bottles were filled with 90 ml of a mixture of mineral and buffer solution and 4 ml of reducing agent (Menke and Steingass, 1988). After adding 10 ml of rumen fluid, eight replications of each treatment were incubated in a pre-heated water bath at 39°C. Additionally, four blanks with buffer solution, reducing agent, and rumen fluid were incubated in each experiment. The mixed rumen content (50:50 solid and liquid phase) was collected at 07:00 h from two rumen cannulated cows (429.5 ± 19.6 kg LW). Cows grazed on a medium-quality pasture dominated by *Cynodon nlemfuensis,* and supplemented daily with 2 kg of concentrate (2.7% N, 7.6% ash, and 37.2% NDF). The serum bottles were sealed by a rubber stopper and gas pressure was set to zero using a transducer and a LED digital readout voltmeter (Red Lion®, Model DP5-1/8 DIN). Gas pressure was recorded at 3, 6, 9, 12, 24, 36, 48, 60, 72, and 96 h after the inoculation. The gas data were fitted to an exponential model according to France *et al.* (1993).

Once readings of gas pressure were done, the bottles were placed in an inverse water bath (4°C). The contents were filtered into glass cresols with porous porcelain plates (pore # 1) previously weighted, to be later dehydrated at 105°C until constant weight. Apparent digestibility (48 and 96 h) was calculated by weight difference between the substrate for the organic matter (OMd) and NDF (NDFd) fractions.

Statistical analysis

The relative frequency of epidermal fragments in feces was analyzed by non parametric statistics, using the Kruskal-Wallis's test, and the chemical composition and nutritive value were examined by a completely randomized design, considering each species as a treatment. ANOVA was conducted on data thorough a linear model. Those variables that showed significant effect by treatments were subjected to Tukey's test. Correlation analyses were used to establish associations between chemical, *in vitro* gas production, and digestibility parameters. The information was processed using the statistical software SAS (2001), with a significance level of $p<0.05$.

RESULTS

The floristic inventory showed the presence of 11 botanical families that grouped 22 identified woody species, 40.9% belonged to the Fabaceae, followed in importance by the Malvaceae, Moraceae and Polygonaceae, which together represented 27.3% of woody species, while 31.8% of the species was distributed in the other six families (Table 1). *Guazuma ulmifolia* (Sterculiaceae), *Pithecellobium lanceolatum* (Fabaceae), *Samanea saman* (Fabaceae) and *Guadua angustifolia* (Poaceae) represented 80.2% of the total species (463 woody plants/ha). *G. ulmifolia* showed high dominance in the community due to its high density and plant development and also the highest IVI, which together with *S. saman* and *P. lanceolatum*, accumulated 71.3% for this index.

A high proportion of the epidermic fragments in the feces corresponded to graminaceous forage, and the other was distributed among five woody species with the highest participation for *S. saman* and *Sida acuta* (Table 2). A small proportion (3.7%) of the fragments did not show morphoanatomic characteristics that could allow its botanical identification. This limitation is due to the integrity degree of the fragments when consumed by the animals, the sensitivity to degradation during its passage through the gastrointestinal tract, and to refringency to colorants used (Bekhuis *et al.*, 2008).

The chemical composition of foliage of selected browse species is shown in Table 3. *I. laurina* and *S. saman* had the highest CP content, and *I. laurina* had the highest NDF. Ca content ranged from 0.84 to 2.21%, while P content did not show significant differences among species. TP and CT varied widely, with the highest values in both cases for *C. caracasana*.

There were a low level of hemicelluloses (15.7 ± 5.4%), and the high proportion of ADF in the cell wall suggests high contents of lignin and cellulose.

The *in vitro* gas parameters, OMd, and NDFd values varied significantly ($P<0.05$) among browsed species (Table 4). The rate of DM degradation was similar among species and the time needed for microorganism to *in vitro* colonize the vegetal tissue (To) was highest for *S. acuta* and *R. spinosa*, and similar among the other species. T½ ranged from 23.1 to 11.6 h, with the lowest value for *R. spinosa* (Table 4). The accumulative gas production varied widely with PEG 8000 addition to the culture medium (Fig. 1), and ranged from 2.1 to 78.1% at 48 h, and 2.0 to 117.1% at 96 h.

The correlations between chemical composition and *in vitro* parameters of the browsed species foliage are shown in Table 5. CT was positively correlated to ADF and negatively correlated CP. TP content was positively correlated to CT and T½, and negatively correlated to a, b, To, and OMd. Additionally, OMd was negatively correlated to T½, and positively correlated to a, b, and To. The a parameter was strongly and positively correlated to To, OMd, and NDFd, and negatively to T½.

DISCUSSION

Compared with other tropical forests with actual or potential nutritional value for silvopastoril systems (Valencia *et al.*, 2004; Zent and Zent, 2004; Mohandass and Davidar, 2009), this semi-deciduous community showed an important floristic diversity. The presence of the Fabaceae is outstanding, which besides its nutritional importance for the ruminant, possesses an aggregate value to the edaphic component because it adds nitrogen and participate in Ca and P recycling (Zent and Zent, 2004). *G. ulmifolia*, *S. saman*, and *P. lanceolatum* showed a high forage potential value because IVI relates to voluntary intake (Mandaluniza *et al.*, 2011) and to species spatial distribution which in turns, it is one of the most influential factors for grazing buffaloes to access a particular plant (Melletti *et al.*, 2007).

Table 1. Botanical composition and Importance Value Index (IVI) of browsed species at a semideciduous tropical forest in Venezuela.

Families	Species	Index			
		Density (%)	Frequency (%)	Dominance (%)	IVI
Arecaceae	*Acrocomia aculeata*	0.4	0.5	0.5	1.4
Boraginaceae	*Rochefortia spinosa*	3.0	0.9	1.6	5.5
Chrysobalanaceae	*Parinari campestris*	0.3	1.9	1.1	3.3
Fabaceae	*Enterolobium cyclocarpus*	0.4	3.9	0.3	4.6
	Erithrina fusca	1.7	1.9	2.4	6.0
	Inga laurina	0.5	1.0	0.2	1.7
	Machaerium humboldtianum	0.7	1.1	0.5	2.3
	Pithecellobium lanceolatum	6.4	11.2	7.2	24.8
	Pterocarpus sp.	1.8	1.9	1.3	5.0
	Samanea saman	6.2	6.8	13.5	26.5
	Senna pallida	1.3	2.9	0.6	4.8
	Senna sp.	0.6	1.8	0.3	2.7
Poaceae	*Guadua angustifolia*	6.0	5.8	1.9	13.7
Malvaceae	*Hibiscus sp.*	1.3	0.2	0.2	1.7
	Sida acuta	1.3	0.9	0.3	2.5
Moraceae	*Morus sp.*	0.4	0.5	0.1	1.0
	Cecropia peltata	1.1	2.9	1.3	5.3
Rubiaceae	*Genipa americana*	0.4	0.3	0.1	0.8
Piperaceae	*Piper sp.*	0.2	0.5	0.1	0.8
Polygonaceae	*Coccoloba caracasana*	0.5	5.8	0.3	6.6
	Triplaris caracasana	0.2	4.8	0.2	5.2
Sterculiaceae	*Guazuma ulmifolia*	61.6	40.6	60.4	162.6
Non identified[2]		3.7	1.9	5.6	11.2

NI, botanically non identified browsed species.

Even though, buffaloes showed an alimentary habit as essentially ungulate grazers, feces analysis showed an important proportion of epidermic fragments of woody species. In similar studies carried out with African buffaloes (*Syncerus caffer nanus*) grazing on a tropical humid forest with open access to savannas, Macandza *et al.* (2004) found that 32.7% of fragments in feces corresponded to woody species, being leaves the fraction with the highest values (26.5%). In these African animals, it has been documented that their grazing habits are strongly related to the quality of the available plant biomass (Melletti *et al.*, 2007).

Table 2. Epidermal fragments in the feces and Ivlev Selectivity Index (ISI) by buffaloes in a silvopastoral system.

Families	Species	Epidermal fragments (%)	ISI
Boraginaceae	*Rochefortia spinosa*	2.01c	0.15c
Fabaceae	*Inga laurina*	0.04c	0.49b
	Samanea saman	18.1b	0.53b
Malvaceae	*Sida acuta*	17.1b	0.78a
Polygonaceae	*Coccoloba caracasana*	0.1c	-0.38d
Poaceae		60.4a	
NI		2.3c	
SEM		17.2*	0.32*
p value		0.03	0.02

NI, non identified species; Poaceae only refers to gramineous plants; [abc]Means followed by different letters within a column are significantly different

Table 3. Chemical composition (% DM) of browse species foliage selected by buffaloes.

Species	DM	CP	Fat	NDF	ADF	Ca	P	TP	CT
R. spinosa	53.2a	15.1b	4.1a	40.6b	23.3b	2.21a	0.13	2.0c	0.33b
I. laurina	40.0b	19.3a	1.7b	45.2a	20.8b	1.03c	0.15	3.7b	0.44b
S. saman	35.9b	20.7a	3.7ab	39.4b	26.9a	0.84c	0.14	4.2b	0.73b
S. acuta	36.3b	14.5b	3.6ab	42.2b	28.3a	1.9b	0.13	3.2b	0.65b
C. caracasana	39.2b	10.4c	1.6b	40.7b	30.1a	1.78b	0.09	6.5a	2.46a
SEM	7.9	2.2	0.6	2.1	3.2	0.3	0.04	1.1	0.8
p value	0.04	0.02	0.001	<0.01	0.04	0.03	0.41	0.01	0.02

DM, dry matter; CP, crude protein; NDF, neutral detergent fiber; ADF, acid detergent fiber; Ca, calcium; P, phosphorous; TP, total phenols; CT, condensed tannins; SEM, standard error of the mean; [abc]Means followed by different letters within a column are significantly different.

The CP content was superior to that of natural and introduced tropical pastures (Minson, 1990), and above of the minimum required for ruminants (7 to 8% DM) for optimum intake and rumen performance (Van Soest, 1994). All evaluated forage showed less than 45% of NDF, so they can be considered as a fibrous resource of good quality (Melaku *et al.*, 2010). The Ca content emphasizes the relative superiority of this foliage of woody species in comparison to tropical grăsses with values adjusted to the estimated requirements of buffaloes with medium levels of production. However, P values indicate the need for supplementation programs for buffaloes grazing in similar forest formations (Habib *et al.*, 2004).

The CT levels were in the range referred as beneficial to ruminant nutrition because of its capacity to reduce the risk of timpanism, to control parasitic infections, and to promote the nitrogen flux to the lower tract (Makkar, 2001), but inferior to expected with regards to the TP content and literature values for non cultivated woody foliage plants (Makkar, 2001; Baldizán *et al.*, 2006; García and Medina, 2006; Melaku *et al.*, 2010). This could be due to the traditional seasonal grazing management of these forests that do not promote an adequate plant-grazer interaction, which would make difficult a genetic fixation of phytochemical strategies of defense, such as synthesis and vacuolar storage of CT (Launchbaugh *et al.*, 2001). In general, bromatological composition, fiber content, tannins, and evaluated minerals showed values similar to those of the edible fraction of other woody species with forage value (Carranza-Montaño *et al.*, 2003; Guimaraes-Beelen *et al.*, 2006; Melaku *et al.*, 2010), and even though there were significant differences ($p<0.05$) among species, all of them can be considered nutrient sources for buffaloes grazing under these conditions.

The *in vitro* gas production parameters for some woody plants mentioned in the literature (*Acacia sp, Erythrina poeppegiana, E. fusta, E. olei, Ficus sp, Flemingia congesta, Gliricidia sepium, Inga marginata, I. ingoides, Leucaena leucocephala, Moringa oleifera and Piptadenia macrocarpa*) show highly variable ranges, for a (22 to 149 ml g^{-1} DM), b (0.02 to 0.06 ml h^{-1}), To (0.8 to 4,5 h), and T½ (11 to 28 h) that reflect the effect of age and physiological development of the plant, secondary metabolites content and environmental conditions (Wood *et al.*, 1993). Compared to the *in vitro* value for a (118.7 to 229.1 ml g^{-1} DM), b (0.01 to 0.04 ml h^{-1}), To (3.1 to 12.4 h) and T½ (19.1 to 48.2 h) of fresh or dried tropical grasses (Bueno *et al.*, 2005; González-Ronquillo *et al.*, 2009), with the exception of *S. acuta* (a= 120.9 ml g^{-1} DM), the foliage of the evaluated woody species presented a lower *in vitro* potential gas production, which could be attributed to an inhibitory effect of abundant lignocelulolitic complexes in the cell wall of semi-deciduous plants (Van Soest, 1994), and the presence of a wide range of secondary metabolites, besides polyphenols, in woody plants of tropical forest, such as saponines, terpenes, cyanogens, steroids, trypsin inhibitors and alkaloids (Baldizán *et al.*, 2006; García and Medina, 2006).

The difference in variation of the cumulated gas production with the addition of PEG 8000 evidenced the presence of CT with different reactivity. In this regard, in spite of *C. caracasana* showed a higher CT content than *S. acuta* ($p=0.02$), in both species PEG generated a similar increment in accumulated gas production (110.8 ± 0.14%). This indicate a higher biological activity of CT of *S. acuta*, which can be linked to variations in molecular weight, tertiary structure, distribution of prodelfidine, procianidine and propelargonidine in the molecule, isoelectric point, and compatibility of the union sites of the secondary compound (Hagerman and Butler, 1980; Guimaraes-Beelen *et al.*, 2006). These characteristics are highly variable among species, and even for the same species according to its phenological stage (Rodriguez *et al*., 2010).

This demonstrated that tannin content is not by itself an indicator of tannins activity and further studies are required to evaluate the potential impact on the productive performance of herbivorous that fed with the foliage of woody plants.

The negative correlation between TP and the *in vitro* gas production parameters and OMd confirm the detrimental effect of these compounds on the rumen microbial activity as previously indicated. However, no correlation was observed between the *in vitro* fermentation parameters and the CT levels, so this could be associated to other non-evaluated polyphenolic compounds, such as flavonoids with proven reactivity towards the PEG 8000 and with negative impact on rumen fermentation (Bodas *et al*., 2009). Gas potential production was highly correlated with the other *in vitro* fermentation parameters and OMd, which indicates its potential as a nutritional indicator for tropical woody foliage evaluation.

CONCLUSIONS

The proximal composition, fiber content, and evaluated minerals from the edible fraction of selected browse species (*Rochefortia spinosa*, *Inga laurina*, *Samanea saman*, *Sida acuta* and *Coccoloba caracasana*) by buffaloes in a silvopastoral system at a semi-deciduous tropical forest showed higher values than tropical grass pastures, so these species could be considered a nutrient source for buffaloes grazing under these conditions. The condensed tannins of this biomass showed different reactivity, and addition of PEG 8000 could be an alternative to reduce the negative ruminal effects of this secondary metabolites of these forage resources.

ACKNOWLEDGEMENT

The authors wish to express their sincere thanks to Consejo de Desarrollo Científico y Humanístico of the Universidad Central de Venezuela for the financial support (CDCH Project N° PG-01-36-4995-2002).

Table 4. *In vitro* gas production and digestibility parameters of browse species foliage selected by buffaloes.

Species	Parameters				Degradability (%)	
	a	b	To	T½	OMd	NDFd
R. spinosa	91.8b	0.06	3.2a	11.6b	64.8a	35.1b
I. laurina	36.3c	0.03	1.1b	23.1a	48.4b	42.4a
S. saman	53.4b	0.04	1.0b	17.3a	50.0b	31.7c
S. acuta	120.9a	0.03	4.4a	18.1a	60.8a	38.7b
C. caracasana	36.5c	0.03	1.0b	23.1a	47.1b	36.6b
SEM	22.1	0.02	1.2	4.9	7.3	4.1
p value	0.03	0.12	0.02	0.03	0.02	<0.01

a, potential gas production (ml g^{-1} DM); b, rate of DM degradation (ml h^{-1}); To, lag time (h); T½, time taken for half the material in the pool described by the exponential component to be degraded (h); OMd, organic matter degradability; NDFd, neutral detergent fiber degradability; SEM, standard error of the mean; [abc]Means followed by different letters within a column are significantly different.

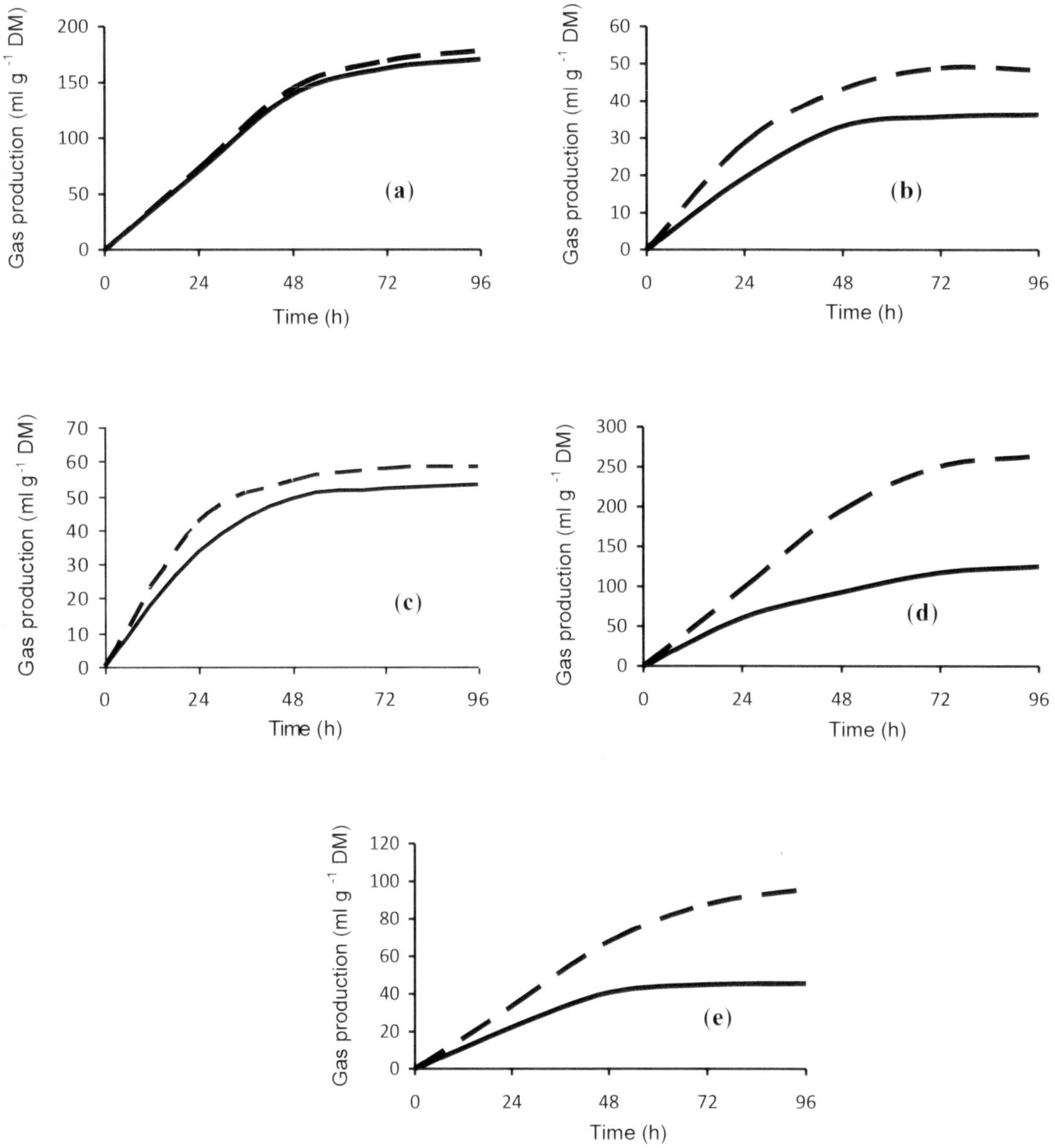

Figure 1. Accumulative *in vitro* gas production (—) and its variation with addition of polyethylene glycol 8000 (– –) in *R. spinosa* (a), *I. laurina* (b), *S. saman* (c), *S. acuta* (d), and *C. caracasana* (e) foliage

Table 5. Correlation coefficients among chemical composition, *in vitro* gas production, and digestibility parameters of browsed species foliage selected by buffaloes.

	CP	NDF	ADF	TP	CT	a	b	To	T½	OMd	NDFd
CP	1										
NDF	0.18	1									
ADF	-0.59	-0.60	1								
TP	-0.41	-0.15	0.62	1							
CT	-0.71*	-0.29	0.73*	0.92**	1						
a	-0.16	-0.14	0.14	-0.66*	-0.45	1					
b	0.09	-0.45	-0.34	-0.63*	-0.43	0.30	1				
To	-0.26	0.01	0.09	-0.64*	-0.41	0.98**	0.25	1			
T½	-0.12	0.51	0.16	0.76*	0.55	-0.64*	-0.90**	-0.56	1		
OMd	-0.14	-0.17	-0.13	-0.81*	-0.55	0.89**	0.66*	0.89*	-0.85**	1	
NDFd	-0.09	0.96**	-0.42	-0.05	-0.11	0.89**	-0.51	0.11	0.55	-0.12	1

CP, crude protein; NDF, neutral detergent fiber; ADF, acid detergent fiber; TP, total phenols; CT, condensed tannins; a, potential gas production; b, rate of DM degradation; To, lag time; T½, time taken for half the material in the pool described by the exponential component to be degraded; OMd, organic matter degradability; NDFd, neutral detergent fiber degradability; *P<0.05; **P<0.01.

REFERENCES

AOAC. 1999. Official methods of analysis. 17th ed. Association of Official Analytical Chemists. Washington, USA.

Abarca, O. 2005. Intensity land use conflicts in the Experimental Stations of the Universidad Central de Venezuela. Spatial analysis with Geographical Information Systems. Agronomía Tropical. 55: 289-313.

Baldizán, A., Domínguez, C., García, D.E., Chacón, E., Aguilar, L. 2006. Secondary metabolites in a tropical deciduous forest vegetation of the Venezuelan central plains. Zootecnia Tropical. 24: 213-232.

Barbour, M.G., Burk, J.H., Pitts, W.D., Shwartz, M.W. 1999. Terrestrial plant ecology. 3rd ed. Addison-Wesley Longman. Menlo Park, USA.

Bekhuis, P.D.B.M, De Jong, C.B., Prins, H.H.T. 2008. Diet selection and density estimates of forest buffalo in Campo-Ma'an National Park, Cameroon. African Journal of Ecology. 46: 668-675

Bodas, R., López, S., Fernández, M., García-González, R., Wallace, R.J., González, J.S. 2009. Phytogenic additives to decrease *in vitro* ruminal methanogenesis. Options Méditerranéennes. 85: 279-283

Borghese, A., Mazzi, M. 2005. Buffalo population and strategies in the world. In: Borghese A (Ed.) Buffalo production and research. FAO, Regional Office for Europe Inter-Regional Cooperative Research Network on Buffalo. Rome, Italy.

Bueno, I.C.S., Cabral, S.L.S., Gobbo, S.P., Louvandini, H., Vitti, D.M.S.S., Abdalla, A.L. 2005. Influence of inoculum source in a gas production method. Animal Feed Science and Technology. 123-124: 95-105.

Carranza-Montaño, M.A., Sánchez-Velásquez, L.R., Pineda-López, M.R., Cuevas-Guzmán, R. 2003. Forage quality and potential of species from the Sierra de Manantlán (México) tropical dry forest. Agrociencia. 37:203-210.

Corrêa, A., Lourenço, J.B., Alves, N.F., Moreira, E.M., Brito M.A., García, A.R. 2008. Sistema silvipastoril na Amazônia: ferramenta para elevar o desempenho produtivo de búfalos. Ciência Rural. 38:2395-2402.

France J., Dhanoa, M.S., Theodorou, M.K., Lister, S.J., Davies, D.R., Isac, D. 1993. A model to interpret gas accumulation profiles associated with *in vitro* degradation of ruminant feeds. Journal of Theoretical Biology. 163: 99-111.

García, D.E, Medina M.G. 2006. Chemical composition, secondary metabolites, nutritive value and relative acceptability of ten fodder trees. Zootecnia Tropical. 24: 233-250.

González-Ronquillo, M., Berchielli, T.T., Beelen, R., Araújo J., de Oliveira, G.S. 2009. Forage production, chemical composition and *in vitro* gas production of the vegetation of a modulated seasonal savanna. Zootecnia Tropical. 27: 407-417.

Greig-Smith, P. 1983. Quantitative plant ecology. 3rd ed. University of California Press, USA.

Guimaraes-Beelen, P.M.M., Berchielli, T.T., Beelen, R., Araújo J., de Oliveira, G.S. 2006. Characterization of condensed tannins from native legumes of the Brazilian northeastern semi-arid. Science Agriculture (Piracicaba, Braz). 63: 522-528.

Habib, G., Jabbar, G., Siddiqui, M.M., Shah, Z. 2004. Paralytic disorders associated with phosphorus deficiency in buffaloes. Pakistan Veterinary Journal. 24: 18-22.

Hagerman, A.E, Butler, L.G. 1980. Choosing the appropriate methods and standards for assaying tannin. Journal of Chemical Ecology. 15:1795-1810.

Holeckek, J. 1982. Sample preparation technique for microhistological analysis. Journal of Range Management. 35: 267-268.

Ichinohe, T., Orden, E.A., Del Barrio, A.N., Lapitan, R.M., Fujihara, T., Cruz, L.C., Kanai, Y. 2004. Comparison of voluntary feed intake, rumen passage and degradation kinetics between crossbred Brahmam cattle (*Bos indicus*) and swamp buffaloes (*Bubalus bubalis*) fed a fattening diet based on corn silage. Animal Science Journal. 75: 533-540.

Launchbaugh, K.L., Provenza, F., Pfister, J. 2001. Herbivore response to anti-quality factors in forages. Journal of Range Management. 54: 431-440.

Lechowicz, M.J. 1982. The sampling characteristics of electivity indices. Oecologia. 52: 22-30.

Macandza, V., Owen-Smith, N., Cross, P.C. 2004. Forage selection by African buffalo (*Syncerus caffer*) through the dry season in two landscapes of the Kruger National Park. South African Journal of Wildlife Research. 34: 113-121.

Makkar, H.P.S. 2001. Quantification of tannins in tree foliage. A laboratory manual for the FAO/IAEA Co-coordinated Research Project on "Use of Nuclear and Related Techniques to Develop Simple Tannin Assays for Predicting and Improving the Safety and Efficiency of Feeding Ruminants on Tanniferous Tree Foliage". FAO/IAEA. Vienna, Austria.

Mandaluniza, N., Aldezabalb, A., Oreguic, L.M. 2011. Diet selection of beef cattle on Atlantic grassland-heathland mosaic: Are heathers more preferred than expected? Livestock Science. 138: 49-55.

Mauricio, R.M., Mould, M.L., Dhanoa, M.S., Owen, E., Channa, K.S., Theodorou, M.K. 1999. A semi-automated *in vitro* gas production technique for ruminant feedstuff evaluation. Animal Feed Science and Technology. 79: 321-330.

Melaku, S., Aregawi, T., Nigatu, L. 2010. Chemical composition, *in vitro* dry matter digestibility and *in sacco* degradability of selected browse species used as animal feeds under semi-arid conditions in Northern Ethiopia. Agroforestry Systems. 80: 173-184.

Melletti, M., Penteriani, V., Mirabile, M., Boitani, L. 2007. Some behavioral aspects of forest buffalo (*Syncerus caffer nanus*): from herd to individual. Journal of Mammalogy. 88: 1312-1318.

Menke, K.H., Steingass, H. 1988. Estimates of the energetic feed value obtained from chemical analyses and *in vitro* gas production using rumen fluid. Animal Research and Development. 28: 7-55.

Minson, D.J. 1990. The chemical composition and nutritive values of tropical grasses. In: Skerman, P.J., Rivers, F. (Eds), Tropical Grasses. FAO. Rome, Italy. pp. 163-180.

Mohandass, D., Davidar, P. 2009. Floristic structure and diversity of a tropical montane evergreen forest (shola) of the Nilgiri Mountains, southern India. Tropical Ecology. 50: 219-229.

Muller-Dumbois, D., Ellemberg, H. 1974. Aims and methods of vegetation ecology. John Wiley & Sons. New York, USA.

Porter, L.J., Hristich, L.N., Chan, B.G. 1986. The conversion of procyanidins and prodelphinidins to cyanidins and delphinidin. Phytochemistry. 25: 223-230.

Rodríguez, R., Mota, M., Castrillo, C., Fondevila, M. 2010. *In vitro* rumen fermentation of the tropical grass *Pennisetum purpureum* and mixtures with browse legumes: effects of tannin contents. Journal of Animal Physiology and Animal Nutrition. 94: 696-705.

SAS, 2001. SAS user's guide. SAS Institute Inc. Cary, USA.

Terril, T.H., Windham, W.R., Evans, J.J., Hoveland, C.S. 1994. Effect of drying method and condenses tannin on detergent fiber analysis of *Sericea lespedeza*. Journal of the Science of Food and Agriculture. 66: 337-343.

Valencia, R., Foster, R.B., Villa, G., Condit, R., Svenning, J.C., Hernández, C., Romoleroux, K., Losos, E., Magård, E., Balslev, H. 2004. Tree species distributions and local habitat variation in the Amazon: large forest plot in eastern Ecuador. Journal of Ecology. 92: 214-229.

Van Soest, P.J. 1994. Nutritional ecology of ruminants, 2nd ed. Cornell University Press. Ithaca, USA. pp. 97.

Van Soest, P.J., Robertson, J.B., Lewis, B.A. 1991. Methods for dietary fiber, neutral detergent fiber and non-starch polysaccharide (NSP) in relation to animal nutrition. Journal of Dairy Science. 74: 3583-3597.

Wood, C.D., Johnson, J., Powell, C. 1993. Evaluation of the Bolivian tree leaves as fodders by an *in vitro* fermentation technique. Agroforestry Forum. 4: 28-34.

Zent, E., Zent, S. 2004. Floristic composition, structure, and diversity of four forest plots in the Sierra Maigualida, Venezuelan Guayana. Biodiversity and Conservation. 13: 2453-2483.

16

Energy use Analysis for Rice Production in Nasarawa State, Nigeria

H.Y. Ibrahim* and H.I. Ibrahim

Department of Agricultural Economics and Extension, Faculty of Agriculture, Nasarawa State University, Keffi. Nigeria

E-Mail: Hussein464@yahoo.com

**Corresponding author*

SUMMARY

The study was conducted to analyze energy use for in rice production in Nasarawa state Nigeria using a sample of 120 randomly selected rice farmers. Energy productivity, energy efficiency and specific energy were computed and simple descriptive statistics was used for data analysis. The energy use pattern shows that, rice production consumed an average total energy of 12906.8 $MJha^{-1}$, with herbicide energy input contributing the largest share (53.55 %). Human labour had the least share (0.74 %) of the total energy input used. The energy productivity, Specific energy and energy efficiency were 0.3 MJ^{-1}, 3.6 MJ^{-1} and 4.1 respectively. A total of 10925.0 MJ of energy was used in the form of indirect energy and 1981.8MJ was in the direct form of energy. Non-renewable energy forms contributed the largest share (80.63 %) of the total energy input used for rice production in the study area. Rice production in the study area was observed to be mainly dependent on non-renewable and indirect energy input especially herbicide. Thus, the study recommends the introduction of integrated weed management system in order to reduce cost and dependence on a non-renewable input for weed control.

Key words: Herbicide; Nigeria; renewable energy; rice; productivity.

INTRODUCTION

Energy is said to be the engine for growth and development in all economies of the world. In all parts of the world today, the demand for energy is increasing almost on a daily basis. According to Pimental (1992), energy is one of the most valuable inputs in agricultural production. Sufficient availability of the right energy and their effective and efficient uses are prerequisites for improved agricultural production (Handan *et al.* 2009). The amount of energy used in agricultural production, processing and distribution is significantly high in order to feed the expanding population and to meet other social and economic goals of a society (Handan *et al.* 2009). It has been realized that crop yield and food supplies are directly linked to energy availability or consumption. Also, increases in yields in the developed countries are as a result of commercial energy inputs, in addition to improved varieties (Tolga *et al.* 2009).

Energy use in agriculture has become more intensive as the Green Revolution led to the increasing use of high yielding seeds, fertilizers and chemicals as well as diesel and electricity. Energy consumption per unit area in agriculture is directly related to the development of the technology in farming and the level of production. Inputs such as fuel, electricity, machinery, seed, fertilizer and chemical take significant share of the energy supplies in the production system of modern agriculture. Thus, the use of intensive inputs in agriculture and access to plentiful fossil energy has provided an increase in food production and standard of living (Selim *et al.* 2005).

Rice is the second most important cereal in the world after wheat in terms of production (Jones 1995). Nigeria ranks the highest as both producer and consumer of rice in the West Africa sub-region (Jones 1995). However, in terms of area of land under food crop production in the country, rice ranks sixth (after sorghum, millet, cowpea, cassava and yam) (Imolehin and Wada 2000). The Federal Ministry of Agriculture (1993) estimated that the annual supply of food crops (including rice) would have to increase at an average annual rate of 5.9 % to meet food demand, and reduced food importation significantly. Studies have shown that aggregate rice production in Nigeria has been growing at about 2.5 % per annum in recent years (Olayemi 1998; Akinbola 2002; Amaza and Olayemi 2002). But the annual rate of population growth has been high (about 3 %) (Akinbola 2002). The reality is that Nigeria has not been able to attain self-sufficiency in rice production despite increasing hectares put into production annually. The constraint to the rapid growth of food production seems to be mainly that of low crop yields and resource productivity. The implication is that there is hope for additional increases of output from existing hectares of rice, if resources are properly harnessed and efficiently allocated (Amaza and Olayemi 2002).
Several studies have outlined a number of factors responsible for the low level of rice production in Nigeria (Kolawole and Scoones 1994; Atala and Voh 1994, Okuneye 2001). However, studies analyzing the relationship between energy inputs and rice yield in Nigeria are not available. There is a dearth of data on energy expenditure and returns in crop production in Nigeria, and other developing countries (Abubakar and Ahmed 2010). Even though, much attention is not given to the knowledge about energy expenditure in crop production in Nigeria, the increasing demand for food production to meet the pressure from an ever-increasing population makes the energy-agriculture relationship very important. For these reasons, energy use pattern, energy efficiency, energy productivity, specific energy and energy inputs-output relationship were determined for rice production in north central Nigeria.

MATERIALS AND METHODS

Nasarawa State is located in north central Nigeria. It lies between North Latitude 7° and 9° and 7° and 10° East Longitude and shares boundary with Benue State to the South, Kogi State to the west, the Federal Capital territory, Abuja to the North – East, Plateau State to the South east. The State covers an area of about 27, 117 Km^2 with an estimated population of 1,863,275 people (National Population Commission 2006). The State has a mean temperature range of 25°C in October to about 36°C in March while annual rainfall varies from 13.73mm in some places to 145mm in others. Alluvial soils are found along the Benue trough and their flood plains. The forest soils, which are rich in humus and laterite, are found in most part of the state. There are also sandy soils in some parts of the state. Solid minerals notable are salt and bauxite. A three (3) stage random sampling technique was used to determine the sample size. In the first stage, a zone from the three ADP zones (Nasarawa North, West and East) was randomly selected. In the selected zone (Nasarawa East), the list of major rice producing communities was obtained, and four communities. (Sabon-Gida, Assakio, Brum-brum and Awe) were randomly sampled. Finally, 30 rice farmers were selected from each community to give a sample size of 120 rice farmers for the study. Structured questionnaire administered to the respondents was used to collect data for the study. Data were collected on the socio-economic characteristics of the respondents as well as production inputs and output in rice production.
Simple descriptive statistics was used for data analysis; energy efficiency, specific energy and energy productivity, for rice crop production were also calculated on per hectare basis using the equations suggested in literature (Canakci *et al.* 2005; Ozkan et al. 2004; Hatirli *et al.* 2005; Singh and Mittal, 1992; Khan *et al.* 2004).

$$\text{Energy efficiency} = \frac{\text{Total energy output (MJ/ha)}}{\text{Total energy input (MJ/ha)}}$$

$$\text{Energy productivity} = \frac{\text{Rice yield/output (kg)}}{\text{Energy input (MJ/ha)}}$$

$$\text{Specific Energy} = \frac{\text{Energy input (MJ/ha)}}{\text{Rice yield/output (kg)}}$$

Each agricultural input and output has its own energy equivalent value. Hence the inputs and output were converted into their equivalent energy units using the conversion factors in Table 1.

Table 1. Energy equivalent for different input and output in rice production.

Input	Energy equivalent (MJ/Unit)	Reference
NPK fertilizer (kg)	11.76*	
Human Labour (Man hours)	01.96	Singh *et al.* (2002)
Chemical Insecticide (L)	199.00	Helsel (1992)
Chemical Herbicides (L)	238.00	Helsel (1992
Machinery (h)	62.70	Singh *et al.* (2002)
Diesel – Oil (L)	56.31	Singh *et al.* (2002)
Seed (kg)	14.7	Singh (2002)
Yield (kg)	14.7	Canackci *et al.* (2005)

*The energy equivalent for a kilogram of NPK fertilizer was derived from the ratio of the elements (N, P and K) in a 50 Kg bag of the fertilizer. The NPK 15:15:15 brand is the most widely used amongst the respondents in the study area. As such, for each element, its quantity in kilogram in a 50 Kg bag was determined as follows;

$$N = \frac{15}{100} \times \frac{50}{1} = 7.5$$

$$P = \frac{15}{100} \times \frac{50}{1} = 7.5$$

$$K = \frac{15}{100} \times \frac{50}{1} = 7.5$$

However, from literature, the energy equivalent for elemental N, P and K are as follows; 60.60 MJ, 11.10 MJ and 6.70 MJ respectively. Thus, the quantity of each element was converted to its energy equivalent as shown below.

$N = 60.60 \times 7.5 = 454.5$ MJ

$P = 11.10 \times 7.5 = 83.25$ MJ

$K = 6.70 \times 7.5 = 50.25$ MJ

Total =588.00 MJ

This implies that a 50 Kg bag of NPK 15:15:15 fertilizer is equivalent to 588 MJ of energy, while a kilogram of the fertilizer is equivalent to 11.76 MJ of energy.

RESULTS AND DISCUSSION

Inputs and outputs level in rice production

The quantities of inputs/ha used and output/ha (paddy) obtained in rice production in the study area are presented in Table 2. To cultivate a hectare of rice in the study area, a total of 163.5 kg of seed, 48.7Man hours of labour, 105.2 kg of fertilizer, 29.0 L of herbicide, 0.5 L of insecticide, 33.5 L of diesel and 6.97 hours of machinery were used. These are equivalent to, 2403.5 MJ, 95.4 MJ, 1237.4 MJ, 9612.4 MJ, 110.4 MJ, 1888.4 MJ and 437 MJ of energy respectively. An average yield of 3597.93 kg/ha of paddy equivalent to 52889.1 MJ of energy was obtained.

The result obtained from this study shows that energy inputs used by the farmers in the study area were manual (human labour), chemical (fertilizer and diesel), mechanical (machinery), and biological (seed). The average quantity of seed, fertilizer and herbicides used per hectare were not in accordance with the recommended rates of 80 – 100 kg/ha for seed, 300 – 400 kg/ha for fertilizer and 16 – 20 L/ha for herbicide for rice production in Nigeria, (Ekeleme *et al.*, 2008).

Energy use pattern in rice production

The pattern of energy usage in rice production is presented in Table 3. The total energy input used per hectare for rice production was 12906.8 MJ. Human labour and herbicide contributed the minimum and maximum energy input values of 95.5 MJ/ha and 6913.9 MJ/ha respectively for rice production, representing 0.7 % and 53.6 % respectively of the total energy used per hectare. The observation for human labour can be attributed to the fact that human labour was only used for planting, fertilizer application and harvesting only. On the other hand, herbicide energy dominated the total energy input used because of its excessive usage for weed control in the study area. Insecticide, diesel and fertilizer inputs contributed 0.9 %, 14.6 % and 9.6% of the total energy inputs per ha respectively. The result indicated that, the energy efficiency and specific energy for rice production were 4.1 and 3.6 MJ^{-1} respectively, while energy productivity was 0.3 MJ^{-1}. The energy efficiency ratio of 4.1 indicates high energy use efficiency. However, the productivity of energy inputs was very low. This can be attributed to the usage of local rice varieties in the study area. The specific energy shows the amount of energy required to produce a kilogram of paddy. Thus, a specific energy of 3.6 means that 3.6 MJ of energy is required to produce a unit of paddy rice.

Energy productivity for various inputs in rice production

The energy productivity of inputs for rice production is presented in Table 4. The value of energy productivity for the inputs ranges from 0.5 to 37.7 for herbicide and labour respectively. Seed, insecticide, diesel, machinery and fertilizer inputs have productivity ratios of 0.9, 30.3, 1.9, 14.4 and 2.9 respectively. The result revealed that, human labour and herbicide were respectively the most productive and the least productive energy inputs for rice production in the study area.

Energy inputs in the forms of direct and indirect, renewable and non-renewable energy for rice production

The results in Table 5 shows that, non-renewable energy forms contributed 80.6 % of the total energy input for rice production, while only 19.4 % comes from the renewable energy forms. On the other hand, 84.7 % of the total energy is also in the form of indirect energy, with the direct energy forms contributing 15.3 % of the total energy. This finding implies that rice production in the study area is mostly dependent on non-renewable and indirect energy forms especially herbicides.

Table 2. Inputs/outputs level in rice production.

Units	Mean	Maximum	Minimum
Yield (kg)	3597.9	16000.0	1166.6
	(52889.1)	(235200.0)	(17150.0)
Seed (kg)	163.5	1000.0	33.3
	(2403.5)	(14700.0)	(490.0)
Human labour (Man hours)	48.7	270.0	7.5
	(95.4)	(529.2)	(14.7)
Fertilizer (kg)	105.2	500.0	35.0
	(1237.4)	(5880.0)	(411.6)
Herbicide (L)	29.0	120.0	09.0
	(6912.4)	(28560.0)	(2142.0)
Insecticide (L)	0.6	10.0	0.0
	(110.4)	(1990.0)	(0.0)
Diesel (L)	33.5	250.0	0.0
	(1888.4)	(14077.5)	(0.0)
Machinery (h)	6.9	60.0	0.0
	(437.03)	(3762.0)	(0.0)

Figures in parenthesis are the energy equivalents of the inputs in MJ

Table 3. Energy use pattern in rice production (ha).

Input	Quantity/unit area (ha)	Total Energy Equivalent	%
Human labour (Man hour)	48.7	95.5	0.7
Machinery (h)	4.0	250.8	1.9
Herbicide (L)	29.05	6913.9	53.6
Insecticide (L)	0.6	119.4	0.9
NPK Fertilizer (kg)	105.2	1237.4	9.6
Diesel (L)	33.5	1885.6	14.6
Seed (kg)	163.5	2403.5	18.6
Total Energy Input (MJ/ha	-	12906.8	100.0
Yield (kg)	3597.9	52889.1	
Energy Efficiency	-	4.1	
Energy Productivity	-	0.3 $KgMJ^{-1}$	
Specific Energy	-	3.6 $MJKg^{-1}$	

Table 4. Energy productivity for various inputs in rice production.

Inputs	Qty/ha	Energy equivalent	Energy productivity	Rank
Human Labour	48.7	95.5	37.7	1st
Machinery	4.0	250.8	14.4	3rd
Herbicides	29.05	6923.9	0.5	7th
Insecticide	0.6	119.4	30.1	2nd
NPK fertilizer	105.2	1237.4	2.9	4th
Diesel	33.5	1885.6	1.9	5th
Seed	163.5	2403.5	0.9	6th
Yield	3597.9	52889.6		

Table 5. Energy inputs in the forms of Direct and Indirect, Renewable and Non-renewable energy for rice production.

Energy forms	Total Energy equivalent(MJ/ha)	%
Direct Energy [a]	1981.8	15.3
Indirect Energy [b]	10925.0	84.7
Total	**12906.8**	**100.0**
Renewable [c]	2499.6	19.4
Non-renewable [d]	10410.2	80.6
Total	**12906.8**	**100.0**

a. Human labour, and Diesel, b. Fertilizer, Insecticides, Herbicides, Machinery and Seed, c. Human labour, Seed, d. Diesel, Fertilizer, Insecticides, Herbicides and Machinery

CONCLUSION

Rice production in the study area was observed to be mainly dependent on non-renewable and indirect energy input especially herbicide. The study recommends the introduction of integrated weed management system to reduce the excessive use of herbicide for weed control. Farmers should be enlightened on the negative effect of excessive use of herbicide on the environment and also the long run impact on climate change. The adoption of high yielding rice varieties such as the Nerica varieties should be promoted in the study area in order to improve the energy productivity in rice production.

REFERENCES

Abubakar, M.S., Ahmed D. 2010. Pattern of energy consumption in Millet production for selected farms in Jigawa, Nigeria. Australian

Journal of Basic and Applied Sciences. 4(4): 665-672.

Akinbola, L. A. 2002. Technology dissemination, agricultural productivity and poverty reduction in rural sector of Nigeria. Poverty reduction and the Nigerian Agricultural Sector, El-Shadai Global Ventures Ltd. Ibadan. pp: 27 – 35.

Amaza, P.S., Olayemi, J.K. 2002. Analysis of technical efficiency in food crop production in Gombe State, Nigeria. Applied Economics Letters. 9: 54- 60.

Atala, T. K., Voh, J. P. 1994. Utilization and sustainability of fadama in northern Nigeria: Input and extension delivery system. Strategies for Sustainable Use of Fadama lands in Northern Nigeria. IIED and CSER, ABU Zaria, Nigeria, p: 147.

Canakci, M., M. Topakci., I. Akinci, A. Ozmerzi. 2005. Energy use pattern of some field crops and vegetable production: case study for Antalya region, Turkey. Energy Conversion Management. 46: 655-666.

Ekeleme, F. Kamara, A.Y., Omoigui, A., Tagabaru, J., Mshelia, J. 2008. Guide to rice production in Borno State, Nigeria. Canadian International Development Agency. (CIDA), 24 pp.

Federal Ministry of Agriculture, 1993: In Amaza, P.S., Olayemi, J.K. 2002. Analysis of technical efficiency in food crop production in Gombe State, Nigeria. Journal of Applied Economics Letters. 9: 51 – 54.

Handan, A. Orhan, O., Hatice, K. 2009. Analysis of energy use for pomegranate production in Turkey. Journal of Food, Agriculture and Environment. 7(2):475-48.

Hatirli S. A., B. Ozkan, Fert, C., 2005. Energy Inputs and Crop Yield Relationships in Greenhouse Tomato Production. Renewable Energy. 31:427-438.

Helsel, Z.R. 1992. Energy and alternatives for fertiliser and pesticide use. In Fluck, R.C. (ed.) Energy in World Agriculture. Vol. 6, Elsevier Science Publishing, pp: 177-210.

Imolehin, E.D., Wada, A.C. 2000.Meeting rice production and consumption demand of Nigeria with improved technologies, International Rice Commission Newsletter; Vol. 49, FAO Rome, pp: 33-44.

Jones, P. M. 1995. The rice plant and its environment; West African Rice Development Association Training Guide 2. pp: 1 – 6.

Khan, M.A., S. Ahmad, Z. Hussain, M. Yasin, M. Aslam, R. Majid. 2004. Efficiency of water and energy use for production of organic wheat. Journal of Science, Technology and Developmen.t 24(1): 25-29.

Kolawole, A., Scoones, I. 1994. Fadama land use and sustainability in northern Nigeria: Strategies for the sustainable use of fadama lands in northern Nigeria. CSER/ABU Zaria and IIED. London, p: 29.

National Population Commission (NPC) 2006. Censu s result. http://www.population.gov.ng/index .php.1st /July/2010.

Okuneye, P. A. 2001. Rising cost of food price and food insecurity in Nigeria and its implication for poverty reduction. Central Bank of Nigeria Economic and Financial Review. 39 (14): 88 – 110.

Olayemi, J.K. 1998. Food security in Nigeria. Research Report NO. 2 Development Policy Centre Ibadan, Nigeria, pp: 1 – 85.

Ozkan, B., Akcaoz, H., Fert, C. 2004. Energy input output analysis in Turkish agriculture. Renewable Energy. 29(1):39-51.

Pimental, D. 1992. Energy inputs in production agriculture. Energy in farm production, Amsterdam: Elsevier, pp: 13-29.

Selim, A. H., Burhan, O., Camel, F. 2005. Energy inputs and crop yield relationship in green house tomato production Journal of Renewable Energy. 31: 427– 438.

Singh, H., Mishra, D., Nahar, N.M. 2002. Energy use pattern in production agriculture of a typical village in Arid Zone India-Part I. Energy

Conversion and Management. 43(16):2275-2286.

Singh, J.M. 2002. On farm energy use pattern in different cropping systems in Haryana, India. MSc. thesis, International Institute of Management University of Flensburg, Germany. 136 pp

Singh, S., Mittal, J.P. 1992. Energy in production agriculture. Mittal Publications, New Delhi, India, 166 pp.

Tolga T., Ç. Bahattin, Vardar A., 2009. An analysis of energy use and input costs for wheat production in Turkey. Journal of Food, Agriculture and Environment. 7(2): 352-356.

17

Technology Transfer Networks on Papaya Production with Transitional Growers

Octavio Cano-Reyes[1], Juan A. Villanueva-Jiménez[1*], Juan L. Reta-Mendiola[1] and Arturo Huerta-De-la-Peña[2]

[1]*Colegio de Postgraduados, Campus Veracruz. Km. 88.5 Carr. Xalapa-Veracruz, Tepetates, Manlio F. Altamirano, Veracruz. México. CP 91690.*

cano.octavio@colpos.mx; jretam@colpos.mx

[2]*Colegio de Postgraduados, Campus Puebla. Cholula, Puebla. México.*

arturohp@colpos.mx

**Corresponding Author: javj@colpos.mx*

SUMMARY

Social networks analysis applied to rural innovation processes becomes a very useful technology transfer tool, since it helps to understand the complexity of social relationships among people and/or institutions in their environment, and it also defines those innovation networks given in specific working groups or regions. This study was conducted from April to May 2011 to determine those networks and key players present in the group of growers associated as "Productora y Comercializadora de Papaya de Cotaxtla S.P.R. de R.L.", that influence the technology transfer process in Cotaxtla, Veracruz, Mexico. Data were analyzed using UCINET 6 software. Three centrality measures were obtained: range, degree of mediation and closeness. Of 32 network players, 27 actively diffuse innovations according to their interests; alliances must be established with them to transfer technology. Four growers stand out as central actors, which along with the Instituto Nacional de Investigaciones Forestales Agricolas y Pecuarias, the Colegio de Postgraduados and the growers' organization itself, could be the most appropriate actors to establish a technology transfer program to accelerate the diffusion and adoption of innovations. Wholesalers, middlemen and credit institutions do not participate in this process, but having capital they could be incorporated in the innovation diffusion process.

Key words: *Carica papaya*; central actors; technological innovation; tacit knowledge; social networks; knowledge diffusion.

INTRODUCTION

Nowadays, a great amount of scientific knowledge is available, even for the rural Mexican society. However, most of this information is out of growers' reach because it is published in scientific journals with limited distribution and written in very technical terminology, not easily understandable by the average rural resident (Reta *et al.*, 2011). Agricultural technology transfer (ATT) is a vital element that carries new knowledge to growers.

Up to the end of 1990s, ATT in Mexico had been

performed in a descendent, lineal and unidirectional way, based on technological packages and under governmental control (Muñoz and Santoyo, 2010a); however, this had limited results in terms of field implementation (Mata, 1997). Innovation is a fundamental condition to reach sustainable economic growth, allowing the reduction of national social disparities and promoting the sustainable use of natural resources (Aguilar *et al.*, 2010). This brings us to a bigger problem.

During the period 1990-2010 (Muñoz and Santoyo, 2010a), Mexican public policy for rural innovation created several extension programs proposing an ascendant paradigm of technology transfer with a greater involvement of producers (Doorman, 1991), supported by the Law of Sustainable Rural Development (SINDER, PESPRO, PRODESCA, PESA and Technical Assistance and Training) (Muñoz and Santoyo, 2010a). However, base participation has not been fully understood by professional service providers (PSP) and by those responsible of several governmental offices, serving only as facilitator to the access of subsidies for agricultural inputs and productive assets (Muñoz and Santoyo, 2010a). Improvements incorporated in the strategy for innovation have been scarce, mainly because ascendant communication mechanisms among producers, extensionists and researchers are not working properly, prevailing the lineal-descendant paradigm of professional technical assistance, according to which "the engineers are those who know" and "they bring all the knowledge to the growers" (Engel, 2004; Aguilar *et al.*, 2005).

Several ATT models presented by social researchers are ascendant in nature, but not all include planning, participative action and reflection in the proposed methodology (Reta *et al.*, 2011). To increase efficiency in the innovation processes, such models require participative research with the rural society to know the social structure complexity and the characteristics of social relationships in a specific agricultural area (Clark, 2006; Núñez, 2008). In those cases, the social network analysis (SNA) is a participative research tool that can be incorporated in ascendant models. Recently, the "Simultaneous Production Growing Groups" (SPGG) technology transfer model has been proposed; it considers participative action-research, previous grower experience, and SNA for planning, as well as group-reflection and action in the diffusion of technological innovations; SPGGs move forward to build significant knowledge that guarantees appropriation and use of the innovation by the participant grower (Abato, 2011; Reta *et al.*, 2011), and improves the way to innovate through collective knowledge and group decision making (Abato-Zarate *et al.*, 2011). Thus, the ascendant SPGG technology transfer model emerges as an alternative to those models implemented in the past, linked to the agricultural product-systems organizations promoted by the Mexican government (Abato, 2011; Reta *et al.*, 2011).

Social networks present among growers in a specific area may be understood through the SNA (Aguilar *et al.*, 2010). The SNA is an ensemble of formal tools for social research that help to determine central actors, that is, growers and/or agents of change that treasure knowledge as a powerful tool, allowing them to enjoy more dominance and influence over all other actors (Muñoz *et al.*, 2004; Navarro and Salazar, 2007). The SNA facilitates establishing and guarantees success of any ATT project, allowing to know which actors of the productive process must establish alliances because of their good acceptance by others in their sector (Clark, 2006; Núñez, 2008). The objective of this study was to determine the networks and central actors present in the growers association "Productora y Comercializadora de Papaya de Cotaxtla S.P.R. de R.L" in Cotaxtla, Veracruz, Mexico, supporting the integration of working groups for the implementation of a technology transfer program based on SPGGs.

MATERIALS AND METHODS

The study to determine technology transfer networks was carried out from April to May 2011 in the municipality of Cotaxtla, Veracruz, Mexico. The participants were the 23 members of the growers association "Productora y Comercializadora de Papaya de Cotaxtla S.P.R. de R.L." Growers affiliated to this group usually plant papaya in areas from 2 to 7 ha, using drip irrigation and having production costs of MX\$100,000 ha^{-1}; their average age is 41.5 years; 50% of them studied only primary school, 17% secondary (3 years), 17% are certified technicians and 17% obtained a bachelor's degree, most of them with a good experience in the crop (5 to 30 years). Consequently, they can be considered in a mature-productive age, with a basic to intermediate education and a good experience in the crop (Abato, 2011). Considering the size of the area cultivated by them, the technology used, their organization and the cash movement, it is classified as a group of "transitional" growers (Abato-Zárate *et al.*, 2011), also called "commercially diversified" growers (Rascón *et al.*, 2006), because they have been unable to consolidate their organization to be considered as "entrepreneurial."

For research purposes, a network was defined as a finite ensemble of actors (persons, groups or organizations) that are associated and connected in a specific system and share common objectives, building a sum of relationships among them (Molina, 2005; Clark, 2006; Núñez, 2008). Thus, it was considered that an innovation and technology transfer network (ITTN) for transitional growers in papaya agroecosystems may operate in relation to the following function: ITTN = f (IG, O, IS, TA, SI, CI, WS, MM), where ITTN = innovation and technology transfer network, IG = innovative growers, O = organizations, IS = input sellers, TA = technical assistants, SI = sectorial institutions, CI = credit institutions, WS = wholeselers and MM = middlemen.

To contrast this theoretical function with local reality, a survey was applied to 20 growers belonging to this group, based on a questionnaire of 21 open-relational questions on actors intervening on the ITTN. The survey asked growers about: a) general information of the grower, b) their organization, c) nearby pesticide chemical companies, d) agrochemical reseller houses working in the area, e) their technical assistants, f) sectorial institutions bringing them attention, g) credit institutions visiting them, h) whole sellers working with them, i) middlemen, and j) growers involved. Questions focused on identifying the actors of each ITTN component and which ones share their knowledge to improve papaya production.

UCINET for Windows v.6.289 software (Borgatti *et al.*, 2002) was used to establish centrality indicators and central actors present on the existing social networks. To define the ITTN actors, an analysis was performed including all network components but growers. For the "growers" component, an analysis apart from the rest of actors was done, using variables such as "inclination to share knowledge" and "use of the best technology adapted to growers conditions". A centrality measure is understood as a group of algorithms calculated for a network, it allows to know the network structure and position of their nodes. Network centrality indicators measure existent relationships among actors of certain social context, and determine the influence respect to the goods, capital and information belonging to each actor (Molina, 2005; Velázquez and Aguilar, 2005; Clark, 2006). "Central actors" are considered those with greater ability to receive and send all type of information to the rest of the network (Zarazúa-Escobar *et al.*, 2011a); to prioritize who might be involved in a project, central actors can help to make that decision (Velázquez and Aguilar, 2005; Clark, 2006). Key and/or central actors are those individuals whose participation is indispensable and required to reach the purpose and goals of the project. They have the power, the ability and the means to decide and influence in vital fields that allow or not the project development, and very often they manifest a direct, explicit and committed interest with the purpose and aims (SEMARNAT, 2011).

In this work, three centrality measurements were determined: nodal degree or range, degree of intermediation and closeness (Molina, 2005; Velázquez and Aguilar, 2005; Clark, 2006). The nodal degree or range is the number of direct ties from an actor or node; it indicates how many other nodes they are directly connected or linked with. The degree of intermediation indicates the appearance frequency of a node in a shorter (or geodesic) section connecting other two, that is, it shows when a person is intermediary between two other persons of the same group that are not familiar to each other (also known as the "bridge person"). The degree of closeness indicates the proximity of a node in relation to the rest of the network and expresses the capacity of that node to provide access for the rest of the actors by means of their direct and indirect relationships with the rest of the nodes; it is calculated taking in consideration all geodesic distances of a node to reach those actors. Due to their structural position, several actors of a network may reach in a faster and easier way a greater amount of actors. Central actors are defined by the greatest nodal degree, greatest closeness and greatest intermediation, in relation to the propensity to share knowledge and use of the best technology judged by growers themselves, making them the most outstanding growers.

The ITTN actors are identified in Figure 3 with the following keys: AS = Agroquímicos Susunaga (agrochemicals company, agchm), FR = Financiera Rural (bank), SA = Servicio Agrotécnico (agchm), FYPA = Fertilizantes y Productos Agropecuarios (agchm), CEPP = Consejo Estatal de Productores de Papaya (State Papaya Growers Council), ST = Secretaría del Trabajo (Work Ministry), PCPC = Productores y Comercializadores de Papaya de Cotaxtla (papaya growers local organization), CP = Colegio de Postgraduados (research institution), BTO = Biotecnología Orgánica (agchm), SE = Secretaría de Economía (Economy Ministry), ASLP = Agroquímicos y Semillas Los Parra (agchm), DASUR = Distribuidora Agrícola del Sureste (agchm), AML = Agroquímicos Mata de Lázaro (agchm), INIFAP = Instituto Nacional de Investigaciones Forestales, Agrícolas y Pecuarias (research institution), SAGARPA = Secretaría de Agricultura, Ganadería, Pesca y Alimentación (Agricultural Ministry), SERAVER = Servicios Agrícolas de Veracruz (agchm), CICY = Centro de Investigaciones Científicas de Yucatán (research institution), AGSB = Agropecuaria Santa Buena Ventura (agchm).

RESULTS AND DISCUSSION

The ITTN among papaya growers of Cotaxtla is shown in Figure 1. The greatest nodal degree was for G. Basurto R. (50%), followed by A. Parra V. and R. Basurto H., both with 36% (Table 1); they showed the greatest number of relationships or links with growers of their organization. These growers share their tacit knowledge with the majoritie of group mates and are open to receive information. It is also shown that a limited number of actors agglutinate in several sub-networks, acting as a powerfull catalizer, generating a high degree of adoption and an elevated propention to establish contacts with the majoritie of actors to share knowledge and for the early and fast adoption of innovations (Muñoz and Santoyo, 2010b). Sub-networks have a central actor (node), who is of great utility to initiate technology innovation processes acting as catalyzers. The three greater nodal degree central actors could work adequately in any technology transfer project, being recognized by their peers in the growers association (Clark, 2006).

Additionally, Table 1 shows G. Basurto R. with the best nodal degree (central actor), besides being the best bridge person or intermediary between two not very familiar group mates, but closer than the rest of the group, allowing a faster and easier way to reach a greater amount of network actors (Molina, 2005). That situation places him in advantage in relation to A. Parra V., R. Basurto H. and A. Basurto H., which present very low intermediary values (bridge person), indicating a lower effectiveness to communicate or mediate situations. However, the three latter persons present good closeness values compared to all other network members, because they are capable of building sub-networks or affinity groups. The ability of G. Basurto R. to communicate with the network allows him to share tacit knowledge that worked for him, besides having the confidence of most of the growers who are members of the group.

The mean and standard deviation of the knowledge exchange among growers of the group (Table 1) indicates that G. Basurto R. outstands the rest of the group, whereas all other central actors are less reliable, besides having centrality values above the group average.

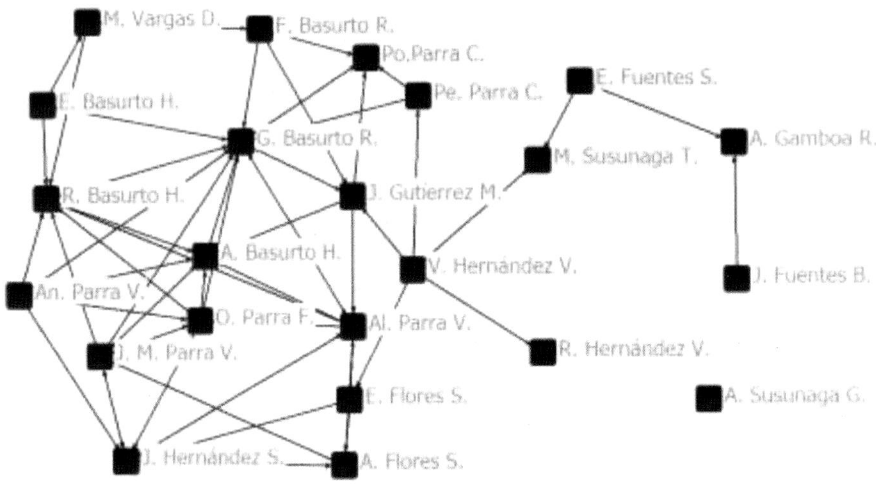

Figure 1. Central actors of the technology transfer network formed by the growers group "Productora y Comercializadora de Papaya de Cotaxtla. S.P.R. de R.L.". 2011.

Table. 1. Extreme centrality measures for actors of the growers group "Productora y Comercializadora de Papaya de Cotaxtla S.P.R. de R.L.", in relation to the variable "knowledge sharing". 2011.

Grower name	Nodal degree	Closeness	Intermediation
Central actors			
G. Basurto R.	50.0	33.8	20.6
A. Parra V.	36.3	32.3	9.0
R. Basurto H.	36.3	29.7	5.6
A. Basurto H.	31.8	31.8	3.0
No central actors			
E. Fuentes S.	9.0	22.2	16.4
R. Hernández V.	4.5	24.7	0.0
A. Susunaga G.	0.0	0.0	0.0
All actors mean (std. dev.)	20.1 (11.9)	28.2 (4.30)	7.5 (9.59)

Innovative growers handling the best technology, indicated by their own organization members, are shown in Figure 2. G. Basurto R., A. Basurto H., A. Parra V. and R. Basurto H. were pointed as those with the highest percentages of nodal relational links (67, 57, 43 and 43%). It was not fortuitous that growers sharing technological innovations and those managing the best technology were the same. This indicates a positive relation between knowledge and avaliability in this network, being probably easy to implement technological innovations with central actors as process facilitators.

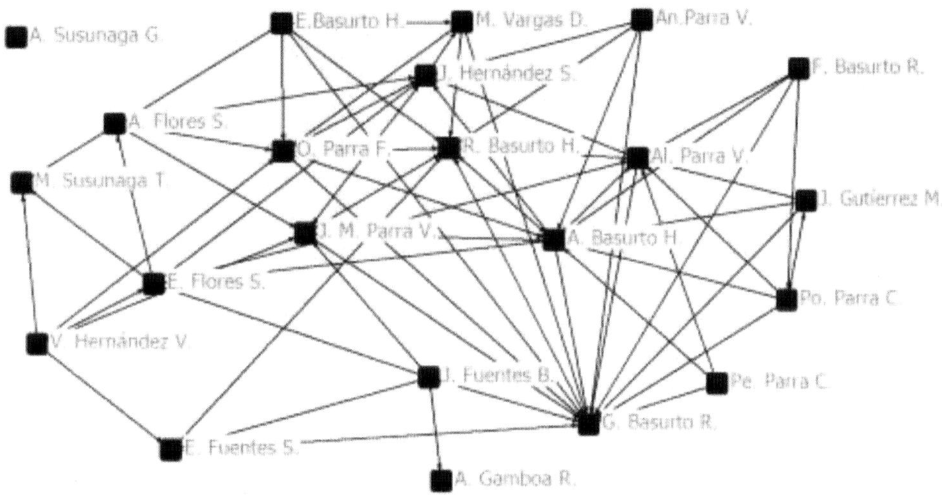

Figure 2. Recognition of growers managing the best technology among members of the network group "Productora y Comercializadora de Papaya de Cotaxta. S.P.R. de R.L.". 2011.

This result differs from what Zarazúa-Escobar *et al.* (2011a) indicated for a strawberry production system in Zamora, Michoacan, Mexico, where a final buyer and a jobber are those who centralize the information, causing low adoption, making necessary to promote organizative innovations to improve competitiveness. On the contrary, in the guava production system (Zarazúa *et al.*, 2011), main actors of the innovation diffusion process are the agrobusinessmen, having an impact on five other actors, whom also have an impact on 12 extra actors.

Núñez (2008) indicates that recruiting first the opinion leaders accelerate the innovation diffusion process On the contrary, early adoption by part of marginal actors yield slow diffusion curves (Clark, 2006; Muñoz and Santoyo, 2010b).

Nodal degree, closeness and intermediary values for central actors in Table 2 are much higher than the mean, which allows trusting them to establish participative research projects due to their level of responsibility conducting their crops.

Sectorial public and private institutions, as well as private consultants providing technical assistance to the group "Productora y Comercializadora de Papaya de Cotaxtla S.P.R. de R.L." are represented in Figure 3. Thirty two network participants were detected, following the ITTN function. However, only 27 participate in the innovation diffusion. In this institutional context, INIFAP, growers organization itself, CP, and the private consultant Ing. E. Sayago are considered network central actors.

In the municipality of Cotaxtla, the only existing grower organization is the one here indicated; however, another organization (Agroproductores de Cotaxtla S.P.R. de R.L.) was active in recent years, but it was unraveled; also "Grupo Agrícola Martínez" is present in the area, but is a growers group that did not originate in this municipality. The last two organizations did not share their technological knowledge with growers of the area, and thus, they are not considered as part of the technological transfer network for small and middle growers. All growers interviewed indicated that wholesale traders, middlemen and credit institutions do not participate in the ATT, being dedicated specifically to their main activity. Financiera Rural arranges credits with several growers of the group, but does not bring any technical assistance service (Figure 3), and it is the only bank that provides individual credits to papaya growers. This funding may be used to establish alliances to distribute at the same time credits and available technological innovations. Funding has decreased in more than 80% in the last decade and it has focused on large growers, meanwhile "ejidos" and communal growers do not have access to it (Morett and Cosío, 2006). This might be a limiting factor for establishing a general program in the area. Although there are some subsidies, as those from the Direct Country Supporting Program (PROCAMPO), that might be used to promote and induct the adoption of knowledge, it has not accomplished the several collateral objectives that are attributed to it, since it only grants direct subsidies (Zarazúa-Escobar *et al.*, 2011b).

Table. 2. Extreme centrality measures for nodes belonging to the group "Productora y Comercializadora de Papaya de Cotaxtla S.P.R. de R.L.", for the variable "growers managing the best technology". 2011.

Grower name	Nodal degree	Closeness	Intermediation
Central actors			
G. Basurto R.	66.6	43.7	24.5
A. Basurto H.	57.1	41.1	10.3
R. Basurto H.	42.8	38.8	4.1
A. Parra V.	42.8	38.8	5.1
No central actors			
J. Gutiérrez M.	19.0	33.8	0.0
P. Parra C.	14.2	33.3	0.0
R. Hernández V.	4.7	27.6	0.0
A. Susunaga G.	0.0	0.0	0.0
All actors mean (std. dev.)	22.27 (15.41)	35.95 (3.40)	3.85 (5.57)

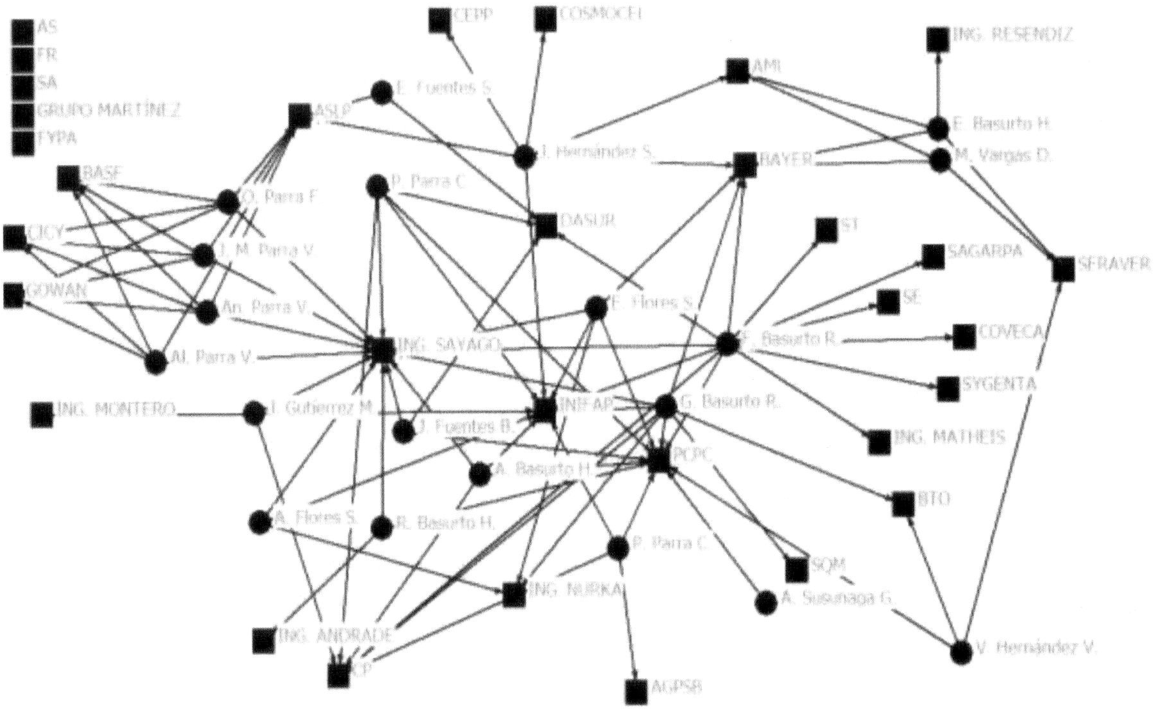

Figure 3. Impact of agricultural agents of change, participating in the technology transfer of the group"Productora y Comercializadora de Papaya de Cotaxtla S.P.R. de R.L.". 2011. Actor codes are indicated in the Materials and Methods section.

The INIFAP and CP are the sectorial institutions participating in the ATT in a signifficant way, both with training programas offered for free. The first one only provides assistance to the group in fertility analysis and vermicompost, meanwhile CP has focused on papaya mite management through participative research, and has proposed a technological transfer model. Ing. E. Sayago emerged as the third central actor in the technology transfer network offering his technical assistance at a monthly cost per grower of MX$500 ha^{-1}, and recommending agrochemicals that he also used to sell. Althoug the general opinion was that his technical assistance enabled yield crop rentability increments, he moved away from the area after not obtainning the expected results with a few associates.

Analysis of network information allowed the detection of central actors indicated in the ITTN function, standing out those that share information with the network and use the best technology, and also two sectorial institutions and an independent technical assistant. The growers G. Basurto R., R. Basurto H., A. Basurto H. and A. Parra V., as well as INIFAP, CP, and their own growers organization might be the catalytic actors in a transdisciplinary and participative research, validation and technology transfer project, improving substantially the quality of papaya production, productivity, profitability and sustainability. This information is of great usefulness to implement a transfer model such as SPGG (Hernández *et al.*, 2002), already proposed to this association. In SPGGs central actors might help the group to grow simultaneously. This model requires working with growers having similar problems, and thus taking advantage of their own experience (tacit knowledge), and their availability to share it; the ascendant and transdisciplinary approach of the model allows an adequate participative research, building their own knowledge (Reta *et al.*, 2011). To be successful in participative papaya projects, substantial modifications are required on the way innovations are diffused. Research and finantial institutions, as well as several professional service providers, must change their descendant and linear paradigm (Muñoz and Santoyo, 2010a) for an ascendant approach (Doorman, 1991), taking central actors as protagonists of their everyday work (Abato-Zarate *et al.*, 2011; Reta *et al.*, 2011).

CONCLUSION

Innovation and technology transfer network of the group "Productora y Comercializadora de Papaya de Cotaxtla S.P.R. de R.L.", in Veracruz, Mexico, is based on 32 participants, where only 27 diffuse technological innovations actively according to their own interest. Agrochemical sellers are an important number of network participants, and this is a market nich disputed among them, mostly because papaya crop demands inputs in large quantities. Eight sectorial institutions converge in this network, all working independently. They could participate in a technology transfer project through "Simultaneous Production Growing Groups", consolidating at the same time this growers organization. Four growers stand out as central actors, along with INIFAP, CP and their own organization; these might be key actors catalyzing innovations diffusion and adoption in papaya. Financiera Rural, wholesellers and middlemen do not participate in the innovation diffusion process, but they have potential as central actors, having the financial capital required to support improvements on rural innovation processes.

ACKNOWLEDGEMENTS

The authors thank "Productora y Comercializadora de Papaya de Cotaxtla S.P.R. de R.L.", for being active participants in this work. Support was provided by CONACYT, Colegio de Postgraduados Campus Veracruz and the "Fideicomiso Revocable de Administración e Inversión No. 167304 del Centro Público Colegio de Postgraduados". This research was conducted as a part of the Priority Research Line (LPI2) - Sustainable Agroecosystems.

REFERENCES

Abato Z., M. 2011. Manejo integrado de la acarofauna del papayo y su transferencia de tecnología. Doctoral Thesis. Colegio de Postgraduados, Campus Veracruz. Tepetates, Manlio F. Altamirano, Veracruz, México. 104 p.

Abato-Zarate, M., Villanueva-Jiménez, J.A., Reta-Mendiola, J.L., Ávila-Reséndiz, C., Otero-Colina, G., Hernández-Castro, E. 2011. Simultaneous productive growth (SPGG) innovation on papaya mite management. Tropical and Subtropical Agroecosystems. 13:397-407.

Aguilar, A.J., Santoyo, C.H., Solleiro, R.J.L., Altamirano, C.J.R., Baca del M., J. 2005. Transferencia e Innovación Tecnológica en la Agricultura: Lecciones y Propuestas. Fundación Produce Michoacán A. C., Universidad Autónoma Chapingo. Michoacán, México.

Aguilar, A.J., Altamirano, C.J.R., Rendón, M.R. 2010. Introducción. In: Aguilar, A.J., Altamirano, C.J.R. Rendón, M.R. (coords.) and Santoyo, C.V.H. (ed.). Del Extensionismo Agrícola a las Redes de Innovación Rural. Universidad Autónoma Chapingo. Chapingo, México. pp. 23-29.

Borgatti, S.P., Everett, M.G., Freeman, L.C. 2002. UCINET 6 for Windows: Software for Social Network Analysis (Version 6.29). Harvard Analytic Technologies. www.mendeley.com. (Consulted: 01/04/2011).

Clark, L. 2006. Manual para el Mapeo de Redes como una Herramienta de Diagnóstico. Centro Internacional de Agricultura Tropical. La Paz, Bolivia. pp. 1-31.

Doorman, F. 1991. La Metodología del Diagnóstico en el Enfoque "Investigación Adaptativa": Guía para la Ejecución de un Diagnóstico con Énfasis en el Análisis de Finca del Pequeño Productor. Universidad Nacional de Heredia. Universidad Estatal San José, Instituto Interamericano de Cooperación para la Agricultura. Costa Rica. pp. 3-4.

Engel, P. 2004. Facilitando el desarrollo sostenible: ¿Hacia una extensión moderna? Boletín InterCambios. 1(10). www.rimisp.org/boletines/bol10. (Consulted: 10/06/2011).

Hernández M., M., Reta M., J., Gallardo L., F., Nava T., M.E. 2002. Tipología de productores de mojarra tilapia (*Oreochromis* spp.): base para la formación de grupos de crecimiento productivo simultáneo (GCPS) en el estado de Veracruz, México. Tropical and Subtropical Agroecosystems. 1:13-19.

Mata G., B. 1997. Avances de una propuesta metodológica para la generación y adopción de tecnología agrícola. In: Mata G., B., Pérez J., G., Sepúlveda G., I. and De León G., F. (coords.). Transferencia de Tecnología Agrícola en México: Crítica y Propuestas. Universidad Autónoma Chapingo. Chapingo, México. pp. 156-171.

Molina, J.L. 2005. Operaciones Básicas con UCINET 6. UAB. Barcelona, España. pp. 1-12.

Morett S., J.C., Cosío R., C. 2006. Impacto de las reformas al Artículo 27 Constitucional en el campo. In: Estudios e Investigaciones. Escenarios y Actores en el Medio Rural. Centro de Estudios para el Desarrollo Rural Sustentable y la Soberanía Alimentaria. México. pp. 153-178.

Muñoz R., M., Rendón M., R., Aguilar A., J., García M., J.G., Altamirano C., J.R. 2004. Redes de Innovación. Un Acercamiento a su Identificación, Análisis y Gestión para el Desarrollo Rural. Fundación Produce Michoacán A. C. Universidad Autónoma Chapingo. Chapingo, México. pp. 11-124.

Muñoz R., M., Santoyo C., V.H. 2010a. Del extensionismo a las redes de innovación. In: Aguilar A., J., Altamirano C., J.R., Rendón M., R. (coords.) and Santoyo C., V.H. (ed.). Del Extensionismo Agrícola a las Redes de Innovación Rural. Universidad Autónoma Chapingo. Chapingo, México. pp. 31-69.

Muñoz R., M., Santoyo C., V.H. 2010b. Pautas para Desarrollar Redes de Innovación Rural. In: Aguilar A., J., Altamirano C., J.R., Rendón M., R. (coords.) and Santoyo C., V.H. (ed.). Del Extensionismo Agrícola a las Redes de Innovación Rural. Universidad Autónoma Chapingo. Chapingo, México. pp. 71-102.

Rascón, F., Hernández, C., Salazar, J. 2006. Tipología de productores. Capítulo IV. In: Escenarios y Actores en el Medio Rural. Estudios e Investigaciones. Centro de Estudios para el Desarrollo Rural Sustentable y la Soberanía Alimentaria. México. pp. 119-152.

Navarro S., L.A., Salazar F., J.P. 2007. Análisis de redes sociales aplicado a redes de investigación en ciencia y tecnología. Síntesis Tecnológica. 3:69-86.

Núñez E., J.F. 2008. Exploración en la modelización de redes sociales de comunicación para el desarrollo rural en zonas marginadas de Latinoamérica. Estudios de caso: Red Nacional de Desarrollo Rural Sustentable (RENDRUS) y Red Iniciativa de Nutrición Humana. Doctoral Thesis. Universitat Politécnica de Catalunya. Barcelona, España. 559 p.

Reta M., J.L., Mena G., J.M., Asiain H., A., Suárez S., C.C.A. 2011. Manual de Procesos de Innovación Rural (PIR) en la Acuacultura. Una Estrategia de Transferencia de Tecnología a través de Grupos de Crecimiento Productivo Simultáneo (GCPS) en el Estado de Morelos. Colegio de Postgraduados-Campus Veracruz. Veracruz, México.

SEMARNAT (Secretaría de Medio Ambiente y Recursos Naturales). 2011. Guía de identificación de actores clave. Serie planeación hidráulica en México. Componente: Planeación Local y Proyectos Emblemáticos. SEMARNAT, México. 25 p.

Velázquez A., O.A., Aguilar G., N. 2005. Manual Introductorio al Análisis de Redes Sociales. Medidas de Centralidad. Ejemplos Prácticos con UCINET 6.85 y Netdraw 1.48. Universidad Autónoma del Estado de México. Toluca, México.

Zarazúa, J.A., Rendón, R., Solleiro, J.L. 2011. Análisis de redes sociales, innovación tecnológica y su transferencia. Estudio de caso en el sistema agroalimentario guayaba del Oriente de Michoacán de Ocampo, México. Editorial Académica Española. Saarbrücken, Alemania. 331 p.

Zarazúa-Escobar, J.A., Almaguer-Vargas, G., Márquez-Berber, S.R. 2011a. Redes de innovación en el sistema productivo fresa en Zamora, Michoacán. Revista Chapingo Serie Horticultura. 17:51-60.

Zarazúa-Escobar, J.A., Almaguer-Vargas, G., Ocampo-Ledesma, J.G. 2011b. El programa de apoyos directos al campo (PROCAMPO) y su impacto sobre la gestión de conocimientos productivo y comercial de la agricultura del estado de México. Revista Agricultura Sociedad y Desarrollo. 8:89-105.

18

Parthenium Hysterophorus L: Distribution, Impact, and Possible Mitigation Measures in Ethiopia

Jemal Abdulkerim-Ute[1]* and Behailu Legesse[2]

[1] *Madda Walabu University, Department of Plant Science, Bale Robe, Ethiopia.*

email: jemiye2003@gmail.com; jemal.abdulkerim@mwu.edu.et

[2]*Madda Walabu University, Department of Rural Development and Agricultural Extension, Bale Robe, Ethiopia*

**Corresponding author*

SUMMARY

The issue of *Parthenium hysterophorus* weed is very important in Ethiopian agriculture. It is spreading rapidly in the highland and lowland areas since its introduction. Currently, most of the regions are invaded by the weed, even the capital city Addis Ababa. There are several factors favoring its rapid distribution in the environment. Its ability to adapt wide range of climatic condition, its allelopathic effect, short life cycle, prolific nature and having large seed bank in the soil environment are among the principal factors. Parthenium has the capacity to decrease the composition and diversity of plant species and it has adverse effects on crop yield, as well as animal and human health. Despite its negative effects, the weed has potential uses such as preparing compost, botanical control of insect pests and bio-herbicide. On the other hand, mechanical control, chemical herbicides, phytochemicals, composting and biological control has been pointed out as an effective management ways and eradication of this weed. There is a need for ecologically balanced management of *Parthenium* weed aiming to enhance the quality of life for farmers and rural communities, especially in pastoral and agro-pastoral areas of Ethiopia. Hence, the objective of this review was to explore the distribution, harmful and beneficial effects as well as the possible controlling practices against *hysterophorus.*

Key words: *Parthenium hysterophorus*; distribution; impact; beneficial effects; allelopathy.

INTRODUCTION

Parthenium weed is an annual herb in the family Asteraceae which is characterized by deep tap root, pale green leaves and an erect stem that becomes woody gradually. At maturity, the plant develops several branches in its top half and may finally reach a height of 1.5-2 meters (EPPO, 2014). It is originated in northern Mexico and southern USA, and spreading in more than 20 countries of Africa, Asia and Oceania (Dhileepan and Strathie, 2009; EPPO, 2014).

Parthenium was probably introduced to Ethiopia through army vehicle during 1976 Ethio-Somalian war or along with contaminated grain in the course of food aid (Tamado and Milberg, 2000; Sushil kumar and Varshney JG, 2010). However, Wise *et al.* (2007) reported Parthenium was first recorded in Ethiopia at the Haramaya University campus in 1968. Since its introduction the weed has rapidly spread throughout agricultural lands, forests, orchards, poorly managed arable crop lands and rangelands in Ethiopia (Tamado and Milberg, 2000; Tefera, 2002).

In the presence of Parthenium the growth and development of crops can be suppressed, and if not controlled on time, it will occupy the land alone. Due to its Aggressive coverage, Bale zone farmers call it 'Anamalee,' in Afaan Oromo- meaning 'Only me' (Personal Communications). Ayele *et al.* (2014) reported species richness and evenness indices of both the above ground vegetation and the soil seed bank significantly decreased at the high level of Parthenium weed infestation and that the decline in species heterogeneity could be due to the strong allelopathic effect of parthenium and/or competition for common resources and thus, suppress the performance of the neighboring plant species. Similarly, Kumari *et al.* (2014) reported that Parthenium weed has the capacity to overwhelm the surrounding weed species and it could absolutely dominate the area inhabits and finally leads to loss of bio diversity. According to O'Donneli and Adkins (2005) parthenium affects agricultural and natural ecosystem production and biodiversity, and on human and animal health. Mirza *et al.* (2013) also said the damage of this weed does not end up with direct competition but also the reduction of the quantity and quality of a crop produced through allelopathic effect (Figures 1 and 2).

The report of case study conducted in South Africa revealed that if parthenium weed left uncontrolled and allowed to spread, returns to small-scale farmers would decline between 26 and 41%, while commercial farmer's yearly total economic returns would reduce between US$38, 818 and US$60, 957 (Wise *et al.*, 2007).

Even though, *P. hysterophurus* is among the top five highly targeted weed in the weed management program of the Research Institutes and Ministry of Rural Development and Agriculture of Ethiopia (G/selase and Getu, 2009) some areas are still left untreated. Ayele *et al.*, (2014) reported that there have been no specific studies on the impact of parthenium weed on the diversity and composition of the standing vegetation and the soil seed bank of rangelands in southeast Ethiopia. Similarly, despite rapid spread and the presence of dense colonies of *Parthenium hysterophorus*, very limited initiative has been taken to investigate the impacts of this dangerous weed and still no visible large scale mapping and quantification of its distribution accomplished in south eastern Ethiopia, especially in Bale Zone. There are also farmers and agricultural experts even who do not know its impact on the eco-system, its way of distribution, and controlling methods. Therefore, the objective of this review was to explore the distribution, harmful and beneficial effects as well as the possible controlling practices against *P. hysterophorus*.

Ecology and causes of rapid distribution of Parthenium weed

Parthenium weed is counted as one of the most serious invasive alien plant on the planet earth, because of its potential for rapid distribution, its impact on ecosystem and socio- economy of the people (Asad and Steve, 2011). So far it was not considered a weed of orchards and forests but now it has spread rapidly into these areas (Kumar, 2012); which propagates itself largely in forest due to the absence of weeding practices in such environments.

The invasion of *parthenium* weed was reported in forest and grazing lands with little or no growth of any other species which results threatening of local biodiversity (Kumar, 2012). Parthenium weed has much more potential to invade bare lands like disturbed road sides and overgrazed pastures than dense pasture (Ayana *et al.*, 2011); it also grows under grass land, and crop field (Netsere and Mendesil, 2011).

It has already become invasive in South Africa and Ethiopia, and appears to be extensively spreading in Uganda and Kenya. Parthenium weed has invaded more than 2 million ha of grazing and crop land, thus becoming a risk for the biodiversity, agriculture and human health in Ethiopia (Asad and Steve, 2011). Based on the estimation of CLIMEX model, south-east and south-west part of Dire Dawa (where *P. hysterophorus* was first recorded in Ethiopia) is to be most Eco climatically suitable area for *Parthenium hysterophorus*. This probably enhanced the distribution of *Parthenium hysterophorus* widely after its introduction in Ethiopia. Rate and extent of spread of this weed since its introduction has been more noticeable in Ethiopia and Swaziland than Kenya, South Africa and Zimbabwe although environmental conditions are also suitable in the latter countries. Higher distribution in Ethiopia and Swaziland could be due to higher levels of disturbance (e.g. overgrazing) and particular land use practices (Mc Connachie *et al.*, 2011).

Figure 1. Range land infestation by Parthenium weed at Ginnir district, Bale Zone (Photo by Jemal A.)

Figure 2: Parthenium infestation near forest land at Madda walabu district, Bale Zone. (Photo by Jemal A.)

According to Rezene *et al.* (unpublished) *Parthenium hysterophorus* is spreading rapidly in various rangeland areas and farm lands of Gambella, Oromia, Afar, Amhara and Somali, national regional states which affecting crop production severely. Hadas and Taye (2015) reported its distribution in to Tigray region particularly,Waja, Alamta town, Bala, kukufto, Zata, Weyrawiha, Bedenoleka, Mohoni town, Maichew town, Kisad Gudo, Adishu, Adigura and Adigudom). It is found in all the Districts but more prominent in Alamata and Raya Azebo. According to the study conducted by Taye (2007) extensive infestation in the central farmlands of east Shewa, Dukem, Bishoftu, Modjo and Koka areas has been prevailed. Gebrehiwot and Berhanu (2015) reported that there has been an urgent need towards the management of parthenium weed in Arba Minch, before it further spread to Nech Sar National Park, which is a home of plants' diversity. Zuberi *et al.* (2014) informed that P. *hysterophorus* is spreading rapidly in the highlands of Ethiopia.

The distribution and spread of parthenium showed that it was not only restricted to the infested Districts but also spread to non-infested Districts like Arero, Bore, Dama and Uraga Districts of Borana and Guji Zones. It is found in Abaya occasionally, present in Bule Hora, abundant in Dugda Dawa, very abundant in Yabello, present in Teltele, frequent in Dire on roadsides, present in Miyo and Moyale, very abundant in Liben, present in Wadera, Adola and Shakiso Districts (Berhanu *et al.*, 2015). Parthenium is widely spread in the range lands and in the cultivable fields of East Showa Zone of Boset district (Belachew and Tessema, 2015).

Causes of rapid distribution

The domination and rapid spread of *P. hysterophorus* in grazing lands with gradual reduction of native plant species could be due to its high invasive capacity, allelopathic properties, short life cycle and prolific character (Dalip *et al.*, 2013**).** The species prefers neutral to alkaline pH soils, but tolerates a wide variety of soil types. *P. hysterophorus* is best suited to areas with an annual summer rainfall greater than 500 mm (Chamberlain and Gittens, 2004). Seeds of parthenium can germinate during any season of the year if the moisture is available. It can keep its viability for a long period of time and can grow under very harsh environmental condition (Williams and Groves, 1980) and at any climatic condition and environment (Netsere and Mendesil, 2011). Tamado *et al.*, (2002) has pointed out that there are no observable climatic conditions that may limit the germination of *Parthenium hysterophorus* in Ethiopia, with exception of high moisture requirement during germination. Such that the only major factor that can limit its germination could be moisture stress during the dry season. However, Ayana *et al.* (2011) concluded that Parthenium weed has the ability to utilize the opportunity of drought prone period in the area to use the chance where the indigenous plants deteriorate and leave much bare ground cover.

Parthenium produced numerous seeds, with reproductive capacity (Lorraine and Lin, 2015), rapid maturity, large quantities of seed production (up to 25,000), easily transported by vehicles, machinery, animals, fodder, pasture seed, stock feed and water. As parthenium weed does not reproduce vegetatively from plant parts, the only method of reproduction and spread is by seed. Large colonies along waterways and drainage floodplains indicate its movement in sheet water. In addition, most long-distance dispersal of seed is by vehicles and farm machinery, as evidenced by the major spread of parthenium along roads. A period of drought followed by rain provides suitable environmental conditions for spread. Drought reduces pasture cover (competition) and increased movement of stock and stock fodder also aids the spread of seed. In particular, flooding after drought is advantageous to the weed, as flood is a dispersal mechanism for parthenium seed. However, seed spread by wind is limited (Agriculture and Resource Management Council of Australia and New Zealand, Australian and New Zealand Environment and Conservation Council and Forestry Ministers, 2001). *Parthenium hysterophorus* is not edible for animals (Hailu, 2010) and this could also be another opportunity for its distribution. Hina and Tahira (2009) suggested that in addition to its capability to withstand wide climatic ranges the weed is germinates and grow in all seasons. The seeds of *Parthenium* germinate in the hottest month of June (mean maximum 38.7ºC) and flourish by producing flowers, fruits and shedding seeds in the coldest months of December and January, when the mean minimum temperature is 3.4 and 2.6 °C respectively.

Impact of Parthenium on biodiversity

Globally, several studies revealed the aggressiveness of parthenium in the ecosystem. The report of Dalip *et al.* (2013) from Mehari Sub-Watershed of Rajouri Forest Range, India, indicated that Parthenium weed occupy new surroundings and often substitute the native plant species, resulting in a serious damage to biodiversity. A study by Ayana *et al.* (2011) in Awash National Park (Ethiopia) showed that Parthenium weed, within a few years from its introduction into Awash National Park, caused a decline (average 69%) in stand density of herbaceous species. Similarly, Asresie (2008) pointed out that an increase in the level of parthenium infestation causing rapid decline in the population and diversity of species in the ecosystem.

Impact of Parthenium on crop production

The impact of parthenium on the yield losses of various crops and orchards has been addressed in the report of earlier works. Crop losses are caused mainly due to allelopathic effects and its ability to compete for common resources like nutrients and moisture and its competitive nature is relatively very much higher than expected from a similar crop weed. Another mechanism by which parthenium affect crop productivity is through its ability to cover crops in pollen, which prevents seed set with resulting losses in yields of up to 40% (Wise *et al.*, 2007). *Parthenium* weed can infest the land where cereals, vegetables and horticultural crops found and reduce agricultural productivity due to its allelopathic effect (Mulatu *et al.*, 2009). The decline in yield due to its highly competitive ability was also reported by Netsere and Mendesil (2011). Tamado *et al.* (2002) reported that if *P. hysterophorus* is not weeded throughout the season the yield of sorghum bicolor can be reduced in the range of 40% and 90% in Ethiopia, this percentage was closer to the report of Wise *et al.* (2007), which was range from 45-80%. According to Nganthoi *et al.* (2014) there was a visible impact on the growth parameters, yield and yield components of *Zea mays* by Parthenium. Accordingly, at high ratio (20:1) population of parthenium the plant height, dry biomass, corn weight, corn length and grain weight per corn were reduced to 21.1%, 42.3%, 50.9%, 51.2%, and 52.7% respectively as compared to control. Furthermore, the finding indicate Parthenium in the form of extract or residue or growing weed can affect the germination and growth by reducing radicle and plumule length of *Zea mays.* Tefera (2002) also reported that 10% leaf aqueous extract of *Parthenium hysterophorus* resulted in total failure of seed germination in *Eragostis tef.* Similarly, Demissie *et al.* (2013) found the presence of allelopathic effect in *Parthenium* extracts which could affect the seed germination and elongation of Onion and Bean. Dangwal *et al.* (2010) investigated that while primary major essential nutrients (NPK fertilizer) supplied, but in the absence of herbicide application and mechanical weeding, parthenium weed along with other weeds were reduced the yield of wheat by 25.35%. Besides reducing the yield they also reduce the quality of germplasm of wheat crop. Raj and Jha (2016) disclosed that higher concentrations of leaf extract have irregularly affected the growth of *Phaseolus mungo* than lower concentrations.

Impact of Parthenium on animal production and human health

Toxic substances found in parthenium are lethal to human beings and animals (Singh *et al.*, 2002). It is considered to be a cause of allergic respiratory problems, contact dermatitis, mutagenicity in human and livestock. In addition, by reducing the species bio diversity it affect the productivity of grazing land and hence reduce feed supply for animals. It releases chemicals that inhibit the germination and growth of pasture grasses and other plants (Dalip *et al.*, 2013). If Parthenium is eaten by animals, the meat gets polluted due to its toxicity problem and these result in

direct economic losses. Thus, domestic animals should avoid eating it (Wise *et al.*, 2007).
When human beings come in contact with this weed, it may cause allergy, dermatitis, eczema, black spots and blisters around eyes, burning rings and blisters over skin, redness of skin and asthma (Handa *et al.*, 2001). Parthenium is spreading at alarming rate, threatening agricultural ecosystem, biodiversity, human and animal health in Ethiopia. The response of the 64 interviewed farmers in Ethiopia showed that all of them have health problems in different nuances. Most frequently they responded to parthenium contact with light allergic symptoms like hay fever or skin prickle on arms and hands. Some farmers had worse health problems: cracks on hand balms, fever, prickle on the whole body, skin irritations, and asthma. In addition to parthenin high concentrations of phenolic acids which might also contribute to health problems (Ulrichs *et al.*, 2007). Studies in Jijiga (Ethiopia) indicated that *Parthenium* causes asthma, bronchitis, dermatitis, and high fever in human (Shashie, 2007).

Potential beneficial uses of *Parthenium hysterophorus*

Parthenium compost helps in moisture conservation which is utilized for better root penetration and crop growth. This enhancement attributed to the higher water holding capacity of the soil due to the influence of organic waste application. The moisture in soil due to application of *Parthenium* compost can be 14.5 and 16.5% at 0-15 and 15-30 cm. depths as compared to 10.7 and 11.6% at 0-15 and 15-30 cm. depths of soil due to application of NPK alone. It was also found that if *Parthenium* compost was prepared following technical recommendation, then it avoid germination of non-dormant seed. It has higher percentage of micronutrient and macronutrient in respect to the other manure. It minimizes the utilization of chemical fertilizers which has bad impact on soil texture and pollutes our whole biosphere. It also reduces the weed growth up to 40% in the maize due to allelopathic effect which minimizes excessive use of herbicides. Examination of *Parthenium* compost was done and found that it has bacteria PSB, *Azotobactor*, Actinomycetes that also improves soil- texture and fertility without affecting soil health (Ambasta and Kumari, 2013). Hailu (2010) conducted research in Tigrai region (Ethiopia) on composting of different weeds including Parthenium and investigated that *Parthenium hysterophorus* contains 38.5, 2.8 and 51 $g.kg^{-1}$ of NPK, respectively, which is relatively high amount. Also, Rajiv *et al.* (2013) prepared various concentration of cow dung and Parthenium mixed compost as well as parthenium alone compost and determined physical properties and analysis of its allelopathic effect on germination and growth of *Arachis hypogeae* L. They investigated that compost formed from a combination of parthenium and cow dung, and compost had enhanced nutrient value and increased germination percentage of *Arachis hypogeae* as compared to their individual controls because high levels of cow dung, induce growth of microbes and reduce the allelopathic potential of parthenium. It can be used as a wealthy source of nutrients because the toxicity due to allelochemicals could be minimized.

Similarly, composted *Parthenium* and poultry manure can be used for preparing organic manures and be used in successfully increasing crop productivity as an alternative source to inorganic fertilizers. Sesame root length was more in composted poultry droppings and *Parthenium* treatments compared to the vermicomposted and fresh forms and NPK. Application of organic manures might have supplied N, P and K nutrients throughout the crop growth period as slow released nutrients. The length of shoots was increased in composted and vermicomposted poultry droppings and vermicomposted *Parthenium* treatments than in fresh *Parthenium*, poultry droppings and NPK (Vijayakumari and Hiranmai, 2012).Addition of different farm and animal wastes helped to degrade the Parthenium and improve the nutrient value (Yadav, 2015). Jelin and Dhanarajan (2013) reported that the combination of parthenium and sawdust in composting takes relatively short time for degradation and for preparing a complete compost of good quality. Similarly, Anbalagan and Manivannan (2012) concluded that *P. hysterophorus* mixed with other organic supplements (Cow dung and Press mud in equal quantities) provides adequate physico-chemical conditions for maximum worm production and large scale vermicompost production.

Apurva *et al.* (2010) revised the use of Parthenium for compost and concluded that ordinary *P. hysterophorus* compost contains allelopathic effects which can affect early growth, development and dry matter yield of both monocot and dicot plants. However, since it is rich in primary major essential nutrients (NPK) and other macro nutrients like Ca, Mg, if properly managed it could be best opportunity to use it for fertilizer. Kishor *et al.* (2010) tested the compost of P. hysterophorus and revealed that its compost increases the soil moisture level more than NPK alone. According to Javaid and Shah (2010) comparing the effect of *P. hysterophorus* green manure and EM (effective microorganisms), a bio fertilizer, on wheat (*Triticum aestivum* L.) cultivation, the highest root biomass was recorded in 3% green manure-amended treatment. Spike length, number of grains per spike and grain yield gradually increased by increasing the quantity of green manure. There was 43–253% increase in grain yield over control due to various green manure treatments as compared with

96% increase due to NPK fertilizers over control. Yadav (2015) concluded that the adoption of vermicomposting to eradicate the weeds and reduce the problems arising due to the weeds in agricultural fields is gaining momentum among farmers. However, Ajayan and Babu (2014) pointed out that traditional use of parthenium as vermicomposting ingredient and fencing material in agricultural field unknowingly affect soil fertility and have potential anthelmintic activity.

Lalitha *et al.* (2012) have utilized the leachate of different plant parts of *Parthenium* plant and observed the effect on several aspects such as seed germination of *Phaseolus mungo*, metal tolerating capacity of the seeds against iron, lead, mercury and nickel of seeds during germination, antimicrobial activity against fungal pathogens and effect on brinjal fruit borer. They investigated that there was a 48.8 % increase in the seeds germination of the seeds treated with flower leachate. The metal tolerating capacity was highest in leaf extract against iron, in stem extract against lead, in flower extract against mercury and leaf extract against nickel. The flower extract exhibited 100% mortality on brinjal fruit borer. Jiten *et al.* (2013) investigated that methanolic extract of *P. hysterophorus* has good acaricidal activity and may find scope in integrated pest management system of *Oligonychus coffeae*. Thus, methanol extract of *P. hysterophorus* is known to be a promising source for the controlling of *Oligonychus coffeae*. Nadir Baloach *et al.* (2014) tested parthenium water extract for its herbicidal effect (for weed control) on wheat crop. The parthenium water extract @ 24 L ha^{-1} combined with reduced doses of Buctril Super 60 EC @ 300, 225, 150 and 75 ml ha^{-1}, respectively, were tested. Individual herbicide (Buctril Super 60 EC) treatments @ 750 ml ha^{-1} full dose and a weedy check were included for comparison. Finally, concluded that water extract @ 24 L ha^{-1} combined with Buctril Super 60 EC@ 150 ml ha^{-1} inhibited total weed density by 38 and 84%, total weeds, fresh weight by 67 and 87% and total weeds dry biomass by 69 and 86% respectively and it increased grain yield by 91% as compared to control. The findings revealed that, allelopathy can be the best option in order to reduce herbicide dose and enhance the wheat yield. Similarly, Khan *et al.* (2008) Comparing the values in control, with the treated plots, there are convincing evidences of the presence of allelochemicals in *Parthenium* extracts that can selectively affect weeds and suggested that *P. hysterophorus* can be used as a bio herbicide.

Control measures of *Parthenium*

The development of effective prevention and control programs are very essential to overcome the problems of parthenium weed. However, no single method of control has been proved satisfactory (Mechanical, chemical and biological control strategies) to tackle the effect of *Parthenium hysterophorus*. Thus, integrated approaches are warranted to restrict the invasion of this weed by combining more than one option (Belachew and Tesema, 2015). To address this problem, public awareness has to be developed and participatory approach to control the invasive weeds should be adopted (Patel, 2011).

Chemical control methods

Effective herbicides are registered for use against parthenium but chemical control requires repeated, regular follow-up treatments (Lorraine and Lin, 2015). Generally, parthenium plants can effectively be controlled with glyphosate tank mixed with low concentrations of urea and common salt (Zelalem, 2013). Thus, treating 3000 ml of glyphosate with 150 ml of urea and 150 ml of common salt solutions and spraying at 6 to 8 leave stage resulted in complete mortality of parthenium weed in short period of time by increasing the phytotoxicity of this herbicide. While spraying this solution at 50 and 75% flowering stages showed poor mortality rates on this weed. The growth of Parthenium can be suppressed using amino acid synthesis and photosynthesis inhibitors as compared to herbicides with other modes of action. In wasteland, non-cropped areas, along railway tracks, water channels and roadsides, the use of glyphosate and metribuzin has been shown promising results. On the other hand, the treatment should be accomplished at rosette stage to be effective. Parthenium weed control at rosette stage is highest with glyphosate (96%) followed by metribuzin 87% at 4 weeks after treatment (WAT) and control is lowest with pendimethalin (42.5%) at 4 WAT. Thus, glyphosate and metribuzin are recommended for the control of parthenium weed in non-cropped areas (Haroon *et al.*, 2012). Bactril Super @ 0.67 ml/L water can control parthenium. However, huge amount of parthenium plants cannot be controlled economically by chemicals (Rezaul Karim, 2012). Reddy *et al.* (2007) suggested that Norflurazon, Clomazone, Fluometuron, Flumioxazin, Halosulfuron, Chlorimuron, and Trifloxysulfuron could provide effective control of parthenium. Fernandez (2013) has shown that under field condition, Saflufenacil + dimethenamid-P and hexazinone were highly effective and rapid in controlling flowering parthenium, providing 100% control at all evaluation timings (3, 6, 9 weeks after treatment). While in greenhouse, aminocyclopyrachlor + chlorsulfuron, aminopyralid, hexazinone, saflufenacil + dimethenamid-P, 2,4 D provided 100% aboveground dry weight reduction of rosette parthenium at 21 DAT.

Biological control

Biological control can play a significant role to check the growth of Parthenium weed because it would be more sustainable and possibly cost effective (Evans, 1997; Lorraine and Lin, 2015). Several effective biological control agents are already available and can be introduced and released with a minimum of additional research required. Six potential agents have been released for biological control, mainly in Australia. The two most important species that have been established are *Zygogramma bicolorata* (leaf feeding beetle) and *Epiblemma strenuana* (stem galling moth) and they have a significant impact on parthenium in Australia. Only *Z. bicolorata* was released in India where it caused widespread defoliation, permitting local vegetation to grow again. In Australia, *E. strenuana* can exert significant control but erratic rainfall has disrupted the moth populations, reducing them to very low levels. Populations take a long time to build-up again, usually too late to have a significant impact on the weed (Wise *et al.*, 2007). Also; according to Kumar *et al.* (2009) Cladosporium sp. (MCPL-461) affects the embryo development which enhances the sterile seeds formation. Spraying of this sp. has deleterious effects on Parthenium weed only, not to other plant species, which living together under same niche. Due to invasive nature and luxuriant growth capacity of Parthenium in different habitat, Cladosporium sp. as a floral and leaf pathogen may be used as a potential myco-herbicide against this weed. Rezaul Karim (2012) reported that *Puccinia abrupta* is appropriate disease that used to control parthenium.Finally, biological control, using natural enemies from the plant in its native range, would be the most sustainable management intervention and requires further research.

Mechanical control

Control operations should focus on preventing spread, the eradication of small and isolated populations (Wise *et al.*, 2007). Khan *et al.* (2013) reported that manual weeding and tillage are the most common control practices used to control Parthenium weed. According to van der Laan (2006) manual removal of *P. hysterophorus* is often not cost-effective and therefore used on a limited basis. Hand-pulling should ensure the removal of the entire crown to prevent regeneration from remaining lateral shoots. Correspondingly, Patel (2011) stated that manual uprooting of Parthenium before flowering and seed setting is the most effective method. This is easily done when the soil is wet. Uprooting the weed after seed setting will increase the area of infestation. Pulling a plant in flower will aid in the dispersal of pollen grains, resulting in allergic reactions. Mulching with *Gliricidia sepium*, or any other suitable and a cost-effective mulch, coupled with manual weeding before land preparation or ploughing would help suppressing the growth and development of the Parthenium weed and enhance yield of tomato (Nishanthan *et al*, 2013). However, burning of *P. hysterophorus* should be avoided in the agricultural field to enhance overall productivity (Kumar and Kumar, 2010). This could be due to the fact that burning requires large quantity of fuel. Moreover, burning destroys all plants and predators in vicinity.

Botanical control methods

Aqueous and methanol extracts of *Nerium oleander* L. leaves inhibited seed germination (Germination percentage, germination speed, germination value and peak value) and early growth (root hair formation, root and shoot lengths of seedlings) of *Parthenium hysterophorus* L. in a concentration dependent manner. Methanol extracts of white flowered variety showed higher inhibition compared to that of pink flowered variety (at 1:40 dilution of the stock, inhibition was 60% and 23.4% with white and pink, respectively) (Rajyalakshmi *et al.*, 2011). Belachew and Tesema (2015) assessed the weed flora composition in Parthenium (*P.hysterophorus L.*) infested area of East Shewa Zone, Ethiopia, and finally investigated three herbaceous species, namely *Cassia tora, Xanthium strumarium* and *Argemone mexicana* which had good association and grow with *P. hysterophrous* in competition and thus, these species are recommended for *Parthenium* eradication. According to Lorraine and Lin (2015) reducing livestock densities to increase grass cover assists to alleviate parthenium infestations.

Root and shoot extracts of the three allelopathic grasses viz., *Dicanthium annulatum* Stapf., *Cenchrus pennisetiformis* Hochest and Sorghum halepense Pers., reduced germination and suppressed early seedling growth of exotic weed Parthenium hysterophorus L. Aqueous extracts of D. annulatum and C. pennisetiformis were more inhibitory than extracts of S. halepense. The highest suppressive ability was exhibited by extracts of C. pennisetiformis where 20% shoot and 25% root extract completely inhibited the germination of *P. hysterophorus.* In general shoot extracts were more inhibitory than the root extracts (Arshad and Tehmina, 2006). In a phytochemical control study, *Cassia occidentalis, Rumex dentatus, Calotropis procera and Withania somnifera* had been evaluated for their herbicidal potential against biochemical activities and mortality percentage of parthenium. The study concluded that the 100%, 9th day aqueous shoot leachates of *Cassia occidentalis* found effective in arresting germination and suppressing seedling growth of parthenium (Jai *et al.*, 2010).Sing *et al.* (2013) studied the effect of leachates of different phenological stages of *Cassia occidentalis* L. on *P.hysterophorus* L. Accordingly, 100% concentration of leaf leachate from vegetative stage completely inhibited the germination while highest concentration of leaf leachate from vegetative, flowering, fruiting and fruit ripening stages caused the death of the plants.

CONCLUSIONS

The distribution and existence of *Parthenium hysterophorus* L. weed in various environments including crop lands, range lands, road sides, forests, watersheds and other economically important ecosystems has been identified. Its dissemination is progressing at an alarming rate which gives the impression of difficulty to control its distribution in the future.

Some general recommendations arise from the present review.

- A well-planned programme that encourage the participation of all stake holders including universities, research centers, individual researchers, government and non-governmental organizations is needed to develop strategies for controlling further distribution or achieve eradication of this weed.
- A single technology in isolation will not give desired results. Thus, adoption of related controlling methods as an integrated package would provide a synergistic effect.
- Creating public awareness especially in farmers' area about the effect of *Parthenium hysterophorus* on agricultural productivity, ecosystem as well as on strategies control methods.
- By using its potential beneficial effects as opportunity there is a possibility to eradicate this weed through time.

REFERENCE

Agriculture & Resource Management Council of Australia & New Zealand, Australian & New Zealand Environment & Conservation Council and Forestry Ministers. 2001. Weeds of National Significance Parthenium weed (Parthenium hysterophorus) Strategic Plan. National Weeds Strategy Executive Committee, Launceston.

Ajayan, K.V and Babu, R.L. 2014. Foliotoxicity of *Parthenium Hysterophorus* L. and *Lantana Camara* L. on earth worm. International Journal of Scientific Research, 3(10):52-55.

Ambasta S.K. and Kumari, S. 2013. A scientific approach of conversion of eco-hazardous parthenium weed into eco-friendly by compost making. International Journal of Geology, Earth and Environmental Sciences, 3 (1): 90-94.

Anbalagan M., Manivannan, S. 2012 Assessment of impact of invasive weed *Parthenium hysterophorus* L. mixed with organic supplements on growth and reproduction preference of *Lampito mauritii* (Kinberg) through vermitechnology. International Journal of Environmental Biology, 2(2): 88-91

Apurva, P., Sinha, S.K., Thakur, P.C. 2010. Composting an obnoxious weed, *Parthenium hysterophorus* L., with the help of a millipede, *Harpaphe haydeniana*. Asian Journal Experimental Biological Sciences 1:337–343

Arshad Javaid and Tehmina Anjum. 2006. Control of *Parthenium Hysterophorus* L., by aqueous extracts of allelopathic grasses. Pakistan Journal Botany, 38(1): 139-145.

Asad Shabbir and Adkins, S.W. 2011. International Parthenium News. Tropical & Sub-Tropical Weed Research Unit, The University of Queensland, Australia.

Asresie. 2008. Impact of Parthenium (*Parthenium Hysterophorus* L.) on herbaceous vegetation and soil seed bank flora in grazing lands and sorghum fields in Eastern Amhara, Ethiopia. M.Sc. Thesis, Haramaya University, Ethiopia.

Ayana Etana, Ensermu Kelbessa and Teshome Soromessa. 2011. Impact of *Parthenium hysterophorus* L. (Asteraceae) on herbaceous plant biodiversity of Awash National Park (Anp), Ethiopia. V.2: 69-80.

Ayele Shashie, Nigatu, Lisanework., Tana, Tamado., and Adkins, S.W. 2014. Impact of parthenium weed (*Parthenium hysterophorus* L.) on the above-ground and soil seed bank communities of rangelands in Southeast Ethiopia. Global Science Research Journals. 2(1): 66-78.

Belachew, K. and Tessema, T. 2015. Assessment of weed flora composition in Parthenium (*Parthenium hysterophorus* L.) infested area of East Shewa zone, Ethiopia. Malaysian Journal of Medical and Biological Research, 2: 63-70.

Berhanu Lemma, Taye Tessema and Rezene Fessehaie. 2015. Distribution, abundance and socio-economic impacts of invasive plant species (IPS) in Borana and Guji zones of Oromia National Regional State, Ethiopia. Basic Research Journal of Agricultural Science and Review, 4(9): 271-279.

Chamberlain, J. and Gittens, A. 2004. Parthenium weed management: challenges, opportunities and strategies. Parthenium Action Group. The State of Queensland (Department of Natural Resources, Mines and Energy), Brisbane (AU), 82.

Ulrichs, C., Wiesner, M., Tessema, T., Hoffmann, A., Wilfried, P., Buettner, C., Mewis, I. 2007. Impact of the pan-tropical weed Parthenium hysterophorus L. on human health in Ethiopia. Humboldt-Universit¨atzu Berlin, Institute of Horticultural Science, Urban Horticulture, Lentzeallee 55, 14129 Berlin, Germany.

Dalip Kumar, Junaid Ahmed, Surat Singh. 2013. Distribution and effect of *Parthenium Hysterophorus L.* In Mehari sub-watershed of Rajouri Forest Range, J&K. International Journal of Scientific Research. 2(6):304-306.

Demissie Abebe G., Ashenafi A., Arega A., Etenash U., Kebede A., Tigist A.2013. Effect of *Parthenium hysterophorus* L. on germination and elongation of onion (*Allium cepa*) and bean (*Phaseolus vulgaris*). Research Journal of Chemical and Environmental Sciences, 1(2): 17-21

Dangwal L.R., Amandeep Singh, Tajinder Singh, Chanchal Sharma. 2010. Effect of weeds on the yield of wheat crop in Tehsil Nowshera. Journal of American Science, 6(10): 405-407.

Dhileepan, K. and L. Strathie.2009. *Parthenium hysterophorus* L. (Asteraceae). In: Biological control of tropical weeds using arthropods (Muniappan, R., Gvp Reddy and Raman, A. eds), pp. 274–318. Cambridge University Press, UK.

EPPO (European and Mediterranean Plant Protection Organization). 2014. Parthenium hysterophorus L. Asteraceae – Parthenium weed. Data sheets on invasive alien plants, OEPP Bulletin. 44(3):474–478.

Fernandez, J.V. 2013. Characterization of glyphosate resistance and management of ragweed Parthenium (*Parthenium hysterophorus* L.) in the everglades agricultural Area of Florida. A Thesis Presented to the Graduate School of the University of Florida in Partial Fulfillment of the Requirements for the Degree of Master of Science. p 32.

Gebrehiwot, N. and Berhanu, L. 2015. Impact of Parthenium on species diversity in GamoGofa, Ethiopia. Scholarly Journal of Agricultural Science, 5(7): 226-231.

G/selase, A and Getu, E. 2009. Evaluation of botanical plants powders against *Zabrotes subfasciatus*(Boheman) (Coleoptera: Bruchidae) in stored haricot beans under laboratory condition. African Journal of Agricultural Research, 4 (10): 1073-1079.

Hadas Beyene and Taye Tessema. 2015. Distribution, abundance and socio-economic impacts of Parthenium (*Parthenium hysterophorus*) in Southern zone of Tigray, Ethiopia, Journal of Poverty, Investment and Development, 19: 22-29.

Handa, S., Sahoo, B. and Sharma, V.K. 2001. Oral hyposensitization in patients with contact dermatitis from *Parthenium hysterophorus* L. Contact Dermatitis, 44(5):279-282.

Hailu Araya Tedla. 2010. The effect of compost on soil fertility enhancement and yield increment under smallholder farming - a case of Tahtai Maichew district - Tigray Region, Ethiopia. PHD dissertation, University of Hohenheim, Germany.

Hina Fatimah and Tahira Ahmad. 2009. Phenology of *Parthenium hysterophorus* – a key factor for the success of its invasion. Advances Environmental Biology, 3(2): 150-156.

Jai Knox, DishaJaggi and Paul, M.S. 2010. allelopathic effect of selected weeds on biochemical activity of *Parthenium hysterophorus*. Current Research Journal of Biological Sciences 2(4): 238-240.

Javaid, A., Shah, M.B.M. 2010. Growth and yield response of wheat to EM (effective microorganisms) and parthenium green manure. African Journal Biotechnology, 9:3373–3381

Jelin, J, Dhanarajan, M.S. 2013. Comparative physicochemical analysis of degrading Parthenium (*Parthenium Hysterophorus*) and saw dust by a new approach to accelerate the composting rate. International Journal of Chemical, Environmental and Biological Sciences, 1(3): 535-537.

Jiten, M., Deka Bhuyan, P., and Bhattacharyya, P.R. 2013. Acaricidal Activities of *Parthenium hysterophorus* L. against Red Spider Mite, *Oligonychus Coffeae Nietner* (Acarina: Tetranychidae) of tea. International Journal of Science and Research, 4(5): 901-904.

Khan Bahadar Marwat, Muhammad Azim Khan, Anwar Nawaz and Anees Amin. 2008. *Parthenium Hysterophorus* L. a potential source of bioherbicide. Pakistan Journal of Botany 40(5): 1933-1942.

Kishor, P., Ghosh, A.K., Singh, S., Maury, B.R. 2010. Potential use of parthenium (*Parthenium hysterophorus* L.) in agriculture. Asian Journal of Agricultural Research, 4:220–225.

Khan, H., Khan, B. M., Hassan, G. and Muhammad, A. K. 2013. Socio-economic impacts of parthenium (*Parthenium hysterophorus* L.) in peshawar valley, Pakistan. Pakistan Journal Weed Science Research. 19(3): 275-293.

Kumar, D. 2012. Current spread, impact and management of Parthenium weed in India. International Parthenium News. Tropical and Sub-Tropical Weed Research Unit, The University of Queensland, Australia.

Kumar and Kumar. 2010. Effect of *Parthenium hysterophorus* ash on growth and biomass of *Phaseolus mungo*. Academia Arena, 2(1):98-102.

Kumar, A., Verma, V.C., Gond, S.K., Kumar, V. and Kharwar, R.N. 2009. Bio-control potential of Cladosporium sp. (MCPL - 461), against a noxious weed *Parthenium hysterophorus* L. Journal of Environmental Biology, 30(2): 307-312.

Kumari, P., Sahu, P.K., Soni, M.Y. and Awasthi, P. 2014. Impact of *Parthenium hysterophorus* L. invasion on species diversity of cultivated fields of Bilaspur (C.G.) India. Agricultural Sciences, 5: 754-764.

Lalitha P., Shivani K., Rama Rao R. 2012. *Parthenium hysterophorus*- An economical tool to increase the agricultural productivity. International Journal Lifesciences Biotechnology Pharma Research. 1 (1): 113-127.

Lorraine Strathie and Lin Besaans. 2015. ARC-PPRI fact sheets on invasive alien plants and their control in South Africa. www.arc.agric.za

Mawal Swapnil S., Shahnawaz, M., Sangale, M.K. and Ade, A.B. 2015. Assessment of allelopathic potential of the roots of *Parthenium hysterophorus* L. on some selected crops. International Journal of Scientific Research in Knowledge, 3(6): 145-152.

Mcconnachie, A.J., Strathie, L.W., Mersie, W., Gebrehiwot, L., Zewdie, K., Abdurehim, A., Abrha, B., Araya, T., Asaregew, F., Assefa, F., Gebre-Tsadik, R., Nigatu, L., Tadesse, B., Tana, T. 2011. Current and potential geographical distribution of the invasive plant *Parthenium Hysterophorus* (Asteraceae) in Eastern and Southern Africa. Weed Research, 51: 71–84.

Mirza Hasanuzzaman, Masum, S.M., Ali, M.H. 2013. Threats of *Parthenium hysterophorus* on agro-ecosystems and its management: a review. International Journal of Agriculture and Crop Sciences. 6 (11):684-697.

Mulatu, Wakjira, Gezahegn, Berecha, Solomon, Tulu. 2009. Allelopathic effects of an invasive alien weed *Parthenium hysterophorus* L. compost on lettuce germination and growth. African Journal of Agricultural Research. 4(11): 1325 - 1330.

Nadir Baloach, Muhammad Yousaf, Shah Fahad, Muhammad Ansar, Bashir Ullah, Wajid Parvez Akhter and Sayed Hussain. 2014. Evaluation of Parthenium (*Parthenium hysterophorus L.*) water extract alone and with low doses of herbicide for weed control in wheat. International Journal of Current Micro Biology and Applied Sciences, 3(6): 354-363.

Nganthoi Devi, Y., Dutta, B.K., Sagolshemcha, R., Singh, N.I. 2014. Allelopathic effect of *Parthenium hysterophorus* L. on growth and productivity of *Zea mays* L. and its phytochemical screening. International Journal of Current Microbiology and Applied Sciences, 3 (7): 837-846.

Netsere A, Mendesil E. 2011. Allelopathic effects of *Parthenium hysterophorus* L. aqueous extracts on soybean (*Glycine max* L.) and haricot bean (*Phaseolus vulgaris* L.) seed germination shoot and root growth and dry matter production. Journal Applied Botany Food Quality, 84: 219-222.

Nishanthan K., Sivachandiran S. and Marambe B. 2013. Control of *Parthenium hysterophorus* L. and its impact on yield performance of tomato (*Solanum lycopersicum* L.) in the northern province of Sri Lanka. Tropical Agricultural Research, 25 (1): 56 – 68.

O'Donnell, C., Adkins, S.W. 2005. Management of parthenium weed through competitive displacement with beneficial plants. Weed Biol. Manage. 5 (2): 77-79.

Patel, S. 2011. Harmful and beneficial aspects of *Parthenium hysterophorus*: an update. 3Biotech, 1:1-9.

Raj Shikha, Jha, A.K. 2016. Evaluation of effect of leaf extract of *Parthenium hysterophorus* L. on seed germination, seedling growth and fresh weight of *Phaseolus mungo*. American Journal of Research Communication, 4(1): 2325-4076.

Rajiv P., Narendhran, S., Subhashkumar, M., Sankar, A., Rajeshwari, S., Rajendran, V. 2013. *Parthenium hysterophorus* L. compost: assessment of its physical properties and allelopathic effect on germination and growth of *Arachis hypogeae* L. International Research Journal of Environment Sciences, 2(2): 1-5.

Rajyalakshmi, M., Amruth Kumar, N., Divyashree, N.R, Kiran, K., Pavithra, G.S., Rohini, B., Sangeeta, A., Sindhu, S. 2011. Inhibitory effects of *Nerium oleander* L. and its compounds, rutin and quercetin, on *Parthenium hysterophorus* L. Journal of Agricultural Science 3(2): 123-137.

Reddy Krishna, N., Bryson, C.T., and Burke, I.C. 2007. Ragweed parthenium (*Parthenium hysterophorus*) control with preemergence and postemergence herbicides. Weed Technology, 21:982–986.

Rezaul Karim S. M. 2012. Ill impacts of parthenium weed on human health, livestock production and environment. The paper has been presented in a seminar at West Virginia University, USA on 04 October 2012. rezaul@umk.edu.my.

Shashie Ayele. 2007. The impact of Parthenium (*Parthenium hysterophorus* L.) on the range ecosystem dynamics of the Jijiga rangeland, Ethiopia. M.Sc. Thesis, Haramaya University, Ethiopia. 50p.

Singh, H.P., Batish, D.R., Kohli, R.K., Saxena, D.B., and Arora, V. 2002. Effect of pathenin-A sesquiterpene lactone from *Parthenium hysterophorus* L. on early growth and physiology of *Ageratum conyzoides*. Journal Chemical Ecology, 28: 2169-2179.

Sing, N.B., Kumar, S., Singh, D., Yadav, K. 2013. Allelopathic effects of different phenological stages of *Cassia occidentalis* L. on *Parthenium hysterophorus* L. Iranian Journal of Plant Physiology 3(4): 817-828.

Subhendu, Datta and Saxena, D.B. 2001. Pesticidal properties of parthenin (from *Parthenium hysterophorus*) and related compounds. Pest Management Science, 57:95-101.

Sushil, Kumar, Varshney, J.G. 2010. Parthenium infestation and its estimated cost management in India. Indian Journal of Weed Science 42: 73–77.

Tamado, Tana, Schütz, W. and Milberg, P. 2002. Germination ecology of the weed *Parthenium hysterophorus* in eastern Ethiopia. Annals Applied Biology, 140:263-270.

Taye, T. 2007. The prospects of biological control of weeds in Ethiopia. Ethiopian Journal of Weed Management, 1(1):63-78.

Tefera, T. 2002. Allelopathic effects of *Parthenium hysterophorus* extracts on seed germination and seedling growth of *Eragrostis tef.* Journal Agronomy Crop Science, 188: 306—310.

Van Der Laan, M. 2006. Allelopathic interference potential of the alien invader plant *Parthenium hysterophorus.* M.Sc. Thesis, University of Pretoria Etd. Page 14.

Vijayakumari, B. Yadav, R Hiranmai, Raja, A. Xavier Vergeese. 2009. Influence of fresh, composted and vermi-composted *Parthenium hysterophorus* and poultry droppings on quality parameters of radish. Journal Applied Science Environmental Management, 13(2): 79 – 82.

Vijayakumari, B. and Hiranmai, Y.R. 2012. Influence of fresh, composted and vermicomposted *Parthenium* and poultry manure on the growth characters of sesame (*Sesamum indicum*). Journal of Organic Systems, 7(1): 14-19.

Williams, J.D. and Groves, R.H. 1980. The influence of temperature and photoperiod on growth and development of *Parthenium hysterophorus*. Weed Research. 20: 47-52.

Wise R.M., B.W. van Wilgen, M.P. Hill, F. Schulthess, D. Tweddle, A. Chabi-Olay, and H.G Zimmermann. 2007. The Economic Impact and Appropriate Management of Selected Invasive Alien Species on the African Continent. CSIR report number: CSIR/NRE/RBSD/ER/2007/0044/C.

Yadav Hiranmai, R. 2015. Assessment of different organic supplements for degradation of *Parthenium hysterophorus* by vermitechnology. Journal of Environmental Health Science Engineering, 13(44): 1-7.

Yibekal Abebe Tessema.2012. Ecological and economic dimensions of the paradoxical invasive species- *Prosopis juliflora* and policy challenges in Ethiopia. Journal of Economics and Sustainable Development, 3 (8): 62-70.

Zelalem, B. 2013. Effect of urea and common salt (NaCl) treated glyphosate on parthenium weed (*Parthenium hysterophorus* L.) at western Hararghe zone, Ethiopia. African Journal of Agricultural Research, 8(23):3036-3041.

Zuberi, M.I., Gosaye, T. and Hossain, S. 2014. Potential threat of alien invasive species: *Parthenium hysterophorus L.* to subsistence agriculture in Ethiopia. Sarhad Journal Agriculture. 30(1): 117-125.

19

Potential of Carbon Storage of Rubber (*Hevea Brasiliensis* Müll. Arg.) Plantations in Monoculture and Agroforestry Systems in the Colombian Amazon

José Alfredo Orjuela-Chaves[1], Hernán J. Andrade C.[2]*, Yeraldine Vargas-Valenzuela[3]

[1]*Universidad de la Amazonia, Colombia.*

[2]*Universidad del Tolima, Colombia. Email: hjandrade@ut.edu.co*

[3]*Universidad de la Amazonia, Colombia.*

**Corresponding author*

SUMMARY

Carbon sequestration potential of rubber (*Hevea brasiliensis*) plantations was estimated in two production systems: monoculture and agroforestry system with copoazú (*Theobroma grandiflorum*), on farms of Florencia, El Doncello and Belén de los Andaquíes, in northeastern Colombian Amazon, department of Caquetá. The plantations were classified into three age classes, according to their productive stage: 1-7, 8-20 and > 20 years. The carbon storage was estimated using the methodology proposed by Andrade and Ibrahim (2003) and recommended by IPCC (2003). Tree carbon sinks were evaluated: above and below ground biomass, and necromass. The highest proportion of carbon storage was found in biomass, with 95 and 92% in monoculture plantations and agroforestry systems, respectively. In both types of production systems, carbon storage is a function of tree age and density. The carbon stored in monoculture plantations was higher than in agroforestry systems, due to a greater density of rubber trees in the first production system. This study confirms that rubber plantations have potential to capture and store atmospheric carbon. With this information, the issue of participating in carbon markets of the rubber production chain can be addressed, and therefore strengthen in the region's competitiveness and sustainability.

Keywords: Productive chain; timber stage; biomass; necromass; roots.

INTRODUCTION

Climate change refers to alterations, directly or indirectly attributed to human activities, affecting the atmosphere composition and which is additional to the climate variability observed over comparable periods of time (IPCC, 2001). This climate change on the planet is mainly due to the increase in the concentration of greenhouse gases (GHG) in the atmosphere, being a large proportion of them emitted by anthropogenic activities associated with land use change, deforestation, fossil fuels use, inadequate agricultural practices and extensive cattle-raising (Acosta *et al.*, 2001a; IPCC, 2008). These GHG cause an overall increase in air and ocean temperature, snow and ice melting, and hence cause sea level to rise (IPCC, 2008). Rainfall periodicity can vary and exceed the usual average. Concentrated droughts may occur which can increase the risk for forest fires and there will be a greater impact on local communities, due to the increasing loss of biodiversity and the reduction of water supply (IDEAM, 2001).

GHG could be diminished by reducing CO_2 emissions and increasing terrestrial sinks (Segura and Andrade, 2008). Some agricultural practices, such as the implementation of agroforestry systems (AFS), are important alternatives for carbon capture and storage, especially in biomass and soil, increasing soil fertility, all of which has positive effects on the ecosystem productivity and sustainability (Orjuela, 2011; Orjuela and Andrade, 2011). AFS, by associating woody plants with crops, are an economical and ecological option for climate change mitigation (Albrecht and Kandji, 2003; Beer *et al.*, 2003; Swamy and Puri, 2005; Andrade, 2007; Segura and Andrade, 2008), which also involves the possibility of recognizing the economic value of carbon capture and storage as an environmental service (Ospina-Ante, 2003).

It has been shown that the rubber plantations (*Hevea brasiliensis* Müll. Arg) accumulate significant amounts of carbon in their biomass (Tsuruta *et al.*, 2000; Duran *et al.*, 2011). Moreover, the agroindustrial transformation process of rubber has great agro-ecological and socio-economic potential in Colombia, emerging as one of the most important productive lines in the Amazon region. This process generates rural employment, building social capital and fulfilling the protective-producer reforestation function in hydrographic basins and is an alternative development culture in illicit crops areas (SENA, 2006).

It is of fundamental importance to estimate with greater precision, the carbon capture and storage potential of *H. brasiliensis*. This information is key for assessing the ecosystemic service of carbon sequestration (Gobbi and Ibrahim, 2004; Andrade *et al.*, 2008), provides access to carbon sequestration economic incentives, generates additional revenue for the producer, and also contributes to the improvement of the production chain competitiveness and, hence, the life quality of local communities (Somarriba *et al.*, 2006). The aim of the study presented here was to estimate the carbon sequestration potential of rubber plantations in monoculture and agroforestry systems in the Colombian Amazon.

MATERIALS AND METHODS

The study was conducted in the Northeastern region of the Colombian Amazon, department of Caquetá, in the municipalities of Florencia, El Doncello and Belén de los Andaquíes (Figure 1). The Caquetá department is located at Southeast of Colombia, on the left bank of Caquetá river, at 2°58' N and 0°40' S, between 71°30' and 76°15' W (IGAC-INPA, 1993). The department has an average rainfall of 3,600 mm year^{-1}, a mean annual temperature of 25.1 °C and the relative humidity varies between 79.5 and 88.6%. Around 20% of the Caquetá is located in the Amazon piedmont at an altitude between 400 and 1,000 m.

Farm selection

The farms were selected in agreement with the *Asociación de Reforestadores y Cultivadores de Caucho del Caquetá* (ASOHECA), taking into account the producers' willingness and the availability of rubber plantations in monoculture and AFS with copoazú, in different stages of formation. Three age ranges were established in the rubber plantations: 1-7 years (stage of growing or unproductive trees); 8-20 years (production peak) and > 20 years (beginning of the decline in rubber production). A productive life of 40 years is estimated. Copoazú (*Theobroma grandiflorum* Willd. Ex Spreng Schum.) is one of the most promising and widespread fruit trees in the Amazonian region and is commonly associated with other crops or productive forest plantations such as rubber. Currently, this fruit tree is widely distributed in the Colombian Amazon departments with crop management and cultural practices similar to those used in cocoa. However, in spite of belonging to the genus *Theobroma*, copoazú has different characteristics in shape, size, internal structure and almond alveolar design (Cohen and Jackix, 2005).

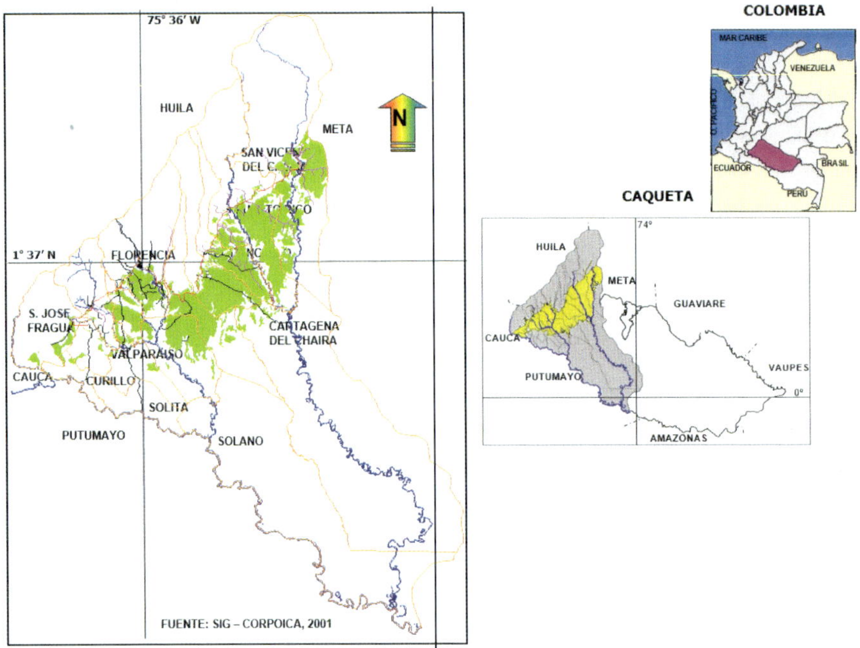

Figure 1. Details of study area.

Sampling techniques and evaluation of samples

Three sinks or carbon components were evaluated:
1) Necromass
2) Aboveground biomass
3) Belowground biomass (fine roots).

This research estimated carbon storage following the methodology proposed by Andrade and Ibrahim (2003) and methods recommended by IPCC (2003).

a) Temporary sampling plots

In each crop system selected, three temporary sampling plots, in rectangular shape, were established and measured, considering age range and location (municipality). A sample plot location map was made with a global positioning system (GPS). The sample plot was 250 m^2 in size considering at least 15 rubber trees (some reached a larger area due to the absence of some trees). The plots were delimited using wooden stakes and colored strings with reflective tape, as recommended by Delgadilla and Soltero (2006). In total, 48 sample plots were established and measured, corresponding to three ages in monoculture for two municipalities (27 plots), while in Florencia and El Doncello only two ages were placed in AFS (21 plots).

b) Carbon and biomass estimation
1. Necromass
Litter

Litter sampling was the first component measured, with the aim of reducing its disturbance. All material, found in 10 frames of 0.25 m^2 each, randomly located within each sample plot, was taken. The samples were fresh weighed and mixed to obtain a subsample of approximately 300 g per plot. In order to estimate dry material content, the subsamples were taken to the laboratory and dried in an oven at 65 °C until a constant weight was obtained.

Standing dead trees and fallen trunks

The standing dead trees were estimated with the method of biomass expansion factor and biomass models with correction according to the absence of part of the biomass. In this case, the diameter of trunk at breast height (dbh) and total height (th) were measured and the proportion of remaining crown, compared to a living tree, was estimated.

The volume of the trunks or fallen dead trees was estimated with the line intersect method (IPCC, 2003), which consists in establishing two perpendicular transects of 50 m each by plot. The diameter of all the pieces (>1 cm) that touched any transect was measured. The volume of fallen timber was estimated with the equation:

$$V = \frac{\pi^2 \, x \, (D_1^2 + D_2^2 + \ldots + D_n^2)}{8 \, x \, L}$$

Where:

V : Volume of fallen timber (m^3 ha^{-1})

D_1, D_2, D_n : Diameter of each piece intercepted by the line (cm)

L : Transects length (m)

The timber volume was converted to dry matter assigning a specific gravity to each piece. In order to do so, the machete method, recommended by IPCC (2003), was used. This method consists on hitting the piece with a machete, defining a class of density: high (consistent), medium (intermediate) and low (rotten), and assigning an specific gravity to each density class: 0.23, 0.42 and 0.60 g cm^{-3}, respectively (Segura, 2005).

2. Above-ground biomass
Herbaceous material

This component was estimated in 10 frames of 0.25 m^2 each, which were randomly distributed within the plot. All herbaceous biomass was collected, cutting the herbaceous material (plants < 1.5 m in height) that was found inside the frame. The collected material was fresh weighed and a subsample of about 250 g was taken to estimate the dry matter content in the laboratory (65 °C until a constant weight was obtained).

Woody components

All rubber and copoazú trees that were found in the temporal sampling plot were measured (th and dbh in rubber and trunk diameter at 30 cm height -D_{30}- in copoazú). Aspects such as the presence of twisted, bifurcation, buttressed, or inclined trunks in sloped terrains were considered when measuring the trees.

The woody component biomass was estimated using allometric models that were developed for these species. The above-ground biomass of 1-7 years trees was estimated with the biomass model developed by Duran *et al.* (2011):

$$AB = -0{,}59 + 0{,}34 \, x \, dbh^2$$

Where:

AB : Aboveground biomass (kg $tree^{-1}$)

dbh : Trunk diameter at breast height (cm)

The allometric model created by Moreno *et al.* (2005) was applied to rubber trees of 8-20 and >20 years age:

$$AB = 0{,}00411 \, x \, CBH^{2{,}596}$$

Where:

AB : Aboveground biomass (kg $tree^{-1}$)

CBH : Trunk circumference at breast height (cm)

Given that no allometric models for copoazú tree have been developed, its aboveground biomass was estimated using an allometric model developed by Andrade et al (2008) for cacao trees (*Theobroma cacao* L.) in Costa Rica:

$$\boldsymbol{AB = 10^{(-1{,}625 + 2{,}63 \, x \, Log(D_{30}))}}$$

Where:

AB : Aboveground biomass (kg $tree^{-1}$)

Log : Base 10 logarithm

D_{30} : Trunk diameter at 30 cm height (cm)

3. Below-ground biomass

In each established plot, three sampling points for fine roots were located, using pits. Fine roots (diameter < 2 mm) biomass was estimated by removing three monoliths (one for each side: south, east and west) of 1000 cm^3 (10 x 10 x 10 cm) at depths of 0-10, 10-20 and 20-30 cm. The roots were extracted from each soil sample manually and were taken to the laboratory for oven drying at 65 °C until constant weight.

4. Carbon storage estimation

Biomass and necromass was converted into carbon by multiplying by 0.5, which is a factor proposed by IPCC (2008).

Information analysis

The results obtained in this study were analyzed with descriptive statistics (mean, standard deviation and standard error) using the Infostat statistical software 2010 version (Di Rienzo *et al.*, 2010).

RESULTS AND DISCUSSION

General Information Regarding Sampled Individuals

In a total of 1.3 ha sampled area, a total of 880 trees was found, of which 774 were *H. brasiliensis* trees and 106 were *T. grandiflorum* individuals The sampled trees were in a good phytosanitary state; some individuals had deciduous behavior due to they were naturally defoliated, which is a usual behavior of this species, according to Torres-Arango (1999). The sampled rubber tree individuals belonged to the clones: FX 25, FX 3864, IAN 710, IAN 873 and FX 4098.

The rubber plantations in monoculture and AFS have a planting distance of 7 x 3 m. AFS older than 7 years were not established with a planned distribution, since only the fruit component between rubber lines was included. However, some young plantations (1-7 years), were established in double rows of rubber, with planting distances of 4 m between rows, 3 m between plants and 10 m between each double line, in the middle of which the copoazú trees were planted. Plantations established in monoculture showed a higher density of rubber trees (Table 1) than in AFS, because in most of the sampled plots, it was found that the producer includes the fruit component when rubber tree losses are obtained due to phytosanitary problems, which allows maintaining its productivity per hectare.

Table 1. Population density of rubber plantations in monoculture and agroforestry systems (AFS) in the Colombian Amazon.

Age (years)	Production System: Monoculture Rubber	AFS Rubber	AFS Copoazú
	Density (trees ha^{-1})		
1 - 7	662	673	232
8 - 20	582	572	126
> 20	563	501	199

Carbon storage in herbaceous plants

Carbon storage in herbs showed no significant differences (p>0.05) between systems, with averages of 1.5 Mg C ha^{-1} for both systems (Table 2). The low carbon content of this component can be attributed to the low amount of photosynthetically active radiation transmitted to the herbaceous layer, where only shade tolerant species can grow (Acosta *et al.*, 2001b). Moreover, in older plantations, herbs are subjected to constant weed control and eventually grazing by cattle.

Table 2. Carbon stored in herbs present in rubber plantations in monoculture and agroforestry systems (AFS) in the Colombian Amazon.

System	Age (years) 1 - 7	8 – 20	> 20
	Mg C ha^{-1}		
Monoculture	1.46 ± 0.10 a	1.56 ± 0.11 a	1.49 ± 0.16 a
AFS	1.65 ± 0.06 a	1.59 ± 0.08 a	1.35 ± 0.08 a

Values correspond to the mean ± standard error. Same letters indicate no statistical differences among systems (p > 0.05).

Carbon storage in necromass

Carbon storage in necromass showed that in AFS more of this element is captured, which could be initially attributed to the extra and constant leaf litter contribution by copoazú. The estimated carbon in this component corresponds to 6.7% of the total stored in the AFS, while in monocultures corresponds to 2.6%.

Litter stored on average 2.6 Mg C ha^{-1} in rubber plantations in agroforestry (Table 3), exceeding reports such as Orjuela (2011) and Orjuela and Andrade (2011), who found that litter stores 1.8 Mg C ha^{-1} in fallows of five years in Caquetá, Colombia. Similarly, Ferreira (2001) found that litter stored 1.8 Mg C ha^{-1} in 1-7 years secondary forests, 1.9 Mg C ha^{-1} in 8-20 years secondary forests, and 2.6 mg C ha^{-1} in >20 years secondary forests, in San Carlos, Nicaragua.

Table 3. Carbon stored in necromass of rubber plantations in monoculture and agroforestry systems (AFS) in the Colombian Amazon.

System	Component	Age (years) 1 - 7	8 – 20	> 20
		Mg C ha^{-1}		
Monoculture	Litter	3.34 ± 0.62 a	1.25 ± 0.16 b	1.18 ± 0.26 b
	Fallen trunks	0.84 ± 0.46 a	1.23 ± 0.59 a	0.29 ± 0.17 a
	Standing dead trees	0.00 (NA) a	0.00 (NA) a	0.03 ± 0.03 a
AFS	Litter	3.25 ± 0.96 a	2.14 ± 0.46 a	2.49 ± 0.40 a
	Fallen trunks	3.80 ± 3.29 a	3.61 ± 1.63 a	2.33 ± 0.95 a
	Standing dead trees	0.00 (NA) a	0.59 ± 0.59 a	0.78 ± 0.78 a

Values correspond to the mean ± standard error. NA: Not Applicable. Same letters indicate no statistical differences among systems (p > 0.05).

Carbon storage in the woody component

As expected, carbon storage increases with the age of the plantations (Table 4), as described by Alegre *et al.* (2001), Callo-Concha *et al.* (2001) y Lapeyre *et al.* (2004). In other words, older plantations showed the highest carbon storage values, which is related to biomass accumulation product of the balance between photosynthetic activity and respiration (Segura and Andrade, 2008).

Younger plantations (1-7 years) stored on average 4.3 Mg C ha^{-1} in monoculture and 4.2 Mg C ha^{-1} in AFS. These findings are similar to those reported by Duran *et al.* (2011) who found that ASOHECA plantations, in the same age range, stored on average 4.2 Mg C ha^{-1}. The 8-20 years age range, in AFS stored 44.8 Mg C ha^{-1}; this result is close to that reported by Yquise-Pérez *et al.* (2009), who found 54.3 Mg C ha^{-1} in forest species of 10 years in AFS with *Paspalum conjugatum* Berg. For the rubber plantations in monoculture included in this study, the amount of carbon in this compartment was 57.2 Mg ha^{-1}. The oldest plantations (> 20 years) contained 121.5 and 97.2 Mg C ha^{-1} in monoculture and AFS, respectively. These same researchers found that *T. cacao* in AFS with forest species of 25 years, stored 111.7 Mg C ha^{-1}.

Carbon storage in belowground biomass

In this compartment, most of the carbon storage was found in the first 10 cm of soil depth (Table 5), which is related to the greater abundance of herbs' root systems. The stored carbon at this depth was 1.2 Mg C ha^{-1} in AFS and 1.5 Mg C ha^{-1} in monoculture; while at 10 to 20 cm depth 0.23 Mg C ha^{-1} in AFS and 0.30 Mg C ha^{-1} in monoculture was found; in the last depth 0.14 Mg C ha^{-1} was estimated for both production systems. In monocultures, more carbon storage was found in fine roots than in AFS, possibly attributed to crop management in the region, where monoculture plantations are subjected to a constant weed control, which in turn stimulates a greater production of roots.

Orjuela (2011) and Orjuela and Andrade (2011) found on five years of formation fallows in the Colombian Amazon, an average of 1.6 Mg C ha^{-1} in the first 10 cm of soil depth and 0.5 Mg C ha^{-1} at depths of 10 to 20, and 20-30 cm. Yquise-Pérez *et al.* (2009) found in rubber plantations of 30 years of age in the department of Huanuco (Peru), a storage of 0.35 Mg C ha^{-1} in the first 20 cm of soil depth, which is a lower result than that found in this study. In the same way, Poroma (2012) found 0.1 Mg C ha^{-1} in the first 20 cm of soil depth, in cacao orchards in Waslala, Nicaragua.

Table 4. Carbon stored in the woody component of rubber plantations in monoculture and agroforestry systems (AFS) in the Colombian Amazon.

System	Species	Age (years) 1 – 7	8 – 20	> 20
		Mg C ha^{-1}		
Monoculture	**Rubber**	4.33 ± 0.48 a	57.16 ± 6.26 a	121.52 ± 19.34 a
AFS	**Rubber**	4.12 ± 1.19 a	44.38 ± 12.32 a	96.62 ± 14.03 a
	Copoazú	0.03 ± 0.02	0.44 ± 0.17	0.60 ± 0.25

Values correspond to the mean ± standard error. Same letters indicate no statistical differences among systems ($p > 0.05$).

Table 5. Carbon stored in fine roots at different depths in rubber plantations in monoculture and agroforestry systems (AFS) in the Colombian Amazon.

System	Depth (cm)	Age (years) 1 - 7	8 - 20	>20
		Mg C ha^{-1}		
Monoculture	**0 – 10**	0.88 ± 0.14 a	2.38 ± 0.19 a	1.28 ± 0.13 a
	10 – 20	0.17 ± 0.03 a	0.42 ± 0.06 a	0.31 ± 0.04 a
	20 – 30	0.08 ± 0.02 a	0.18 ± 0.03 a	0.17 ± 0.03 a
AFS	**0 – 10**	0.70 ± 0.10 a	1.60 ± 0.15 b	1.20 ± 0.17 a
	10 – 20	0.14 ± 0.03 a	0.32 ± 0.05 a	0.22 ± 0.03 a
	20 – 30	0.14 ± 0.06 a	0.16 ± 0.02 a	0.12 ± 0.02 a

Values correspond to the mean ± standard error. Same letters indicate no statistical differences among systems ($p > 0.05$).

Total carbon storage

The total carbon storage of the studied plantations, specifically in the aboveground biomass compartment, is higher in rubber plantations in monoculture than in AFS with copoazú (Table 2), which is mainly due to differences in tree density, being it 31% higher in monoculture plantations compared with the AFS studied. In relation to this, Salgado and Flores (2004) emphasize that the ability to capture and store carbon depends on the species used, tree density, silvicultural management, ecological conditions, site quality and possible leakages.

The plant aboveground sub-compartment was the largest reservoir of carbon: 95.4 and 91.7% of the total carbon stored in monoculture plantations and AFS, respectively (Figure 2); which is explained by its high content of timber biomass. The necromass sub-compartment, held 2.6 and 6.7% of the total carbon stored in monoculture and AFS, respectively. The fine root component held 1.9 and 1.6% of the total carbon stored in monoculture and AFS, respectively.

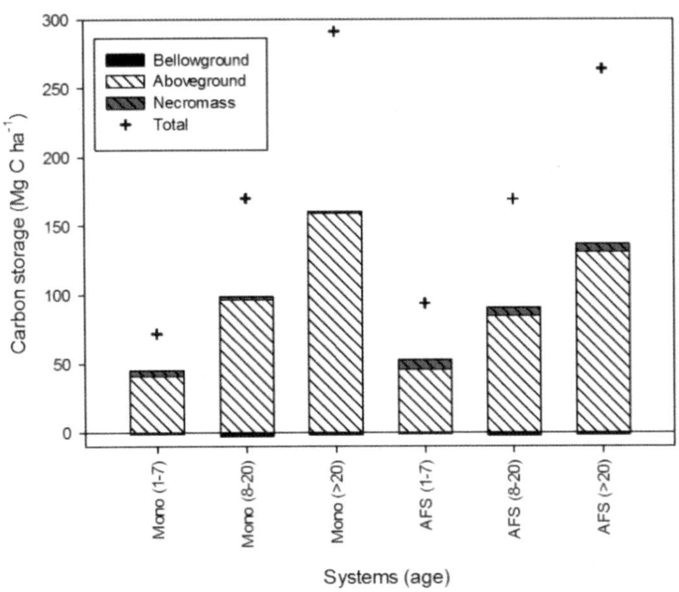

Figure 2. Carbon stored in rubber plantations in monoculture (Mono) and agroforestry systems (AFS) in the Colombian Amazon.

CONCLUSIONS

Rubber production systems in monoculture fixed 3.7% more atmospheric carbon than those in AFS with copoazú. The major carbon storage is found aboveground in the plant sub-compartment. However the difference between the densities of rubber trees sampled in monoculture and AFS showed, in the same way, the difference in total carbon storage. This is related to the fact that in the region rubber plantations are very heterogeneous and present tree loss because of various reasons, such as damages in the trunk by termites and fungi.

It is confirmed that plantations of *Hevea brasiliensis* in the Colombian Amazon, have great potential for carbon capture and storage, making it possible to find options to strengthen the competitiveness and sustainability of the rubber production chain of the in the region. This can be a first step to access to compensation mechanisms (payment for ecosystem services: PES), which could generate additional revenue to the producer and hence help to improve the life quality of local communities.

Acknowledgements

The authors express their gratitude to the Administrative Department of Science, Technology and Innovation of Colombia (COLCIENCIAS), the University of Amazonia, the Reforestation and Rubber Growers Association of Caquetá (ASOHECA), and the Project to Strengthen the Rubber Production Chain, through the biophysical and economic assessment of carbon sequestration in rubber systems in the Colombian Amazon.

REFERENCES

Acosta, M., Etchevers, J.D., Monreal, C., Quednow, K., Hidalgo, C., 2001a. Un método para la medición del carbono en los compartimientos subterráneos (raíces y suelo) de sistemas forestales y agrícolas en terrenos de ladera en México. Simposio Internacional Medición y Monitoreo de la Captura de Carbono en Ecosistemas Forestales. INIFAP, Colegio de Postgraduados. Montecillo, México, Valdivia, Chile. Accessed on 13-Nov-2013 at http://www.uach.cl/procarbono/pdf/simposio_carbono/10_Acosta.PDF

Acosta, M., Quednow, K., Etchevers, J., Monreal, C., 2001b. Un método para la medición del carbono almacenado en la parte aérea de sistemas de vegetación natural e inducida en terrenos de ladera en México. Simposio

Internacional Medición y Monitoreo de la Captura de Carbono en Ecosistemas Forestales. Valdivia, Chile. Accessed on 13-Nov-2013 at http://procarbono.site88.net/pdf/simposio_carbono/08_Acosta.PDF

Albrecht, A., Kandji, S.T., 2003. Carbon sequestration in tropical agroforestry systems. Agriculture, Ecosystems and Environment 99 (1): 15-27.

Andrade, H.J., 2007. Growth and inter-specific interactions in young silvopastoral systems with native timber trees in the dry tropics of Costa Rica. University of Walles Bangor (UWB)- Tropical Agricultural Research and Higher Education Center (CATIE), Turrialba, Costa Rica, p. 224. Accessed on 13-Nov-2013 at http://bit.ly/1aTnxu7

Andrade, H.J., Brook, R., Ibrahim, M., 2008. Growth, production and carbon sequestration of silvopastoral systems with native timber species in the dry lowlands of Costa Rica. Plant and Soil 308 (1-2): 11-22.

Andrade, H; Segura, M; Somarriba, E; Villalobos, M. 2008. Valoración biofísica y financiera de la fijación de carbono por uso del suelo en fincas cacaoteras indígenas de Talamanca, Costa Rica. Revista Agroforestería en las Américas 46: 45-50.

Andrade, H.J., Ibrahim, M., 2003. ¿Cómo monitorear el secuestro de carbono en los sistemas silvopastoriles? Agroforestería en las Américas (CATIE-Costa Rica) 10 (39-40): 109-116.

Beer, J., Harvey, C., Ibrahim, M., Harmand, J.M., Somarriba, E., Jiménez, F., 2003. Servicios ambientales de los sistemas agroforestales. Agroforestería en las Américas (CATIE-Costa Rica) 10 (37-38): 80-87.

Callo-Concha, D., Krishnamurthy, L., Alegre, J., 2001. Cuantificación del carbono secuestrado por algunos SAFs y testigos, en tres pisos ecológicos de la amazonia del Perú. Simposio Nacional de Medición y Monitoreo de la Captura de Carbono en Ecosistemas Forestales, Valdivia-Chile, pp. 18-20. Accessed on 22-Nov-2013 at http://www.uach.cl/procarbono/pdf/simposio_carbono/53_Callo_Concha.PDF

Cohen, K.d.O., Jackix, M.d.N.H., 2005. Estudo do liquor de cupuaçu. Ciência e Tecnologia de Alimentos (Food Science and Technology - Sociedade Brasileira de Ciência e Tecnologia de Alimentos) 25 (1): 182-190. Accessed on 26-Nov-2013 at http://bit.ly/Imj1bF

Delgadilla, M., Soltero, Q., 2006. Manual de monitoreo de carbono en Sistemas Agroforestales. Comisión Nacional Forestal CONAFOR y AMBIOS S.C. DE R.L. Méx. Accessed on 13-Nov-2013 at http://bit.ly/1hFGcxh

Di Rienzo, J.A., Casanoves, F., Balzarini, M.G., Gonzalez, L., M, T., Robledo, C.W., 2010. InfoStat versión 2010. Grupo InfoStat, FCA, Universidad Nacional de Córdoba, Argentina.

Ferreira, C.M., 2001. Almacenamiento de carbono en bosques secundarios en el Municipio de San Carlos, Nicaragua. CATIE. CATIE, Turrialba, Costa Rica, p. 100. Accessed on 13-Nov-2013 at http://bit.ly/17pDTEC

Gobbi, J., Ibrahim, M., 2004. Creating win-win situations: The Strategy of Paying for Environmental Services to Promote Adoption of Silvopastoral Systems. The Importance of Silvopastoral Systems in Rural Livehoods to Provide Ecosystem Services. In: Mannetje, L.T., Ramírez, L., Igrahim, M., Sandoval, C., Ojeda, N., Ku, J. (Eds.), International Symposium Silvopastoral Systems, Mérida, Yucatán, México. Univerisada Autónoma de Yucatán, México, pp. 98-101.

IDEAM, 2001. Colombia Primera Comunicación Nacional ante la Convención Marco de las Naciones Unidas sobre el Cambio Climático. Ministerio del Medio Ambiente, Instituto de Hidrología, Meteorología y Estudios Ambientales de Colombia (IDEAM), Programa de las Naciones Unidas para el Desarrollo. Accessed on 19-Nov-2013 at http://unfccc.int/resource/docs/natc/colnc1.pdf

IGAC-INPA, 1993. Aspectos ambientales para el ordenamiento territorial del Occidente del departamento del Caquetá. Tomo I. In: Saldarriaga, J.G., van der Hammen, T. (Eds.). Instituto Geográfico Agustín Codazzi, Colombia - Comité Editorial Programa TROPENBOS.

IPCC, 2001. Climate change 2001: Synthesis Report. In: Watson, R.T., Albritton, D.L., Dokken, D.J., Core-Writing-Team (Eds.). World Meteorological Organization, p. 148.

Accessed on 19-Nov-2013 at http://www.ess.uci.edu/researchgrp/prather/files/2001IPCC_SyR-Watson.pdf

IPCC, 2003. Guía de Buenas Prácticas del Uso de la Tierra, Cambio del Uso de la Tierra y Silvicultura (GBPUTCUTS). Métodos Complementarios y Orientación Sobre las Buenas Prácticas que Emanan del Protocolo de Kyoto., Intergovernmental Panel on Climate Change (IPCC). Accessed on 13-Nov-2013 at http://www.ipcc-nggip.iges.or.jp/public/gpglulucf/gpglulucf_languages.html

Lapeyre, T., Alegre, J., Arévalo, L., 2004. Determinación de las reservas de carbono de la biomasa aérea, en diferentes sistemas de uso de la tierra en San Martín, Perú. Ecología Aplicada (Universidad Nacional Agraria La Molina, Perú) 3 (1-2): 35-44. Accessed on 26-Nov-2013 at http://www.lamolina.edu.pe/ecolapl/Articulo6vol3.pdf

Moreno, J.A., Burgos, J.D., Nieves, H.E., Buitrago, C.E., 2005. Modelo alométrico general para la estimación del secuestro de carbono por plantaciones de caucho Hevea brasiliensis Müll arg. en Colombia. Revista Colombia Forestal (Universidad Distrital, Bogotá-Colombia) 9 (18): 5-21.

Orjuela, J., 2011. Almacenamiento de carbono en rastrojos de paisajes ganaderos en la Amazonía intervenida Colombiana. Universidad de la Amazonia, Florencia – Colombia, p. 92.

Orjuela, J.A., Andrade, H.J., 2011. Almacenamiento de carbono en áreas de regeneración natural en paisajes ganaderos de la Amazonia Colombiana. III Seminario Internacional en Ambiente, Biodiversidad y Desarrollo y I Simposio Nacional de Química Aplicada. Vicerectoria de Investigaciones y Posgrados - Facultad de Ciencias Básicas - Universidad de la Amazonía, Colombia, p. 68.

Ospina-Ante, A., 2003. Agroforesteria: aportes conceptuales, metodologicos y practicos para el estudio agroforestal. Asociacion del Colectivo de Agroecologia del Suroccidente Colombiano (ACASOC) - Serie Agroforestería.

Poroma, D., 2012. Estrategias de reducción de la huella de carbono en la producción del grano de cacao (*Theobroma cacao* L.) para la cooperativa CACAONICA en Waslala, Nicaragua. Centro Agronómico Tropical de Investigación y Enseñanza (CATIE), Turrialba, Costa Rica, p. 79. Accessed on 22-Nov-2013 at http://biblioteca.catie.ac.cr:5151/repositoriomap/bitstream/123456789/204/1/Poroma_Estrategias.pdf

Segura, M., 2005. Estimación del carbono almacenado y fijado en sistemas agroforestales indígenas con cacao en la zona de Talamanca, Costa Rica, Informe de Consultoría. Proyecto: Captura de carbono y desarrollo de mercados ambientales en sistemas agroforestales indígenas con cacao en Costa Rica (TF-052118).

Segura, M., Andrade, H.J., 2008. ¿Cómo construir modelos alométricos de volumen, biomasa o carbono de especies leñosas perennes? Agroforestería en las Américas (CATIE-Costa Rica) (46): 89-96.

SENA, 2006. Estudio de Caracterizacion Ocupacional del Sector del Caucho Natural en Colombia: Mesa de Sectorial del Caucho. Servicio Nacional de Aprendizaje (SENA). Bogotá-Colombia. Accessed on 13-Nov-2013 at http://observatorio.sena.edu.co/mesas/01/CAUCHO.pdf

Somarriba, E., Quesada, F., Villalobos, M., 2006. La captura de carbono: un servicio ambiental en fincas cacaoteras indígenas. . Serie Técnica. Manual Técnico CATIE (ST MT-64), Turrialba, Costa Rica, p. 28.

Swamy, S., Puri, S., 2005. Biomass production and C-sequestration of *Gmelina arborea* in plantation and agroforestry system in India. Agroforestry systems 64 (3): 181-195. Accessed on 26-Nov-2013 at http://bit.ly/IhcNJv

Torres-Arango, C.H., 1999. Manual para el cultivo del Caucho en la Amazonia Colombiana. Plan Nacional De Desarrollo Alternativo (PLANTE). Universidad de la Amazonia, Colombia. Accessed on 26-Nov-2013 at http://bit.ly/1cVXZy0

Tsuruta, H., Ishizuka, S., Ueda, S., Murdiyarso, D., 2000. Seasonal and spatial variations of

CO_2, CH_4, and N_2O fluxes from the surface soils in different forms of land-use/cover in Jambi, Sumatra. In: Murdiyarso, D., Tsuruta, H. (Eds.), The impacts of Land–use/Cover Change on Greenhouse Gas Emissions in Tropical Asia, IC-SEA, Bogor, Indonesia and NIAES, Tsukuba, Japan, pp. 7-30.

Yquise-Pérez, A.R., Pocomucha, V., Vargas, Y., 2009. Carbono Almacenado en Diferentes Sistemas de uso de la Tierra del Distrito de José Crespo y Castillo, Huánuco, Perú. Accessed on 13-Nov-2013 at http://www.inteligentesite.com.br/arquivos/floagri/File/ANEXO20_CarbonoPeru.pdf

Diversty and Abundance of Arbuscular Mycorrhizal Fungi across Different Land use Types in a Humid Low Land Area of Ethiopia

Zerihun Belay[1*], Mauritz Vestberg[2] and Fassil Assefa[1]

[1]*Department of Microbial, Cellular and Molecular Biology, College of Natural Sciences, Addis Ababa University, Ethiopia. Email: zebelay2009@gmail.com*

[2]*Natural Resources Institute Finland (Luke), Laukaa, Finland,*

Email: mauritz.vestberg@mtt.fi

[*]*Corresponding author*

SUMMARY

The aim of this work was to study the effect of different land use types on arbuscular mycorrhizal (AM) fungal populations in soil and trap cultures from Showa robit, Ethiopia. Seven land use types were selected. There were low-input arable systems, either having a mixture of crops (Arable1) or monocropped with sorghum (Arable2) or maize (Arable3). Arable4 was relatively a high-input system with monocropped sorghum. A fruit cropping area (FC) managed with composts and plant residues, a natural forest (NF) and an acacia plantation (AP) were also studied. AMF spore abundance, species richness, diversity indices and mycorrhizal inoculum potential (MIP) were studied. In field soil, significantly higher spore numbers were recorded from FC, Arable1 and Arable3 (5.8-6.1 spores g^{-1} soil) than in Arable4, NF and AP (2.8-3.9 spores g^{-1} soil). In trap cultures, AP, FC, and Arable2 had the highest spore numbers (9.8-11.1 g^{-1} soil) and Arable4 and NF the lowest (2.5-3.8 g^{-1} soil). Slightly different MIP patterns also occurred with Arable1 (53.7%) and FC (52.6%), having significantly higher hyphal colonization, 53.7% and 52.6%, respectively, compared to the other land use types that fell within percentage colonization of 19.9-25.8 %. A total of 42 and 33 morphospecies of AMF were identified in field soil and trap culture soil, respectively. Trap culturing increased spore numbers but caused a loss of AMF species richness. Higher species richness was obtained in FC and Arable1 compared with the other systems. *Claroideoglomus* and *Funneliformis* were the dominant genera in all land use types in both trap culture and field soil. The results clearly imply that organic management and diversification of crops enhances AMF diversity of low-input agricultural systems.

Keywords: *Claroideoglomus; Funneliformis; Glomus;* mixed cropping; monocropping; mycorrhizal inoculum potential.

INTRODUCTION

It is a well-known fact that rapid land use change as a result of deforestation, cropland and pasture expansion, dry land degradation, urbanization, and agricultural intensification results in reduction in soil fertility (Hartemink *et al.*, 2008). The conversion of natural ecosystems to agriculture affects the above-ground plant and animal biodiversity (Wardle and Lavelle, 1997), which, in turn affects below-ground macro and microbial community structure and their functions (Bossio *et al.*, 2005).

Arbuscular Mycorrhizal Fungi (AMF) are one of the below-ground microbial groups that are severely affected by changes in vegetation cover and physical and chemical characteristics due to deforestation and land degradation (Smith and Read, 2008). These fungi are associated with more than 80 % of terrestrial plants. They enhance host growth and survival by facilitating nutrient uptake (Smith and Read, 2008) and improving tolerance to drought (Yamato *et al*., 2009) and to some root pathogens and nematodes (Azcón-Aguilar and Barea, 1997).

The influence of land use changes on AM fungal diversity and abundance has recently received considerable attention in both temperate (Oehl *et al.*, 2003; Verbruggen *et al.*, 2010) and tropical areas (Li *et al.*, 2007; Jefwa *et al.*, 2009; Stürmer and Siqueira, 2011; Jefwa *et al.*, 2012; Muchane *et al.*, 2012). It is reported that land use changes (Tchabi *et al*., 2008; Jewfa *et al.*, 2012; Muchane *et al*., 2012), ploughing, fertilizer and fungicide application (Douds *et al.*, 1993; Helgason *et al.*, 1998) land use intensity (Oehl *et al.*, 2003, 2004; Verbruggen *et al*., 2010) and tillage intensity (Jansa *et al.*, 2002) decrease AMF species richness, spore abundance and root colonization in different parts of Central Europe and tropical ecosystems in Africa.

Verbruggen *et al.* (2010) showed a decrease in mycorrhizal fungal communities in relation to land use intensity in that the average number of AMF taxa identified was highest in grasslands (8.8), intermediate in organically managed fields (6.4) and significantly lower in conventionally managed fields (3.9) in the agricultural soils of the Netherlands. It is also shown that AMF diversity and species richness in organically managed fields and natural vegetation fields was higher than that of conventionally managed fields (Oehl *et al.*, 2004; Tchabi *et al.*, 2008).

In Ethiopia, the land cover has been changed from natural forest to farmland, open grazing and fast growing plantation forests for several decades. According to FAO (2007), the country lost an average of 141,000 ha, or 1.1% per year, of its forest covers between 1990 and 2005, due to deforestation. Another study in the Central Ethiopian Rift Valley indicated that woodland cover declined from 40% to 9% at one site, while another site lost 54% of its woodland cover due to rapid deforestation (Garedew *et al.*, 2009). In all these years, deforestation has resulted in massive soil degradation with a decline in soil organic matter (SOM) and available nitrogen in the highlands of Ethiopia (Lemenih *et al*., 2005; Girmay *et al.*, 2008). A recent study also showed drastic changes in several of the physical and chemical properties of soils from different parts of the country due to rapid land use changes (Getachew *et al.*, 2012).

The drastic change in deforestation and land use in Ethiopia decimated the large biodiversity and plant community structure of the country. This is also presumed to affect the underground microbial composition including the AMF, because several studies have showed that plant community structure affects diversity and community composition and species richness of AMF (Burrows and Pfleger, 2002; Vandenkoornhuyse *et al.*, 2002; Johnson *et al.*, 2003: Scheublin *et al.*, 2004; Sýkorová *et al.*, 2007).

It was also reported that changing the vegetation cover from tree-based intercrops to mono-cropping systems can reduce AMF fungal richness (Chifflot *et al*., 2009). Lower AMF species richness was found in arable fields, compared to different natural ecosystems and perennial communities such as tropical forests (Snoeck *et al*., 2010). It may well be that the intensive land use change in Ethiopia for several decades has brought a reduction and/or shift in abundance and diversity of AMF under monocropping and intercropping systems.

The hypothesis of this study was that AMF abundance and diversity may have been affected by land management practices such as monocropping cultivation and use of fertilizer in the agricultural systems of the country. It is also equally important that understanding the role of AMF over a broad range of land use systems is essential for land rehabilitation and effective management for sustainable production through AMF technology in the future (Estaún *et al.*, 1997; Oehl *et al.*, 2003).

The objectives of this study were; (1) to compare AM fungal diversity and community composition among different land-use systems within a single agroecosystem; (2) to determine mycorrhizal inoculum potential in soil of these systems (3) and to determine whether AM fungal species richness, mycorrhizal inoculum potential (MIP) and spore abundance are influenced by land use changes.

MATERIAL AND METHODS

Study site description

The study site is located in Showa robit (10^0 06' 650" - 09^0 57'957" N, 039^0 54'37"-$039^0$56'579" E), (**Figure 1**) in north Showa Zone of Amhara Regional State, 225 km north of Addis Ababa, Ethiopia. The agro-ecology of the study site is low-land or Erteb Kola (sub-moist warm) with altitude ranging between 1120 and 1350 m a.s.l. The climate data of the study area recorded for the last ten years shows average annual maximum and minimum temperature and precipitation of 32.1 and 16.1 ^{0}C, and 968 mm, respectively (NMA, 2002-2010).

The land use of the area is mainly characterized by agroforestry practices such as agrisilvicultural (crops including and shrub/tree crops-trees) and agropastoral systems (trees+ crops+ pasture/animals) (Nair, 1993). The vegetation cover of the area is wood-land dominated by trees such as *Acacia*, *Erythrina*, *Cordia* and, *Ficus* species. The main grain crops of the area are sorghum, teff, finger millet and maize, whereas horticultural and commercial crops such as mango, banana, sugar cane, coffee, orange, tobacco, onion, tomato and cabbage are also grown in a mixed and/or rotation cropping system.

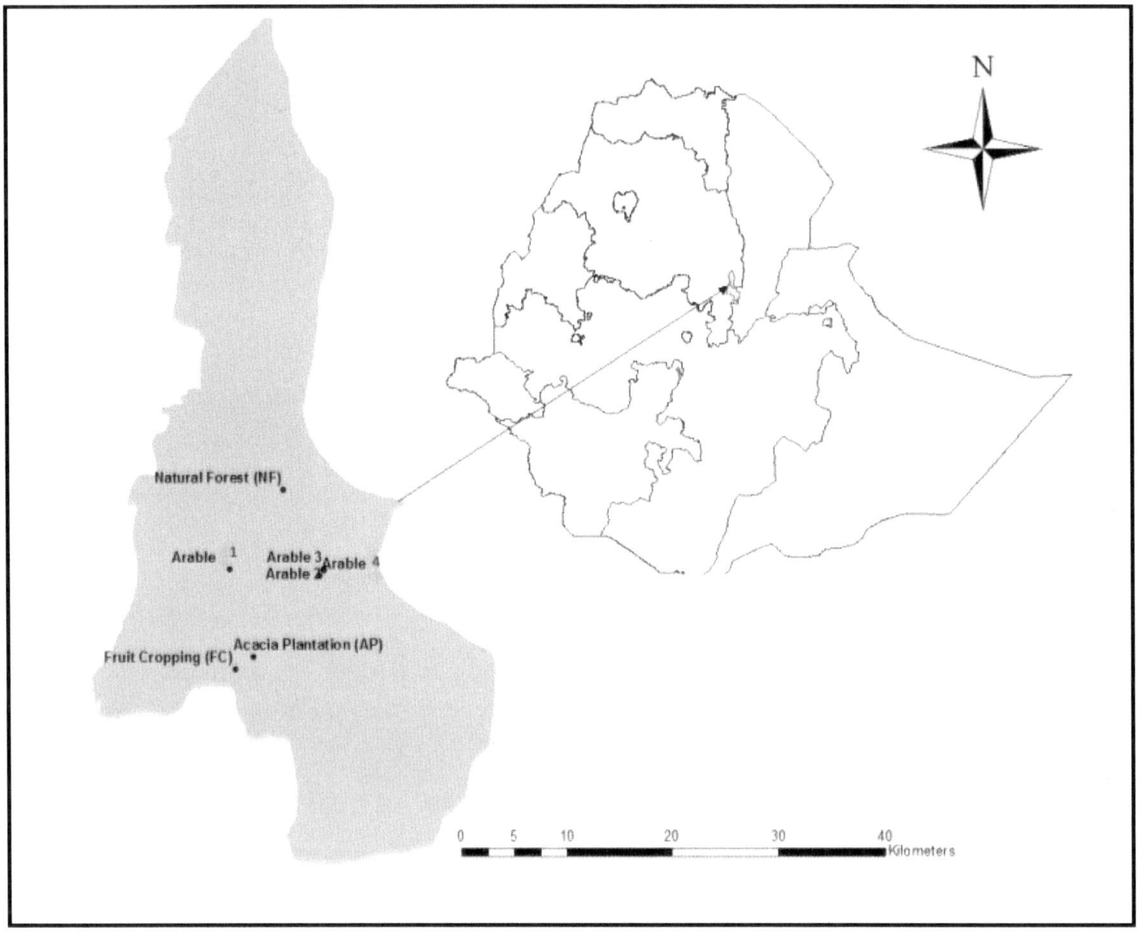

Figure 1. Map of the study site and sampling location.

Characteristics of the sampling site

The sampling area included seven different land use types and vegetation covers. There were four arable lands (Arable1-Arable4), one fruit cropping area (FC), one natural forest (NF) and one acacia plantation (AP). Arable1 was a low-input, mixed subsistence cropping (teff, sunflower and sesame). Arable2 and Arable3 were low-input sorghum and maize monocrop fields, respectively. Arable4 was relatively a high-input, conventional field cropped with sorghum. In the fruit cropping area (FC), mainly fruits, vegetables and garden cash crops were grown in an intercropping system with banana, papaya, mango, lemon, avocado, tomato and coffee. Two forest relic sites were also included. The natural forest site was dominated by acacia, fig trees and red stinkwood. The other forest relic site was an acacia plantation dominated by two acacia species (*Acacia seyal* and *A. nilotica*). A description and agricultural management practices of the study sites are given in Table 1.

Sampling was conducted during the dry season from November to December 2011. Three replicate sampling locations (approximately 100 m^2) were established for each land use type and dominant plant species. Three replicates of each dominant plant species were randomly selected.

From each sampling site, 500 g of rhizosphere soil samples were taken from a depth of 0-30 cm and subsequently pooled into one composite sample per location. A total of 54 samples, 3×3 from the three arable monocrop fields, 3×3 each from the arable mixed crop field, 3×7 from the fruit crop field, 3×3 from the natural forest and 2×3 from the acacia plantation were collected. The samples were collected in alcohol sterilized plastic containers, air dried and stored at room temperature for further analysis. Subsamples were used as inoculum for the trap culture system and for the spore extraction.

Soil physical and chemical characteristics

Soil particle size of samples from all land use types was determined using the hydrometer method (Gee and Bauder, 1986). Soil organic carbon (OC) was determined by the Walkley-Black dichromate oxidation procedure (Walkley and Black, 1934). Total nitrogen (TN) was determined by the Kjeldahl method (Hinds and Lowe, 1980). Soil available phosphorus was measured according to the method described by Olsen et al. (1954). The soil sample analyses were done at the Addis Ababa city administration environmental protection authority. Soil pH was measured at the Department of Microbial Cellular and Molecular Biology, Addis Ababa University, following 1:2.5 (v/v) soil: water suspension with a digital pH meter (HD8602) (Table 2).

Establishment of trap cultures

Trap cultures in pots were set up in triplicates for all seven land use types in a greenhouse to obtain fresh spores for identification of AM fungi and induce sporulation of species present only as hyphae in field samples. Pots with the size of 15cm were filled with 250 g of soil sub-sample containing root sections from each plot of the field and were thoroughly homogenized with sterile sand (1:1; v/v) according to Morton *et al.* (1993). Maize (*Zea mays* L.), a mycotrophic crop, was selected as an appropriate trap plant for its ability to induce high spore density, diversity and species richness (Yao *et al.*, 2010).

Seeds of maize were surface sterilized by immersing them in a 0.5% sodium hypochlorite solution for 15 minutes. After washing the seeds with sterile water they were sown at 2 cm depth in each plastic pot and covered with sterilized sand. Pots were irrigated daily as needed. No fertilizer was added during the growing period. All seedlings were grown in the greenhouse under natural ambient light and temperature conditions (about 29 °C day/18°C night).

The maize roots were checked for AMF colonization after 45 days. Pots supporting successful mycorrhization were maintained for six months. Watering was reduced during the final three weeks to maximize spore production. At the end of 6 months the plants were cut near the base, and the cultures were air-dried and checked for the presence and identification of spores.

Estimation of mychorrhizal inoculum potential (MIP)

MIP (mycorrhizal inoculum potential) of the different land use types was assessed in a greenhouse bioassay according to Sieverding (1991). A 300g soil sample from each land use type was placed in 450-ml sterile plastic pots. Five seeds of maize were sown per pot, and the seedlings were thinned down to three per pot after emergence. The pots were arranged in a completely randomized design with three replicates. After five weeks, the trap plants were harvested and their roots were cut into 1-cm sections and stained according to Brundrett *et al.* (1996). The percentage of colonized roots was quantified using the magnified intersection method (McGonigle *et al.*, 1990).

Table 1: Charactersitcs of study sites; management practices, standing crops or dominant plant species, fertilization and plant protection system. (Sources:Debere Brehan Agricultural Research Centere and Kewit Wereda Agriculture Office).

	Cropping system, land history	Fertilization (kg ha^{-1})	Plant protection	Standing crops or dominant plant species during sampling	Common name
Arable1	Crop rotation; three crops in mixture	Low-input urea (50) DAP (100)	Chemical and mechanical	*Eragrostis teff* (Zucc.) Trotter *Sesamum indicum* L. *Helianthus annuus* L.	Teff Sesame Sunflower
Arable2	Continuous sorghum monocropping field for 2 yr.	Low-input, Urea (50) DAP (50)	Chemical and mechanical	*Sorghum bicolor* L.	Sorghum
Arable3	Continuous maize monocropping field for 3 yr.	Low-input, Urea (100) DAP (100)	Chemical and mechanical	*Zea may* L.	Maize
Arable4	High-input; Sorghum monocropping field site used by DBRC	Mineral high-input ; Urea (100), DAP (150)	Chemical	*Sorghum bicolor* L.	Sorghum
Fruit cropping (FC)	Mixed fruit crops field, adjacent to river and irrigated regularly	Manure, compost and crop residues	Mechanical	*Persea americana* Mill. *Mangifera indica* L. *Coffee arabica* L. *Carica papaya* L. *Musa acuminata* Colla *Lycopersicon esculentum* Mill. *Citrus limonum* Risso.	Avocado Mango Coffee Papaya Banana Tomato Lemon
Natural forest (NF)	Mature forest (30 yr.), mixture of different trees, protected	None	None	*Acacia nilotica* (L.) Delile. *Ficus vasta* Forssk. *Prunus africana* (Hook.f.) kalkman	Acacia Fig tree Red Stinkwood
Acacia Plantation (AP)	Community managed acacia dominated forest, not protected	None	None	*Acacia seyal* Del. *Acacia nilotica* (L.) Delile.	Acacia Acacia

Table 2: Physical and chemical properties of soil samples from seven land use types at Showa robit, Ethiopia.

Land use type	pH	P (ppm)	T.N %	O.C %	Sand %	Clay %	Silt %	Texture class
Arable1	7.4	9.42	0.1	1.6	56	17	27	Sandy loam
Arable2	7.3	11.92	0.1	1.9	45	27	28	Clay loam
Arable3	7.4	5.98	0.2	1.6	52	22	26	Sandy clay loam
Arable 4	7.3	11.02	0.1	1.7	52	20	28	Loam
Fruit cropping(FC)	7.8	7.5	0.3	2.5	55	16	29	sandy loam
Natural forest (NF)	7.7	34.7	0.26	1.1	61	7	32	Sandy loam
Acacia plantation (AP)	7.9	5.4	0.1	1.3	50	25	25	Sandy loam

P: available phosphorus; T.N: Total nitrogen; O.C: organic carbon; Arable1: Low -input mixed cropping; Arable2: low-input monocropping, sorghum; Arable3: low-input monocropping, maize; Arable4: high-input monocropping, sorghum.

Staining of mycorrhizal roots

Staining of mycorrhizal roots was made according to Brundrett *et al.,* (1996). The root samples were carefully washed several times with tap water. About 0.5 g of root segments were cleared in 10 % (w/v) KOH at 90^{0}C in a water-bath for 2h, after which they were bleached with alkaline hydrogen peroxide at 10% for 3 minutes at room temperature. The roots were then treated with 1% HCl (v/v) for 15-20 minutes at room temperature and finally stained with 0.05% w/v trypan blue in lactoglycerol (1:1:1; lactic acid, glycerol and water) at 90^{0}C for 30 minutes in a water-bath.

With the exception of the HCl treatment, samples were drained and washed thoroughly with distilled water at the end of every step. The root samples were then left overnight in the lactoglycerol destaining solution (1:1:1; lactic acid, glycerol and water) in a dark room to remove coloration from root cells. Finally, roots were mounted in PVLG mountant on microscopic slides and covered with 40×22 mm coverslips.

Quantification of AMF root colonization

AM colonization was assessed from cleared and stained roots according to McGonigle *et al.* (1990). A total of 100 intersections were taken for each subsample to estimate percent AM root colonization under a compound microscope (OLYMPUS-BX51) at a magnification of ×200.

The presence of arbuscular colonization (AC) and vesicular colonization (VC) were calculated by dividing the count for the 'arbuscular' and 'vesicles' categories, respectively by the total number of intersections. Hyphal colonization (HC) was calculated as the proportion of non-negative intersections (McGonigle *et al.,* 1990).

Spore extraction

Soil and trap culture samples were air-dried and sieved through a 2-mm sieve to remove coarse debris before extracting, counting and identifying AM fungal spores. One hundred gram of dry soil was taken to extract the spores using the wet sieving and decanting method (Gerdemann and Nicolson, 1963), followed by centrifugation in water and in 50% sucrose solution (Brundrett *et al.,* 1996).

Each soil sample was mixed in a substantial volume of water and decanted through a series of sieves of 500, 250 and 50 μm. The contents from 250 and 50 μm sieves were mixed with water and centrifuged (Wagtech International) for 5 minutes at 2000 RPM. After having discarded the supernatant, the pellets were re-suspended in 50% sucrose solution and centrifuged for 1 minute as before. The supernatant was carefully poured through a 50 μm sieve and carefully washed with water to remove the sucrose.

Finally spores, spore clusters and sporocarps were carefully washed and transferred to a Petri dish to prepare for spore counting under the dissecting microscope (ISO 1006) at ×4. Enumeration of spore numbers per gram of dry soil was undertaken according to INVAM, *http://invam.caf.wvu.edu.*

Identification and characterization of spores

The AMF spores were morphologically identified at the Department of Microbial, Cellular and Molecular Biology, Addis Ababa University, Ethiopia and Natural Resources Institute Finland (Luke), Laukaa, Finland. About 50-70% of healthy looking spores were picked with forceps and mounted on slides in polyvinyl-lactic acid-glycerol (PVLG), (Omar *et al.,* 1979) or in PVLG mixed with Melzer's reagent (1:1 v/v) (Morton, 1991). Spores were examined under a compound microscope (OLYMPUS-BX51) at a magnification of ×400 and identified to the species level or to a specific morphotype based on (Schenck and Perez, 1990), online references of species description INVAM *http://invam.caf.wvu.edu,* West Virginia University, USA , University of Agriculture in Szczecin, Poland http://www.zor.zut.edu.pl/Glomermycota/, Schüßler and Walker (2010) and the Schüßler AMF phylogeny website http://www.lrz.de/~schuessler/amphylo/.

Determination of AMF diversity and spore density

The AMF communities on different land use types were detected and calculated based on the following parameters: Spore density (SD) was expressed as the number of AMF spores g^{-1}soil. Species richness (S) was measured as the total number of morphospecies. The Shannon–Wiener index (H') of diversity was calculated using the formula: $H' = -\sum ((n_i/n) \ln (n_i/n))$ where: n_i = number of individuals of species i and n = number of all individuals of all species. The Simpson's dominance index (D) was calculated using the formula $D = \sum (n_i/n)^2$; Evenness (E) was calculated by dividing Shannon–Wiener diversity value by the logarithm of the species richness. These analyses were conducted using the software PAST3 (ver. 3.0).

Isolation frequency (IF) was calculated as (the number of samples in which a given species was isolated/ the total number of samples) ×100%. Relative abundance of spores (RA) was calculated as (the number of spores in a given species / total number of spores) ×100%. The importance value (IV) was used to evaluate the dominance of AMF species based on IF and RA and was calculated as IV = (IF + RA)/2. An IV ≥ 50% indicates that a genus or species is dominant; 10% < IV < 50% applies to common genera or species; an IV ≤ 10% indicates that a genus or species is rare (Chen *et al.*, 2012).

Statistical analysis

Spore abundance data were log(x) transformed and the proportion of root colonization values were arcsine (the inverse sine of the square root of the proportion) transformed prior to analysis to meet assumptions of ANOVA such as normality and homogeneity of variance, but values were expressed as number of spores g^{-1}soil and percentage of root colonization, respectively. ANOVA and correlation analyses were carried out with the SPSS software package (version 21.0).

Significance of differences in AM fungal spore abundance and inoculum potential was tested using Fisher's least significant difference (LSD) at $p < 0.05$ after one-way ANOVA. The relationships between AMF parameters and soil chemical properties (pH, OC, available P, and TN) were determined using Pearson's correlation analysis. The same statistical tests were applied for initial mycorrhizal root colonization and spore formation for the trap cultures inoculated soils from the different sites.

RESULTS

AMF spore abundance in the soil and in trap cultures

The AMF spore densities of the different land use types recovered directly from the soil and trap cultures is shown in (**Table 3**). Spore densities in trap cultures were up to twice as high as those recovered from soil. The data also showed differences in the spore counts between soil and trap culture. Accordingly, FC (fruit cropping) and Arable3 land use types showed the highest spore count of 6.1 spores g^{-1} soil from field soil whereas FC and AP displayed the highest spore count of more than 11 spores g^{-1} soil in trap cultures. In general, NF (2.5) and Arable4 (3.8) showed the lowest number of spores per gram of soil in trap culture.

Data are reported as averages and standard errors for the three replicates per land use type. Values followed by different letters denote significant differences between land use types according to Fisher's LSD test at the 5% level after a one-way ANOVA. N: number of replicates; Arable1: Low-input mixed cropping; Arable2: low-input monocropping, sorghum; Arable3: low-input monocropping, maize; Arable4: high-input monocropping, sorghum.

Table 3. AMF spore abundance in soil and trap cultures of the different land use types in a humid lowland sampling area at Showa robit, Ethiopia

Land use type	N	AMF spores g^{-1} of soil	
		Field soil culture	Trap
Arable1	9	5.8 ± 0.8c	7.2 ± 1.7bc
Arable2	3	5.5 ± 1.5bc	9.8± 2c
Arable3	3	6.1 ± 1.4c	6.6± 1.4bc
Arable4	3	3.9 ± 0.5ab	3.8± 0.1ab
Fruit cropping (FC)	21	6.1 ± 0.7c	11.4± 1.4c
Natural forest (NF)	9	3.5 ± 0.2ab	2.5 ± 0.2a
Acacia plantation (AP)	6	2.8 ± 0.5a	11.1± 0.7c

Spore density correlated significantly with species richness and VC%, ($r=0.84$, $P<0.05$; $r=0.94$, $P<0.01$, respectively). Species richness correlated significantly with TN% and O.C %,($r=0.79$, $P<0.05$ and $r=0.76$, $P<0.05$, respectively).

Mycorrhizal inoculum potential

The AM fungal colonization patterns within the roots of maize plants showed that there was considerable heterogeneity between the land use types (**Table 4**). Root colonization occurred with typical structures (arbuscules, vesicles and hyphae) in almost all land use types except in Arable2, where vesicles were not observed.

The highest hyphal colonization of 53.7% and 52.6% were recorded from low-input mixed cropping (Arable1) and fruit cropping systems (FC), respectively, compared with the other land use types that showed mycorrhization rate ranging from 19.9% to 25.8% ($P=0.011$).

Table 4. Percentage of mycorrhizal root colonization in maize after five weeks of growth in soil from seven land use types at Showa robit, Ethiopia.

	AM colonization (%)		
Land use type	Arbuscular Colonization	Vesicular Colonization	Hyphal Colonization
Arable1	8.6 ±3ab	4.1 ±0.6ab	53.7±10.5c
Arable2	2 ± 1.5a	0	21±7.5ab
Arable3	7.4 ±0.3ab	0.9±0.5ab	36±8abc
Arable4	3.3 ±0.1ab	1.2±0.2ab	19.9±1.2ab
Fruit crops	12.6 ± 1.8b	7.5±1.4b	52.6±5.8c
Natural forest	2.6 ± 0.8a	2.2±0.9a	25.8±7.7ab
Acacia plantation	5.5± 1.7ab	4.7±2.1ab	30.2±2.1abc

Data are reported as averages and standard errors for three replicates per land use types. Values followed by different letters denote significant differences among land use types according to Fisher's LSD test at the 5% level after a one-way ANOVA. Arable1: Low-input mixed cropping; Arable2: low-input monocropping, sorghum; Arable3: low-input monocropping, maize; Arable4: high-input monocropping, sorghum.

AMF community composition

A total of 42 and 33 AMF morphospecies, were identified from field soil and trap culture, respectively (**Fig. 2**). Four morphospecies, from *Glomus, Acaulospora* and *Gigaspora,* from field soil samples and five morphospecies, from *Glomus*, *Acaulospora, Gigaspora* and *Ambispora* were not unidentified (data not shown). With a few exceptions of Arable 2, 3, and 4 land use types, field soil revealed more species than trap culture, and mixed cropping (Arable1), fruit cropping (FC), and natural forest (NF) and acacia plantation (AP) harbored more species than did monocrops (Arable2, Arable 3, and Arable 4).

A total of fourteen species, from *Acaulospora, Funneliformis, Glomus* and *Scutellospora* and *Claroideoglomus* and *Gigaspora* were detected in field soil samples but not from trap cultures; whereas four species from *Sclerocystis, Racocetra, Gigaspora* and *Ambispora* were identified from trap cultures but not detected from soil samples (data not shown).

The genera *Glomus* and *Acaulospora*, were the most diversified group represented by the highest number of morpho-species (9 species each), followed by the genera *Funneliformis* and *Gigaspora* (**Table 5).** More species from the dominant genera *Glomus, Funneliformis,* and *Acaulospora* were retrieved from field soil than from trap cultures, but the other genera showed no significant difference in their distribution between the two methods.

Isolation frequency, relative abundance and dominant AMF species

The Isolation frequency (IF) and relative abundance (RA) of AMF species varied greatly among land use types **(Table 5)**. The genera *Claroideoglomus* (all species), *Funneliformis* (*F. mossae*, and *F. geosporum*), *Paraglomus* (*P. occultum*), *Rhizophagus* (*R. diaphanus*) were distributed across all land use types. The genera *Entrophospora* and *Ambispora* were recovered only in two of the land use types. *Gigaspora* species were more limited to arable lands (Arable1, 2, 3 & 4) both in field and trap culture soils. Although the genus *Acaulospora* included a relatively large number of species, their distribution was more limited to perennial cropping systems (FC, NF, and AP).

On the basis of IV (important values), the genera *Claroideoglomus* and *Funneliformis* were categorized into the dominant genera with IV 63 and 56, respectively (Chen *et al.*, 2012). However, the different species under the different AMF genera were categorized into "common" and "Rare" groups with 10<IV< 50 and IV<10), respectively.

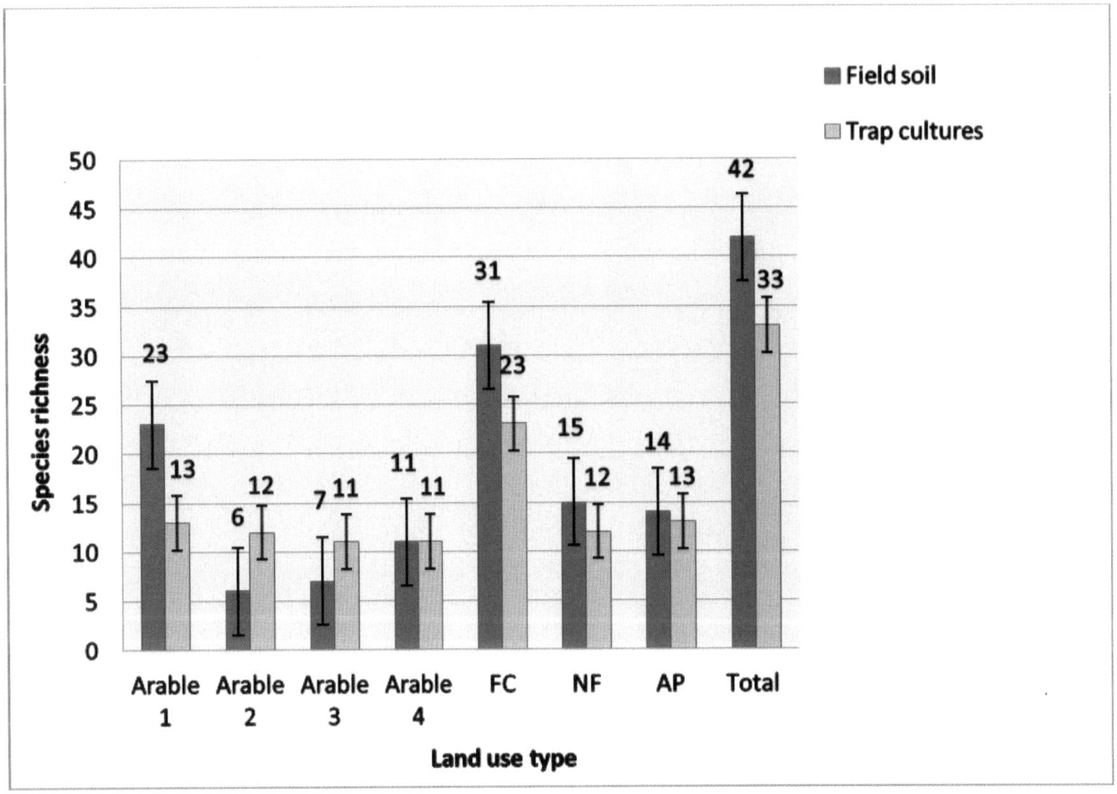

Fig. 2 AMF species richness (numbers on top of bars) in field soil and trap cultures for different land use types at Showa robit, Ethiopia. Vertical bars indicate ± standard errors of means, N= 3. Arable1: Low-input mixed cropping; Arable2: low-input monocropping, sorghum; Arable3: low-input monocropping, maize; Arable4: high-input monocropping, sorghum; FC: fruit cropping; NF: natural forest; AP: acacia plantation.

AMF species richness and diversity across land use types

AMF species richness varied among different land use types (**Fig. 2**). It ranged from 6-31 species in field soil samples and from 11-23 species in trap culture, respectively. Species richness was the highest in FC (31) followed by Arable1 (23), NF (15), AP (14) and the others (6-11) in the field soil. Likewise, the greatest species richness from trap cultures was observed in FC (23) but the others were not significantly different from one another (13-11). On average, the species richness recorded in FC, Arable1, and NF in field soil was five, four and three times greater than in the monocropped arable fields Arable2 and Arable3.

AMF diversity, expressed by the Shannon-Weaver diversity index also varied among different land use types (**Table 6**). Values for Shannon–Wiener diversity index, species dominance and species evenness were 2-2.64, 0.1-0.16 and 0.53-0.88, respectively. The highest Shannon-Wiener diversity, the lowest dominance and the lowest evenness were recorded from fruit crops, whereas the lowest diversity index, the highest dominance and the highest evenness were recorded from the monocropped fields (Arable 2, 3 & 4). In general, AMF diversity was lower in the trap cultures than in the field samples, whereas there was no significant difference in species dominance (Simpson's index) between the soil and the trap culture samples (**Table 6**).

Table 5: Community structure, isolation frequency (IF) and relative abundance (RA) and importance values (IV) of AMF species in soil and trap culture on different land use types from Showa robit, Ethiopia.

AMF genera	No of species	IF (%)	RA (%)	IV (%)	Status	AMF species	IF (%)	RA (%)	IV (%)	Status	Occurrence in land use types
Claroideoglomus	4	92.9	34.7	63.8	Dominant	*Cl. claroideum*	80.5	17.5	49	Common	All
						Cl. etunicatum	61.5	11	36.2	Common	All
						Cl. luteum	58	7.5	32.8	Common	All
Funneliformis	6	91.3	21.6	56.4	Dominant	*Funneliformis mossae*	71.5	10.5	41	Common	All
						*F. caledonium**	49	9.7	29.4	Common	All except AP
						F. geosporum	47	5.9	26.5	Common	All
Glomus	9	80.2	16.8	48.5	Common	*Glomus sp2*	42.5	7.2	24.9	Common	All except A3 &A4
						G. aggregatum	45	5.2	25.1	Common	All except A4
Paraglomus	*1*	27.3	4.06	15.7	Common	*Paraglomus occultum*	27.5	4	15.7	Common	All
Rhizophagus	*2*	34	3.8	18.9	Common	*Rhizophagus diapahanus*	30.8	3.4	17.1	Common	All
Acaulospora	*9*	21	2.8	11.6	Common	*A. scrobiculata*	6.8	0.7	3.7	Rare	A1, FC, NF, AP
Gigaspora	*5*	35	6.5	20.8	common	*Gigaspora gigantea*	25	4.8	14.9		A1, A2,A3,A4,
Diversispora	*1*	6.8	0.87	3.83	Rare	*Diversipora epigaea*	6.9	0.8	3.8	Rare	A1, FC, NF, AP
Septoglomus	*1*	16.5	3	9.7	Rare	*Septoglomus constrictum*	16.6	3	9.8	Rare	A1, FC, NF, AP
Pacispora	*1*	7.7	0.8	4.3	Rare	*Pacispora scintillans*	8	0.8	4.4	Rare	A1, FC, NF, AP
*Scutellospora**	*2*	12.7	2.8	7.75	Rare	*S. pellucida**	10.3	1.7	6	Rare	A1, NF, AP
Racocetra	*2*	8.3	1.5	4.9	Rare	*Racocetra gregaria*	3.6	0.5	2	Rare	A4, FC, NF
*Sclerocystis***	*1*	15.6	2	8.8	Rare	*Sclerocystis sinuosa***	17.2	2	9.6	Rare	A1, FC, NF
Entrophospora	*1*	8	0.6	4.4	Rare	*E. nevadensis*	8.4	0.65	4.5	Rare	A2, FC
Ambispora **	*2*	12.9	1.4	7.15	Rare	*Ambispora fennica***	11	1.3	6.15	Rare	A3, FC

*Only in field soil; **only in trap culture

Table 6. Diversity indices of AMF community in different land use types of Showa robit, Ethiopia.

	Shannon_H			Dominance _D			Evenness_ e^H/S		
Land use type	Field soil	Trap culture	Mean	Field soil	Trap culture	Mean	Field soil	Trap culture	Mean
Arable1	2.54	2.27	2.41	0.12	0.13	0.12	0.55	0.74	0.65
Arable2	1.74	2.28	2.01	0.19	0.12	0.15	0.94	0.81	0.88
Arable3	1.8	2.19	2	0.18	0.14	0.16	0.87	0.81	0.84
Arable4	2.25	2.14	2.2	0.13	0.15	0.14	0.86	0.77	0.82
Fruit cropping	2.81	2.46	2.64	0.09	0.11	0.1	0.54	0.51	0.53
Natural forest	2.61	2.1	2.36	0.08	0.15	0.11	0.91	0.68	0.8
Acacia plantation	2.38	2.21	2.3	0.11	0.13	0.12	0.77	0.7	0.74

Arable1: low-input mixed cropping; Arable2: low-input monocropping, sorghum; Arable3: low-input monocropping, maize; Arable4: high-input monocropping, sorghum.

DISCUSSION

Spore abundance

AMF species diversity and spore abundance were studied in soil from different land use types in a low-land (sub-moist warm) agro-ecosystem, Ethiopia. The spore densities recovered through direct count from soils of all land use types varied between 2.8 spores g^{-1} and 6.1 spores g^{-1} of soil (**Table 3**). Trap cultures established from the same land use types showed higher spore numbers, 2.5-11.4 spores g^{-1} trap culture soil. The numbers of spores recovered from fruit crops and acacia plantation were 2-4 times higher in the trap cultures than when counting directly from the soil. In general, trap culturing enhanced spore abundance but decreased AMF species richness. The monocroped land use types, where the number of species recovered in the trap culture were higher or similar, were an exception from this main rule.

There was also an inverse relation between soil P content and spore density in the different land use types. The highest spore numbers in both field and trap cultures were obtained from FC associated with low P content and the lowest from natural forest and arable 4 characterized by high soil P content. Similarly, other studies in Finnish and Swedish soil showed negative relationship between spore density and P content. This indicates that certain AMF species are induced to sporulate abundantly under low P availability in the soil (Mårtensson and Carlgren, 1994; Kahiluoto *et al.*, 2001).

Mycorrhizal inoculum potential

The MIP bioassay showed that fruit cropping (FC) and mixed cropping in Arable1 were colonized by higher percentage of AMF than the relatively high-input sorghum monocroped field (Arable4), low-input sorghum monocropped field (Arable2) and natural forest (NF) (**Table 4**). In general, there was a slight, but not significant positive correlation between spore density and hyphal colonization both in the soil and trap culture. However, vesicular colonization was strongly correlated with spore density ($r=0.94$, $P<0.01$). This result is consistent with studies from southeast Spain (Azcón-Aguilar *et al.*, 2003) where it was found that the numbers of spores of AM fungal species are the propagule sources which were best correlated with the total mycorrhizal potential in the rhizosphere of the target plant species from Mediterranean shrublands.

The highest values of MIP in low-input mixed cropping and organically managed fruit cropping can be related to higher plant species diversity compared to the monocrops indicating that AMF colonized roots of different plants species are the major sources of propagules that would result in higher MIP values. Several studies also show higher levels of AMF root colonization under organic management and low input mixed cropping system than in monocropping with maize and other crops (Gosling *et al.*, 2010; Verbruggen *et al.*, 2010; Bedini *et al.*, 2013). However, contrary to this result, Purin *et al.* (2006) obtained no differences in MIP values between conventional and organic apple orchards in Brazil. A study conducted among different cropping systems and land use types in Kenya showed a significantly higher AMF inoculum potential in maize-bean intercropping systems than in maize or wheat monocrops in both dry and wet regions (Muchane *et al.*, 2012).

AMF community composition

A total of 42 and 33 AMF morphospecies belonging to 15 genera and 8 families were identified from soil and trap cultures, respectively (**Fig. 2**). This result is quite similar to a study of different cropping systems in Sudan (Abdelhalim *et al.*, 2012), in which 42 AMF species belonging to 12 genera in 8 families were discovered. The genera *Glomus*, *Funneliformis*, *Septoglomus*, *Claroideoglomus*, *Entrophospora*, *Acaulospora*, *Paraglomus*, *Diversispora*, *Pacispora*, and *Ambispora* were commonly detected in both studies. However, the genera *Gigaspora*, *Rhizophagus*, *Racocetra*, *Sclerocystis*, and *Scutellospora* were not identified from Sudan, and the genera *Archaeospora* and *Kuklospora*, were not detected in this study.
The AMF species diversity observed in this study was much higher than the 17 species identified in *Acaulosporaceae* (5), *Glomeraceae* (4), *Gigasporaceae* (5) and others (3) from different land use types in Kenya (Jewfa *et al.*, 2009). This may be related to the diversity and the type of plants sampled from the land use types. Other studies have also showed that coexisting plant species within a habitat are associated with divergent AMF communities, showing that host preference has a strong influence on AMF community composition in soil (Vandenkoornhuyse *et al.*, 2002; Scheublin *et al.*, 2004).

Although trap culturing enhanced spore abundance it reduced AMF species richness compared with the field soil samples. Similarly, Chaturvedi *et al.* (2012) found that AMF diversity in trap cultures of one year was decreased from 50 to 21, although spore abundance was higher. Tchabi *et al.* (2008) reported that out of a total of 59 AMF species detected in soils of different ecological zones of West Africa, only seven had sporulated after 10 and 24 months of trap culturing.

Isolation frequency, relative abundance and dominant AMF species

Claroideoglomus and *Funneliformis* were dominant genera according to Chen *et al.* (2012), because they were found in all land use types. The genera *Glomus*, *Paraglomus*, *Rhizophagus*, *Acaulospora* and *Gigaspora* were categorized as common. It is interesting to note that more than 50% of the genera were classified as rare. Previous reports have also shown that *Glomus* was dominant in other agroecological regions of Ethiopia (Muleta *et al.*, 2008; Birhane *et al.*, 2010).

The genera *Glomus*, *Funneliformis*, and *Claroideoglomus* were also reported to be dominant in Cameroon (Snoeck *et al.*, 2010) and other sub-Saharan regions, in North Côte d'Ivoire (Nandjui *et al.*, 2013), in different land use types of Kenya (Jefwa *et al.*, 2009, 2012), in the Namibia desert (Stutz *et al.*, 2000), in natural and cultivated savannas of Benin, West Africa (Tchabi *et al.*, 2008), in selected crops in the White Nile State, Central Sudan (Abdelhalim *et al.*, 2013) and in temperate agroecosystems in Europe (Oehl *et al.*, 2003). The high incidence of *Glomus* and *Funneliformis* spp. has been associated with their capacity to produce more spores in a shorter time than genera such as *Gigaspora* and *Scutellospora* (Bever *et al.*, 1996: Oehl *et al.*, 2009). These species could therefore, be selected for future studies as AMF inocula after testing their compatibility with different crops and checking their persistence in the field.

AMF species richness and diversity

The number of AMF morphospecies recovered from the fruit cropping system and the mixed cropping system (Arable1) was almost double that of the number of morphospecies collected from each of the other land use types from monocrop fields (Arable 2, Arable 3, and arable 4) and woody vegetation (natural forest, and acacia plantation) (**Fig. 2**). In general, lower AMF species diversity was recorded in high–input (Arable 4) and low-input monocropped fields compared to organically managed fruit crops or low-input mixed cropping (Arable1) **(Fig. 2).**

Our study also showed that the AMF species diversity (23 species) of the mixed cropping system (Arable 1) was much higher than the 12 AMF species reported from similar maize and sesbania intercrops from Southern Malawi (Jefwa *et al.*, 2006). However, the AMF diversity of monocrops (Arable 2 and Arable 3) was almost similar to the 12 AMF species collected from indigenous forest to croplands in Southern Kenya (Jefwa *et al.*, 2012) and from maize monocrops in Southern Malawi (Jefwa *et al.*, 2006). Similar pattern of AMF diversity (15-17morphospecies) was also reported from crop land, fallow land, natural forest and tree plantations in the high altitude regions of Kenya (Jefwa *et al.*, 2009) and from grassland, woodland and intensified monocropping systems in Maasai Mara ecosystems in Kenya (Muchane *et al.*, 2012).

This study showed no clear impact of soil P on the diversity of AM fungi. Accordingly, Gosling *et al.* (2013) suggested that host species is more important than soil P for determining AM diversity, except at the highest P concentration. We observed a strong positive correlation between AMF species richness and spore density ($r=0.84$, $P<0.05$). AMF species richness also correlated strongly positively with soil organic carbon and total nitrogen both in field soil and trap cultures ($P<0.05$). Other studies have also shown that spore density and species richness are usually positively correlated with soil organic carbon contents and soil pH ($P<0.05$) (Tchabi *et al.*, 2008).

AMF diversity indices

The highest diversity index value was recorded from the FC land use type both in field soil and trap cultures. This result is similar to previous reports which show that organic systems have higher AMF community diversity indices than conventional or monocropping systems (Helgason *et al.*, 1998; Oehl *et al.*, 2003; Verbruggen *et al.*, 2010). It has been suggested that conventional farming systems may select for a small set of generalist AMF species (Helgason *et al.*, 2007; Verbruggen *et al.*, 2010), while organic farming systems are characterized by high species diversity (Mäder *et al.*, 2002).

CONCLUSION AND RECOMMENDATION

This study showed high AM fungal diversity, but also high variation in AM fungal community composition among seven land use types in the humid and semi-arid soil of Showa robit, Ethiopia. The difference in AMF community structure was most closely related to a variety of biotic and abiotic factors, including various aspects of agricultural management practice and land use. Markedly higher numbers of AMF species and higher rate of mycorrhizal infectivity potential were obtained in FC and Arable1 compared with the other land use types. This clearly implies that organic farming and diversification of crops in agriculture is a more sustainable land use system for enhancing biological soil qualities, including maintenance of AMF diversity, than highly fertilized monocropping systems.

The study also showed that *Claroideoglomus* and *Funneliformis* were the dominant genera in all land use types in both trap culture and field soil. It also showed that some AMF species could be missed when studying either soil or trap cultures implying that there is a need to use both methods for getting a full picture of the AMF species diversity in a study area. Future studies should be focused on the dominant species for further selection of AMF inocula for enhancing productivity in different cropping system.

In this study, the AM fungal diversity and community composition analyses relied on an assessment of spore morphotypes for identification. If it is complemented with molecular identification of AMF species directly from plant roots it can fully show the heterogeneity of the organisms in relation to land use types. Except the AMF genera and species that dominated all land use types and plants, the occurrence of some rare species in specific land use types should also be studied to fully realize their role in nutrient and water uptake and protection against plant pathogens and in improvement of crop productivity.

Acknowledgments

This work was supported by the School of Graduate study, Addis Ababa University. We thank Rocío Vega-Frutis for her support in translating the English summary of this work to Spanish. We also wish to thank Dr Jonathan Robinson for valuable revision of the English manuscript.

REFERENCES

Abdelhalim, T. S., Finckh, M. R., Babiker, A. G., Oehl, F. 2013. Species composition and diversity of arbuscular mycorrhizal fungi in White Nile state, Central Sudan. Archives of Agronomy and Soil Science. 60:377-391.

Azcón-Aguilar, C., Palenzuela, J., Roldán, A., Bautista, S., Vallejo, R., Barea, J. M. 2003. Analysis of the mycorrhizal potential in the rhizosphere of representative plant species from desertification-threatened Mediterranean shrublands. Applied Soil Ecology. 22:29-37.

Bedini, S., Avio, L., Sbrana, C., Turrini, A., Migliorini, P., Vazzana, C., Giovannetti, M. 2013. Mycorrhizal activity and diversity in a long-term organic Mediterranean agroecosystem. Biology and Fertility of Soils. 49:781-790.

Bever, J. D., Morton, J. B., Antonovics, J., Schultz, P. A. 1996. Host-dependent sporulation and species diversity of arbuscular mycorrhizal fungi in mown grassland. The Journal of Ecology. 84:71-82.

Birhane, E., Kuyper, T. W., Sterck, F. J., Bongers, F. 2010. Arbuscular mycorrhizal associations in *Boswellia papyrifera* (frankincense-tree) dominated dry deciduous woodlands of Northern Ethiopia. Forest Ecology and Management. 260:2160-2169.

Bossio, D. A., Girvan, M. S., Verchot, L., Bullimore, J., Borelli, T., Albrecht, A., Scow, K. M., Ball, A. S., Pretty, J. N., Osborn, A. M. 2005. Soil microbial community response to land use change in an agricultural landscape of western Kenya. Microbial Ecology. 49:50-62.

Brundrett, M., Bougher, N., Dell, B., Grove, T., Malajczuk, N. 1996. Working with mycorrhizas in forestry and agriculture. ACIAR Monograph 32. Australian Centre for International Agricultural Research, Canberra.

Burrows, R. L., Pfleger, F. L. 2002. Arbuscular mycorrhizal fungi respond to increasing plant diversity. Canadian Journal of Botany. 80:120-130.

Chaturvedi, S., Tewari, V., Sharma, S., Oehl, F., Wiemken, A., Prakash, A., Sharma, A. K. 2012. Diversity of arbuscular mycorrhizal fungi in oak-pine forests and agricultural land prevalent in the Kumaon Himalayan hills, Uttarakhand, India. British Microbiology Research Journal. 2:82-96.

Chen, K., Liu, W., Guo, S., Liu, R., Li, M. 2012. Diversity of arbuscular mycorrhizal fungi in continuous cropping soils used for pepper production. African Journal of Microbiology Research 6:2469 – 2474

Chifflot, V., Rivest, D., Olivier, A., Cogliastro, A., Khasa, D. 2009. Molecular analysis of arbuscular mycorrhizal community structure and spores distribution in tree-based intercropping and forest systems. Agriculture, Ecosystem and Environment. 131:32-39.

Douds, D. D., Janke, R. R., Peters, S. E. 1993. VAM fungus spore populations and colonization of roots of maize and soybean under conventional and low-input sustainable agriculture. Agriculture, Ecosystem and Environment. 43:325-335.

Estaún, V., Savé, R., Biel, C. 1997. AM inoculation as a biological tool to improve plant revegetation of a disturbed soil with *Rosmarinus officinalis* under semi-arid conditions. Applied Soil Ecology. 6:223-229.

FAO. 2007. State of the World's Forests Report.

Garedew, E., Sandewall, M., Söderberg, U., Campbell, B. M. 2009. Land-use and land-cover dynamics in the central Rift Valley of Ethiopia. Environmental Management. 44:683-694.

Gee, G. W., Bauder, J. W. 1986. Particle size analysis. In: Klute, A. (ed),. Method of soil analysis, part 1: Physical and mineralogical methods, Soil Science Society of America, Madison, Wisconsin USA. 383- 411.

Gerdemann, J. W., Nicolson, T. H. 1963. Spores of mycorrhizal *Endogone* species extracted from soil by wet sieving and decanting. Transactions of the British Mycological Society. 46:235-244.

Getachew, F., Abdulkadir, A., Lemenih, M., Fetene, A. 2012. Effects of different land uses on soil physical and chemical properties in Wondo Genet area, Ethiopia. New York Science Journal. 5:110-118.

Girmay, G., Singh, B. R., Mitiku, H., Borresen T., Lal, R. 2008. Carbon stocks in Ethiopian soils in relation to land use and soil management. Land Degradation & Development. 19:351-367.

Gosling, P., Mead, A., Proctor, M., Hammond, J. P., Bending, G. D. 2013. Contrasting arbuscular mycorrhizal communities colonizing different host plants show a similar response to a soil phosphorus concentration gradient. New Phytologist. 198:546-556.

Gosling, P., Ozaki, A., Jones, J., Turner, M., Rayns, F., Bending, G. D. 2010. Organic management of agricultural systems results in a rapid increase in colonization potential and spore populations of arbuscular mycorrhizal fungi. Agriculture, Ecosystem and Environment. 139:273-279.

Hartemink, A. E., Veldkamp, T., Bai, Z. 2008. Land cover change and soil fertility decline in tropical regions. Turkish Journal of Agriculture & Forestry. 32:195-213.

Helgason, T., Daniell, T. J., Husband, R., Fitter, A. H., Young, J. P. W. 1998. Ploughing up the wood-wide web? Nature. 394:431.

Helgason, T., Merryweather, J. W., Young, J. P. W., Fitter, A. H. 2007. Specificity and resilience in the arbuscular mycorrhizal fungi of a natural woodland community. Journal of Ecology. 95:623 -630.

Hinds, A., Lowe, L. E. 1980. Ammonium-N determination. Soil nitrogen. Berthelot reaction. Soil Science and Plant Analysis 11, 469– 475.

Jansa, J., Mozafar, A., Anken, T., Ruh, R., Sanders, I. R., Frossard, E. 2002. Diversity and structure of AMF communities as affected by tillage in a temperate soil. Mycorrhiza. 12:225-234.

Jefwa, J. M., Mung'atu, J., Okoth, P., Muya, E., Roimen, H., Njuguini, S. 2009. Influence of land use types on occurrence of arbuscular mycorrhiza fungi in the high altitude regions of Mt. Kenya. Tropical and Subtropical Agroecosystems. 11:277-290.

Jefwa, J. M., Okoth, S., Wachira, P., Karanja, N., Kahindi, J., Njuguini, S., Ichami, S., Mung'atu, J., Okoth, P., Huising, J. 2012. Impact of land use types and farming practices on occurrence of arbuscular mycorrhizal fungi (AMF) Taita-Taveta district in Kenya. Agriculture, Ecosystem and Environment. 157:32-39.

Jefwa, J. M., Sinclair, R., Maghembe, J. A. 2006. Diversity of glomale mycorrhizal fungi in maize/sesbania intercrops and maize monocrop systems in southern Malawi. Agroforestry System. 67:107-114.

Johnson, D., Vandenkoornhuyse, P. J., Leake, J. R., Gilbert, L., Booth, R. E., Grime, J. P., Young J. P. W., Read, D. J. 2003. Plant communities affect arbuscular mycorrhizal fungal diversity and community composition in grassland microcosms. New Phytologist. 161:503-515.

Kahiluoto, H., Ketoja, E., Vestberg, M., Saarela, I. 2001. Promotion of AM utilization through reduced P fertilization 2. Field studies. Plant and Soil. 231:65-79.

Lemenih, M., Karltun, E., Olsson, M. 2005. Soil organic matter dynamics after deforestation along a farm field chronosequence in southern highlands of Ethiopia. Agriculture, Ecosystems and Environment. 109:9-19.

Li, L.-F., Li, T., Zhao, Z.W. 2007. Differences of arbuscular mycorrhizal fungal diversity and community between a cultivated land, an old field, and a never-cultivated field in a hot and arid ecosystem of southwest China. Mycorrhiza. 17:655-665.

Mäder, P., Flieβbach, A., Dubois, D., Gunst, L., Fried, P., Niggli, U. 2002. Soil fertility and biodiversity in organic farming. Science. 296:1694-1697.

Mårtensson, A. M., Carlgren, K. 1994. Impact of phosphorus fertilization on VAM diaspores in 2 Swedish long-term field experiment. Agriculture, Ecosystem and Environment. 47:327-334.

McGonigle, T. P., Miller, M. H., Evans, D. G., Fairchild, G. L., Swan, J. A. 1990. A new method which gives an objective measure of colonization of roots by vesicular-arbuscular mycorrhizal fungi. New Phytologist. 115:495-501.

Morton, J. B. 1991. INVAM newsletters. vol 1-5. West Virginia University, Morgantown.

Morton, J. B., Bentivenga, S. P.,Wheeler, W. W. 1993. Germ plasm in the International Collection of Arbuscular and Vesicular-arbuscular Mycorrhizal Fungi (INVAM) and procedures for culture development, documentation and storage. Mycotaxon. 48:491-528.

Muchane, M. N., Muchane, M., Mugoya, C., Masiga, C. W. 2012. Effect of land use system on arbuscular mycorrhiza fungi in Maasai Mara ecosystem, Kenya. African Journal of Microbiology Research. 6:3904-3916.

Muleta, D., Assefa, F., Nemomissa, S., Granhall, U. 2008. Distribution of arbuscular mycorrhizal fungi spores in soil of southwestern Ethiopia. Biology and Fertility of Soils. 44:653-659.

Nair, P. K. R. 1993. An introduction to agroforestry. Kluwer Academic Publishers, pp.21

Nandjui, J., Voko, D. R. R., Kouadio, A. N. M.-S., Fotso, B., Tano, Y., Zeze, A. 2013. Assessment of the occurrence and abundance of mycorrhizal fungal communities in soils from yam (*Dioscorea* spp.) cropping fields in Dabakala, North Côte d'Ivoire. African Journal of Agricultural Research. 8:5572-5584.

NMA (National Metrological Agency, Ethiopia). 2013. Annual climate bulletin for year 2002-2010.

Oehl, F., Sieverding, E., Ineichen, K., Mäder, P., Boller, T., Wiemken, A. 2003. Impact of land use intensity on the species diversity of arbuscular mycorrhizal fungi in agroecosystems of central Europe. Applied and Environmental Microbiology. 69:2816-2824.

Oehl, F., Sieverding, E., Ineichen, K., Mäder, P., Wiemken, A., Boller, T. 2009. Distinct sporulation dynamics of arbuscular mycorrhizal fungal communities from different agroecosystems in long term microcosms. Agriculture, Ecosystems and Environment. 134:257-268.

Oehl, F., Sieverding, E., Mäder, P., Dubois, D., Ineichen, K., Boller, T., Wiemken, A. 2004. Impact of long-term conventional and organic farming on the diversity of arbuscular mycorrhizal fungi. Oecologia. 138:574-583.

Olsen, S.R., Cole, C.V., Watanabe, F.S., Dean, L. 1954. Estimation of available phosphorus in soils by extraction with sodium bicarbonate. USDA Circular No. 939.

Omar, M. B., Bolland, L., Heather, W. A. 1979. PVA (polyvinyl alcohol). A permanent mounting medium for fungi. Bulletin of the British Mycological Society. 13:31-32.

Purin, S., Filho, O. K., Stürmer, S. L. 2006. Mycorrhizae activity and diversity in conventional and organic apple orchards from Brazil. Soil Biology and Biochemistry. 38:1831-1839.

Schenck, N. C., Perez, Y. 1990. Manual for the identification of VA mycorrhizal Fungi, 3rd edn. Synergistic publications, Gainesville, Fla.

Scheublin, T. R., Ridgway, K. P., Young, J. P. W., van der Heijden, M. G. A. 2004. Non legumes, legumes, and root nodules harbor different arbuscular mycorrhizal fungal communities. Applied and Environmental Microbiology. 70:6240-6246.

Schüßler, A., Walker, C. 2010. The Glomeromycota. A species list with new families and new genera. Published in December 2010 in libraries at The Royal Botanic Garden Edinburgh, The Royal Botanic Garden Kew, Botanische Staatssammlung Munich, and Oregon State University. Electronic version freely available at www.amf-phylogeny.com.

Sieverding, E. 1991. Vesicular-arbuscular mycorrhiza management in tropical agrosystems. GTZ, Eschborn and Hartmut Bremer Verlag, Friedland, 365 pp.

Smith, S. E., Read, D. J. 2008. Mycorrhial symbiosis, 3rd Ed. Elsevier Ltd.

Snoeck, D., Abolo, D., Jagoret, P. 2010. Temporal changes in VAM fungi in the cocoa agroforestry systems of central Cameroon. Agroforestry System. 78:323-328.

Stürmer, S. L., Siqueira, J. O. 2011. Species richness and spore abundance of arbuscular mycorrhizal fungi across distinct land uses in Western Brazilian Amazon. Mycorrhiza. 21:255-267.

Stutz, J. C., Copeman, R., Martin, C. A., Morton, J. B. 2000. Patterns of species composition and distribution of arbuscular mycorrhizal fungi in arid regions of southwestern North America and Namibia, Africa. Canadian Journal of Botany. 78:237-245.

Sýkorová, Z., Wiemken, A., Redecker, D. 2007. Co-occurring *Gentiana verna* and *Gentiana acaulis* and their neighboring plants in two Swiss upper montane meadows harbor distinct arbuscular mycorrhizal fungal communities. Applied and Environmental Microbiology. 73:5426-5434.

Tchabi, A., Coyne, D., Hountondji, F., Lawouin, L., Wiemken, A., Oehl, F. 2008. Arbuscular mycorrhizal fungal communities in sub-Saharan savannas of Benin, West Africa, as affected by agricultural land use intensity and ecological zone. Mycorrhiza. 18:181-195.

Vandenkoornhuyse, P., Husband, R., Daniell, T. J., Watson, I. J., Duck, J. M., Fitter, A. H., Young, J. P. W. 2002. Arbuscular mycorrhizal community composition associated with two plant species in a grassland ecosystem. Molecular Ecology.11:1555-1564.

Verbruggen, E., Röling, W. F. M., Gamper, H. A., Kowalchuk, G. A., Verhoef, H. A., van der Heijden, M. G. A. 2010. Positive effects of organic farming on below-ground mutualists: large-scale comparison of mycorrhizal fungal communities in agricultural soils. New Phytologist. 186:968-979.

Walkley, A., Black, I. A. 1934. An examination of the Degtareff method for determining soil organic matter, and a proposed modification of the chromic acid titration method. Soil Science 37:29–38.

Wardle, D. A., Lavelle, P. 1997. Linkages between soil biota, plant litter quality and decomposition. In Cadisch, G., Giller, K. E, (eds.). Driven by nature: Plant litter quality and decomposition. CAB International, Wallingford (UK). pp.107-125.

Yamato, M., Ikeda, S., Iwase, K. 2009. Community of arbuscular mycorrhizal fungi in drought-resistant plants, *Moringa* spp., in semiarid regions in Madagascar and Uganda. Mycoscience. 50:100 – 105.

Yao, Q., Gao, J.-L., Zhu, H.-H., Long, L.-K., Xing, Q.-X., Chen, J.-Z. 2010. Evaluation of the potential of trap plants to detect arbuscular

Appendices: List, isolation frequency (IF %), relative spore abundance (RA) and important value (IV) of the AMF species identified in field soil and trap culture

Appendix A. field soil

AMF genera and species	IF%								RA%									
	A1	A2	A3	A4	FC	NF	AP	Mean	A1	A2	A3	A4	FC	NF	AP	Mean	IV	Status
Acaulospora	33.3	-	-	33	71.4	11.1	50	28.4	3	-	-	5	9.4	3.1	8.6	4.16	16	Common
A. denticulata Sieverd. & S. Toro	11.1	-	-	33	9.5	-	-	7.7	1	-	-	5	0.9	-	-	0.99	4.3	Rare
A. faveata Trappe & Janos	-	-	-	-	4.7	-	-	0.67	-	-	-	-	0.4	-	-	0.06	0.4	Rare
A. kentinesis (Wu & Liu) Kaonongbua, Morton & Bever	-	-	-	-	9.5	-	-	1.36	-	-	-	-	0.9	-	-	0.13	0.7	Rare
A. rehmii Sieverd. & S. Toro	11.1	-	-	-	4.7	11.1	-	3.84	1	-	-	-	0.4	3.1	-	0.64	2.2	Rare
A. scrobiculata Trappe	-	-	-	-	28.6	-	16.7	6.47	-	-	-	-	2.9	-	2.8	0.81	3.6	Rare
A. spinosa Walker & Trappe	-	-	-	-	19	-	-	2.71	-	-	-	-	1.9	-	-	0.27	1.5	Rare
A. splendida Sieverd, Chaverri & Rojas	11.1	-	-	-	-	-	-	1.59	1	-	-	-	-	-	-	0.14	0.9	Rare
A. tuberculata Janos & Trappe	-	-	-	-	4.7	-	33.3	5.43	-	-	-	-	0.4	-	5.7	0.87	3.2	Rare
Acaulospora sp1	-	-	-	-	9.5	-	-	1.36	--	-	-	-	0.9	-	-	0.9	1.1	Rare
Claroideoglomus	100	66.7	33.3	100	100	100	100	85.7	26.3	22.2	7.6	35	31	28.1	25.7	25.1	55	Dominant
C. claroideum (Schenck & Sm.) Walker & Schuessler	100	33.3	33.3	100	100	44.4	16.7	61.1	15.2	16.6	7.6	25	20.7	12.5	2.8	14.3	38	Common
C. etunicatum (Becker & Gerd.) Walker & Schuessler	22.2	-	-	67	42.8	33.3	83.3	35.5	2	-	-	10	4.4	9.3	14.3	5.7	21	Common
C. lamellosum (Dalpé, Koske & Tews) Walker & Schuessler	22.2	-	-	-	-	-	-	3.17	2	-	-	-	-	-	-	0.29	1.7	Rare
C. luteum (Kenn, Stutz & Morton) Walker & Schuessler	66.6	33.3	-	-	57.1	22.2	50	32.7	6	16.6	-	-	5.9	6.2	8.6	6.19	19	Common
Diversispora	11.1	-	-	-	28.6	22.2	-	8.84	1	-	-	-	2.9	6.2	-	1.44	5.1	Rare
D. epigaea (Daniels & Trappe) Walker & Schuessler	11.1	-	-	-	28.6	22.2	-	8.84	1	-	-	-	2.9	6.2	-	1.44	5.1	Rare
Entrophospora	-	-	-	-	47.6	-	-	6.8	-	-	-	-	4.9	-	-	0.7	3.8	Rare
E. nevadensis J. Palenzuela, N. Ferrol, Azcón-Aguilar & Oehl	-	-	-	-	47.6	-	-	6.8	-	-	-	-	4.9	-	-	0.7	3.8	Rare
Funneliformis	100	100	100	100	100	66.7	50	88.1	39.4	33.3	38.5	25	30	18.7	11.4	28	58	Dominant

F. badium (Oehl, Redecker & Sieverd.) Walker & Schuessler	-	-	-	-	4.7	-	-	0.67	-	-	-	-	0.4	-	-	0.06	0.4	Rare
*F. caledonium (*Nicolson & Gerd.) Walker & Schuessler	55.5	66.6	100	67	42.8	11.1	-	49	5	22.2	23	10	4.4	3.1	-	9.67	29	Common
F. coronatum (Giovann.) Walker & Schuessler	-	-	-	-	14.3	-	-	2.04	-	-	-	-	1.5	-	-	0.21	1.1	Rare
F. geosporum (Nicolson & Gerd.) Walker & Schuessler	88.8	-	33.3	33	80.9	22.2	50	44.1	8.1	-	7.6	5	8.4	6.2	11.4	6.67	25	Common
F. mosseae (Oehl, Redecker & Sieverd.)Walker & Schuessler	100	33.3	33.3	67	100	33.3	-	52.4	26.3	11.1	7.6	10	14.8	9.3	-	11.3	32	Common
F. verruculosusm (Błaszk.) C. Walker & Schuessler	-	-	-	-	4.7	-	-	0.67	-	-	-	-	0.4	-	-	0.06	0.4	Rare
Glomus	100	66.7	66.7	67	100	55.6	100	79.4	17.2	22.2	23	15	10.8	15.6	25.7	18.5	49	Common
Gl. aggregatum Schenck & Sm.	44.4	-	-	-	57.1	11.1	16.7	18.5	4	-	-	-	5.9	3.1	2.8	2.26	10	Common
Gl. albidum N.C. Schenck & G.S. Sm.	-	-	-	-	9.5	-	-	1.36	-	-	-	-	0.9	-	-	0.13	0.7	Rare
Gl. hoi Berch & Trappe	11.1	-	-	-	-	-	-	1.59	1	-	-	-	-	-	-	0.14	0.9	Rare
Gl. microaggregatum Koske, Gemma & Olexia	-	-	-	-	-	-	16.7	2.39	-	-	-	-	-	-	2.8	0.4	1.4	Rare
Gl. microcarpum Tul. & Tul.	-	-	-	-	19	-	-	2.71	-	-	-	-	1.9	-	-	0.27	1.5	Rare
Gl. monosporum Gerd. & Trappe	11.1	-	-	33	-	-	-	6.34	1	-	-	5	-	-	-	0.86	3.6	Rare
Gl. tortuosum N.C. Schenck & G.S. Sm.	11.1	-	-	-	-	-	-	1.59	1	-	-	-	-	-	-	0.14	0.9	Rare
Glomus sp1(#2) sporocarpic , thick wall , smooth (80-110μm)	-	-	-	-	4.7	-	100	15	-	-	-	-	0.4	-	17.1	2.5	8.8	Rare
Glomus sp2(#3) red brown geosporum like	100	66.7	66.7	67	14.3	44.4	16.7	53.6	10	22.2	23	10	1.5	12.5	2.8	11.71	33	Common
Rhizophagus	11.1	-	-	67	14.3	22.2	33.3	21.1	1	-	-	10	1.5	6.2	5.7	3.49	12	Common
R. diaphanus (Morton & Walker) Walker & Schuessler	11.1	-	-	67	4.5	22.2	33.3	19.7	1	-	-	10	0.4	6.2	5.7	3.33	12	Common
R. fasciculatus (Thaxt.) Walker & Schuessler	-	-	-	-	9.5	-	-	1.36	-	-	-	-	0.9	-	-	0.13	0.7	Rare
Septoglomus	22.2	-	-	-	46.7	11.1	50	18.6	2	-	-	-	4.9	6.2	17.1	4.31	11	Common
S. constrictum (Trappe) Sieverd., Silva& Oehl	22.2	-	-	-	46.6	11.1	50	18.6	2	-	-	-	4.9	6.2	17.1	4.31	11	Common

Racocetra	-	-	-	-	4.7	11.1	-	2.26	-	-	-	-	0.4	3.1	-	0.5	1.4	Rare
R. gregaria (Schenck & Nicolson) Oehl, Souza & Sieverd.	-	-	-	-	4.7	11.1	-	2.26	-	-	-	-	0.4	3.1	-	0.5	1.4	Rare
Gigaspora	66.7	66.7	-	67	14.3	-	-	30.6	6	22.2	-	10	1.5	-	-	5.67	18	Common
Gi. albida Schenck & Sm	22.2	-	-	-	-	-	-	3.17	2	-	-	-	-	-	-	0.29	1.7	Rare
Gi. gigantea (Nicolson & Gerd.) Gerd. & Trappe	44.4	66.7	-	33	-	-	-	20.6	4	22.2	-	5	-	-	-	4.46	13	Common
Gi. margarita Becker & Hall	-	-	-	33	4.7	-	-	5.43	-	-	-	5	0.4	-	-	0.77	3.1	Rare
Gigaspora sp.	-	-	-	-	9.5	-	-	1.36	-	-	-	-	0.9	-	-	0.13	0.7	Rare
Scutellospora	33.3	-	33.3	-	-	22.2	16.7	12.7	3	-	7.6	-	-	6.2	2.8	2.8	7.8	Rare
S. cerradensis Spain & Miranda	-	-	33.3	-	-	-	-	4.7	-	-	7.6	-	-	-	-	1.09	2.9	Rare
S. pellucida (Nicolson & Schenck) Walker & Sanders	33.3	-	-	-	-	22.2	16.7	10.3	3	-	-	-	-	6.2	2.8	1.71	6	Rare
Pacispora	11.1	-	-	-	23.8	-	-	4.99	1	-	-	-	2.5	-	-	0.5	2.7	Rare
Pacispora scintillans (Rose & Trappe) Walker, Vestberg & Schuessler	11.1	-	-	-	23.8	-	-	4.99	1	-	-	-	2.5	-	-	0.5	2.7	Rare
Paraglomus	11.1	-	66.7	-	-	22.2	16.7	16.7	1	-	23	-	-	6.2	2.8	4.71	11	Common
Paraglomus occultum (Walker) Morton & Redecker	11.1	-	66.7	-	-	22.2	16.7	16.7	1	-	23	-	-	6.2	2.8	4.71	11	Common

Appendix B. Trap culture

(A1)Arable1: low-input mixed cropping; (A2) Arable2: low-input monocropping, sorghum; (A3)Arable3: low-input monocropping, maize; (A4)Arable4: high-input monocropping, sorghum; FC: fruit cropping; NF: natural forest; AP: acacia plantation

AMF genera and species	IF%								RA%									
	A1	A2	A3	A4	FC	NF	AP	Mean	A1	A2	A3	A4	FC	NF	AP	Mean	IV	Status
Acaulospora **Trappe & Gerd.**	33.3	-	33.3	-	14.3	11.1	-	**13.1**	2.4	-	4.2	-	1	1.8	-	**1.343**	**7.2**	**Rare**
A. denticulata Sieverd. & S. Toro	-	-	33.3	-	-	-	-	**4.76**	-	-	4.2	-	-	-	-	**0.6**	**2.7**	**Rare**
A. rehmii Sieverd. & S. Toro	-	-	-	-	4.7	-	-	**0.67**	-	-	-	-	0.3	-	-	**0.043**	**0.4**	**Rare**
A. scrobiculata Trappe	33.3	-	-	-	4.7	11.1	-	**7.01**	2.4	-	-	-	0.3	1.8	-	**0.643**	**3.8**	**Rare**
Acaulospora sp.	-	-	-	-	4.7	-	-	**0.67**	-	-	-	-	0.3	-	-	**0.043**	**0.4**	**Rare**
Claroideoglomus	100	100	100	100	100	100	100	**100**	51.2	45.6	56.5	20.7	47.1	46.3	42.7	**44.3**	**72**	**Dominant**
C. claroideum (Schenck & Sm.) Walker & Schuessler	100	100	100	100	100	100	100	**100**	21.4	17.5	25	13.8	21.6	25.5	21.3	**20.87**	**60**	**Dominant**
C. etunicatum (Becker & Gerd.) Walker & Schuessler	100	100	100	66.7	100	44.4	100	**87.3**	21.4	17.5	16.7	6.9	17	12.7	17.9	**15.73**	**52**	**Dominant**
C. luteum (Kenn, Stutz & Morton) Walker & Schuessler	100	100	100	100	100	33.3	50	**83.3**	7.1	10.5	12.5	10.3	8	7.3	3.4	**8.443**	**46**	**Common**
Diversispora	-	-	-	-	-	-	33.3	**4.76**	-	-	-	-	-	-	2.2	**0.314**	**2.5**	**Rare**
D. epigaea (Daniels & Trappe) Walker & Schuessler	-	-	-	-	-	-	33.3	**4.76**	-	-	-	-	-	-	2.2	**0.314**	**2.5**	**Rare**
Entrophospora	-	33.3	-	-	33.3	-	-	**9.51**	-	1.7	-	-	2.6	-	-	**0.614**	**5.1**	**Rare**
E. nevadensis Błaszk., Madej & Tadych; J. Palenzuela, . N. Ferrol, Azcón-Aguilar & Oehl	-	33.3	-	-	33.3	-	-	**9.51**	-	1.7	-	-	2.6	-	-	**0.614**	**5.1**	**Rare**
Funneliformis	100	100	100	100	100	88.8	100	**98.4**	9.7	15.8	13	13.8	14.5	14.8	24.7	**15.19**	**57**	**Dominant**
F. badium (Oehl, Redecker & Sieverd.) Walker & Schuessler	-	-	-	-	-	11.1	-	**1.59**	-	-	-	-	-	1.8	-	**0.257**	**0.9**	**Rare**
F. geosporum (Nicolson & Gerd.) Walker & Schuessler	33.3	100	33.3	33.3	38.1	11.1	100	**49.9**	2.4	5.3	4.2	3.4	2.6	1.8	15.7	**5.057**	**27**	**Common**

F. mosseae (Oehl, Redecker & Sieverd.)Walker & Schuessler	100	100	66.7	100	100	66.6	100	**90.5**	7.1	10.5	8.3	10.3	11.7	10.9	8.9	**9.671**	**50**	**Common**
Glomus	100	100	33.3	33.3	100	100	100	**80.9**	14.6	21	4.2	3.4	19.5	27.7	14.6	**15**	**48**	**Common**
Gl. aggregatum Schenck & Sm.	100	66.6	33.3	-	100	100	100	**71.4**	7.1	3.5	4.2	-	11	21.8	8.9	**8.071**	**40**	**Common**
Gl. hoi Berch & Trappe	-	100	-	33.3	52.4	-	-	**26.5**	-	12.3	-	3.4	4.3	-	-	**2.857**	**15**	**Common**
Gl. microaggregatum Koske, Gemma & Olexia	-	-	-	-	4.7	-	-	**0.67**	-	-	-	-	0.3	-	-	**0.043**	**0.4**	**Rare**
Gl. microcarpum Tul. & Tul.	-	-	-	-	-	-	16.7	**2.39**	-	-	-	-	-	-	1.1	**0.157**	**1.3**	**Rare**
Gl. tortuosum N.C. Schenck & G.S. Sm.	66.7	-	-	-	19	-	-	**12.2**	4.8	-	-	-	1.3	-	-	**0.871**	**6.5**	**Rare**
Glomus sp1 sporocarpic , thick wall , smooth (80-110 µm)	-	-	-	-	-	-	50	**8.33**	-	-	-	-	-	-	3.4	**0.486**	**4.4**	**Rare**
Glomus sp2 red brown geosporum like	33.3	100	-	-	33.3	33.3	16.7	**30.9**	2.4	5.3	-	-	2.3	5.5	1.1	**2.371**	**17**	**Common**
Rhizophagus	-	100	100	-	28.6	-	100	**46.9**	-	5.3	13	-	2	-	8.9	**4.171**	**26**	**Common**
R. diaphanus (Morton & Walker) Walker & Schuessler	-	100	66.7	-	28.6	-	100	**42.2**	-	5.3	8.3	-	2	-	8.9	**3.5**	**23**	**Common**
R. fasciculatus (Thaxt.) Walker & Schuessler	-	-	33.3	-	-	-	-	**4.76**	-	-	4.2	-	-	-	-	**0.6**	**2.7**	**Rare**
Septoglomus	-	-	-	-	33.3	-	66.6	**14.3**	-	-	-	-	6	-	5.6	**1.657**	**8**	**Rare**
S. constrictum (Trappe) Sieverd., Silva& Oehl	-	-	-	-	33.3	-	66.6	**14.3**	-	-	-	-	6	-	5.6	**1.657**	**8**	**Rare**
Sclerocystis	100	-	-	-	9.5	-	-	**15.6**	11.9	-	-	-	0.7	1.8	-	**2.057**	**8.8**	**Rare**
S. sinuosa (Gerd. & Bakshi)	100	-	-	-	9.5	11.1	-	**17.2**	11.9	-	-	-	0.7	1.8	-	**2.057**	**9.6**	**Rare**
Racocetra	-	-	-	100	-	-	-	**14.3**	-	-	-	17.2	-	-	-	**2.457**	**8.4**	**Rare**
R. alborosea (Ferrer & Herrera) Oehl, Souza & Sieverd.	-	-	-	100	-	-	-	**14.3**	-	-	-	13.8	-	-	-	**1.971**	**8.1**	**Rare**
R. gregaria (Schenck & Nicolson) Oehl, Souza & Sieverd.	-	-	-	33.3	-	-	-	**4.76**	-	-	-	3.4	-	-	-	**0.486**	**2.6**	**Rare**
Gigaspora	100	33.3	33.3	100	9.5	-	-	**39.4**	7.3	1.7	4.2	31	0.7	-	-	**7.367**	**23**	**Rare**
Gi. albida Schenck & Sm	-	-	-	33.3	-	-	-	**4.76**	-	-	-	3.4	-	-	-	**0.486**	**2.6**	**Rare**
Gi. gigantea (Nicolson & Gerd.) Gerd. & Trappe	33.3	33.3	33.3	100	-	-	-	**28.6**	2.4	1.7	4.2	27.6	-	-	-	**5.129**	**17**	**Common**

Gi. rosa Nicolson & Schenck	66.6	-	-	-	4.7	-	-	**10.2**	4.8	-	-	-	0.3	-	-	**0.729**	**5.5**	**Rare**
Gigaspora sp (unidentified)	-	-	-	-	4.7	-	-	**0.67**	-	-	-	-	0.3	-	-	**0.043**	**0.4**	**Rare**
Ambispora	-	-	66.7	-	23.5	-	-	**12.9**	-	-	8.3	-	1.7	-	-	**1.429**	**7.2**	**Rare**
A. fennica Walker, Vestberg & Schuessler	-	-	66.7	-	9.5	-	-	**10.9**	-	-	8.3	-	0.7	-	-	**1.286**	**6.1**	**Rare**
Ambispora sp.	-	-	-	-	14.3	-	-	**2.04**	-	-	-	-	1	-	-	**0.143**	**1.1**	**Rare**
Pacispora	-	-	-	-	23.8	33.3	16.7	**10.5**	-	-	-	-	1.6	5.5	1.1	**1.171**	**5.8**	**Rare**
P. scintillans Oehl & Sieverd	-	-	-	-	23.8	33.3	16.7	**10.5**	-	-	-	-	1.6	5.5	1.1	**1.171**	**5.8**	**Rare**
Paraglomus	66.6	100	-	33.3	42.8	22.2	-	**37.8**	4.8	8.8	-	3.4	3.3	3.7	-	**3.429**	**21**	**Common**
P. occultum (Walker) Morton & Redecker	66.6	100	-	33.3	42.8	22.2	-	**37.8**	4.8	8.8	-	3.4	3.3	3.7	-	**3.429**	**21**	**Common**

Permissions

All chapters in this book were first published in TSA, by Universidad Autónoma de Yucatá¡n; hereby published with permission under the Creative Commons Attribution License or equivalent. Every chapter published in this book has been scrutinized by our experts. Their significance has been extensively debated. The topics covered herein carry significant findings which will fuel the growth of the discipline. They may even be implemented as practical applications or may be referred to as a beginning point for another development.

The contributors of this book come from diverse backgrounds, making this book a truly international effort. This book will bring forth new frontiers with its revolutionizing research information and detailed analysis of the nascent developments around the world.

We would like to thank all the contributing authors for lending their expertise to make the book truly unique. They have played a crucial role in the development of this book. Without their invaluable contributions this book wouldn't have been possible. They have made vital efforts to compile up to date information on the varied aspects of this subject to make this book a valuable addition to the collection of many professionals and students.

This book was conceptualized with the vision of imparting up-to-date information and advanced data in this field. To ensure the same, a matchless editorial board was set up. Every individual on the board went through rigorous rounds of assessment to prove their worth. After which they invested a large part of their time researching and compiling the most relevant data for our readers.

The editorial board has been involved in producing this book since its inception. They have spent rigorous hours researching and exploring the diverse topics which have resulted in the successful publishing of this book. They have passed on their knowledge of decades through this book. To expedite this challenging task, the publisher supported the team at every step. A small team of assistant editors was also appointed to further simplify the editing procedure and attain best results for the readers.

Apart from the editorial board, the designing team has also invested a significant amount of their time in understanding the subject and creating the most relevant covers. They scrutinized every image to scout for the most suitable representation of the subject and create an appropriate cover for the book.

The publishing team has been an ardent support to the editorial, designing and production team. Their endless efforts to recruit the best for this project, has resulted in the accomplishment of this book. They are a veteran in the field of academics and their pool of knowledge is as vast as their experience in printing. Their expertise and guidance has proved useful at every step. Their uncompromising quality standards have made this book an exceptional effort. Their encouragement from time to time has been an inspiration for everyone.

The publisher and the editorial board hope that this book will prove to be a valuable piece of knowledge for researchers, students, practitioners and scholars across the globe.

List of Contributors

Ajao Olajide A
Agricultural economics and extension dept Ladoke akintola university of tech Ogbomoso-nigeria

Fernando López-Morales
Programa de Maestría en Estrategias para el Desarrollo Agrícola Regional. Colegio de Postgraduados Campus Puebla, México

Oswaldo Rey Taboada-Gaytán, Abel Gil-Muñoz and Pedro Antonio López
Colegio de Postgraduados, Campus Puebla. Km. 125.5 Carretera Federal México-Puebla, Santiago Momoxpan, San Pedro Cholula, Puebla. CP. 72760

Delfino Reyes-López
Facultad de Ingeniería Agrohidráulica, Benemérita Universidad Autónoma de Puebla. CP. 73695. Teziutlán, Puebla

Alexandre Giesel
UDESC-Santa Catarina State University, Centro de Ciências Agroveterinárias, 88520-000. Lages, SC. Brazil

Mari. I. C. Boff
UDESC-Santa Catarina State University, Centro de Ciências Agroveterinárias, 88520-000. Lages, SC. Brazil

Paulo. A. S. Gonçalves
EPAGRI-Agricultural Research and Rural Extension Agency of Santa Catarina State, CEP 88400-000. Ituporanga, SC, Brazil

Pedro Boff
EPAGRI-Agricultural Research and Rural Extension Agency of Santa Catarina State, CEP 88502-970 Lages, SC, Brazil

N.G. Cortés-López, S. del Moral and J. Abad-Zavaleta
Universidad del Papaloapan. Instituto de Biotecnología. Calle Circuito Central No. 200, Col. Parque Industrial. CP. 68301. Tuxtepec, Oaxaca, México

José A. Rueda
Universidad del Papaloapan. Licenciatura en Zootecnia. Av. Ferrocarril S/N, Col. Ciudad Universitaria, C.P. 68400. Loma Bonita, Oaxaca, México

C. Luna-Palomera
Universidad Juárez Autónoma de Tabasco. División Académica de Ciencias Agropecuarias. Av. Universidad S/N, Col. Magisterial, Villahermosa, Tabasco, México

C. A. Meza-Herrera
Universidad Autónoma Chapingo. Unidad Regional Universitaria de Zonas Áridas. Bermejillo, Durango, México

Abebe Nigussie, Gebremedihin Ambaw and Endalkachew Kissi
Jimma University College of Agriculture and Veterinary Medicine Department of Natural Resource Management, Jimma, Ethiopia

E. U. Mbah
Department of Crop Production Technology, Federal College of Agriculture, P.M.B. 7008, Ishiagu, Ebonyi State, Nigeria

E. Ogidi
Department of Biological Sciences, Michael Okpara University of Agriculture, Umudike, Abia State, Nigeria

Edgar Castro Alvarado, Ana Tztzqui Chávez Bárcenas, Pedro Antonio García Saucedo, Leovigilda Reyes Ramírez and Ana Elizabeth Bárcenas Ortega
Facultad de Agrobiología "Presidente Juárez", Universidad Michoacana de San Nicolás de Hidalgo. Lázaro Cárdenas esq. Berlín S/N, CP 60190, Uruapan, Michoacán, México

Guadalupe Lopez-Puc
Unidad Sureste de Centro de Investigación y Asistencia en Tecnología y Diseño del Estado de Jalisco A.C. Av. Normalistas # 800, Colinas de la Normal, Guadalajara Jalisco, México C.P. 44270

H. M Lawal
Department of Soil Science, Ahmadu Bello University, Zaria Nigeria

A. Tolera
Department of Animal and Range Sciences, Hawassa University, Awassa, Ethiopia

G. Oba
Noragric, Department of International Environment and Development Studies, Norwegian University of Life Sciences, N-1432 Ås, Norway,

A. Angassa
Department of Animal and Range Sciences, Hawassa University, Awassa, Ethiopia
Botanical Institute, University of Cologne

Gisele do Socorro Santos Pompeu
Universidade Federal do Pará, Brazil

Leonilde Santos Rosa
Universidade Federal Rural da Amazônia, Brazil

Mônica Mota Santos
Secretaria Municipal da Gestão do Meio Ambiente e Turismo de Altamira, Brazil

Rozi Silva Modesto
Prefeitura Municipal de Ananindeua, Brazil

Thiago Almeida Vieira
Universidade Federal do Oeste do Pará, Brazil

María del Carmen Rivera-Cruz
Colegio de Postgraduados Campus Tabasco. Periférico Carlos A. Molina s/n km. 3.5 CP. 86570. H. Cárdenas, Tabasco, México

Eduardo Maldonado-Chávez
Universidad Popular de la Chontalpa. Carretera Cárdenas-Huimanguillo km 2.0 CP. 86500. H. Cárdenas, Tabasco, México

Antonio Trujillo-Narcía
Colegio de Postgraduados Campus Tabasco. Periférico Carlos A. Molina s/n km. 3.5 CP. 86570. H. Cárdenas, Tabasco, México
Universidad Popular de la Chontalpa. Carretera Cárdenas-Huimanguillo km 2.0 CP. 86500. H. Cárdenas, Tabasco, México

Solomon T. Beyene
University of Fort Hare, Faculty of Science and Agriculture, Department of Livestock and Pasture, Alice, 5700 South Africa

Victor Mlambo
Department of Food Production, Faculty of Science and Agriculture, University of the West Indies, St. Augustine, Trinidad and Tobago

Chinnamadasamy Kalidass and Veerabahu Ramasamy Mohan
Ethnopharmacology unit, Research Department of Botany, V.O.Chidambaram College, Tuticorin-628008, Tamil Nadu

Álvaro Ojeda, Carlos Canelones and Dorian Muñoz
Facultad de Agronomía, Universidad Central de Venezuela. CP 2101, Maracay, Venezuela

José L. Gil
Instituto Nacional de Investigaciones Agrícolas, Centro Nacional de Investigaciones Agropecuarias. Maracay, Venezuela

H. Y. Ibrahim and H. I. Ibrahim
Department of Agricultural Economics and Extension, Faculty of Agriculture, Nasarawa State University, Keffi. Nigeria

Octavio Cano-Reyes, Juan A. Villanueva-Jiménez and Juan L. Reta-Mendiola
Colegio de Postgraduados, Campus Veracruz. Km. 88.5 Carr. Xalapa-Veracruz, Tepetates, Manlio F. Altamirano, Veracruz. México. CP 91690

Arturo Huerta-De-la-Peña
Colegio de Postgraduados, Campus Puebla. Cholula, Puebla. México

Jemal Abdulkerim-Ute
Madda Walabu University, Department of Plant Science, Bale Robe, Ethiopia

Behailu Legesse
Madda Walabu University, Department of Rural Development and Agricultural Extension, Bale Robe, Ethiopia

José Alfredo Orjuela-Chaves
Universidad de la Amazonia, Colombia

Hernán J. Andrade C.
Universidad del Tolima, Colombia

Yeraldine Vargas-Valenzuela
Universidad de la Amazonia, Colombia

Zerihun Belay and Fassil Assefa
Department of Microbial, Cellular and Molecular Biology, College of Natural Sciences, Addis Ababa University, Ethiopia

Mauritz Vestberg
Natural Resources Institute Finland (Luke), Laukaa, Finland

Index

A

Acacia Senegal, 71-72, 74-77
Acaulospora Scrobiculata, 56-57
Acid Detergent Fibre, 111, 113
Acromyrmex, 23
Agroforestry Systems, 93-94, 96-99, 135, 174-175, 178, 180, 182, 199
Allelic Frequencies, 31-34
Arbuscular Mycorrhizal Fungi, 56-57, 62, 184-185, 196-199
Azadirachta Indica, 71-72, 74-77

B

Belladonna Homeopathy, 22, 24, 26, 28
Bos Indicus, 32, 35-36, 144
Bos Taurus, 32
Bothriochloa Radicans, 83, 85-87, 113-114, 116
Brachiaria Mutica, 101-102, 104-108

C

Carica Papaya, 153
Cation Exchange Capacity, 40, 43-45
Cenchrus Ciliaris, 82-87
Chloris Roxburghiana, 83-87
Claroideoglomus, 184, 191, 193, 195-196, 200, 203
Coccoloba Caracasana, 139

D

Data Envelopment Analysis, 1, 3
Decision-making Units, 2
Digitaria Milanjiana, 83, 85-87
Double Purpose Cattle, 31-32, 34-35
Drosophyllum Lusitanicum, 64-65
Drymifolia, 56-57

E

Elettaria Cardamomum, 64, 70
Enteropogon Somalensis, 83-87
Epidendrum Chlorocorymbos, 63-65, 68
Eucalyptus Camadulensis, 71-72, 74-77
Eutric Nitisol, 39-41, 43-44

F

Fresh Tuber Yield, 48, 51
Funneliformis, 184, 191, 193, 195-196, 200, 203

G

Glomus Fasciculatum, 56-57
Glycine Max, 48, 55, 125, 171
Guazuma Ulmifolia, 135, 138-139

H

Harpachne Schimperi, 83, 85-87
Herbaceous Biomass, 80-81, 83-90, 92, 177
Heteropogon Contortus, 83-87
Hevea Brasiliensis, 174, 180-182

I

In Vitro Protein Digestibility, 121, 124, 130
Intercropping Cassava, 48, 51

L

Leaf Litter, 71, 73, 75, 178
Leptothrium Senegalense, 83-87
Leucaena Leucocephala, 101-102, 104-107, 140
Leucoagaricus Gongilophorus, 23

M

Malmquist Productivity Index, 3
Manihot Esculenta Crantz, 48
Mannitol, 63-66, 69
Mexican Landrace Avocado, 56-57
Morphological Diversity, 9-10, 20
Mycorrhizal Plants, 56, 59, 62

N

Neutral Detergent Fibre, 111, 113
Nutrient Management Decision Support System, 39

P

Panicum Coloratum, 83-87
Parthenium Hysterophorus, 162-173
Pennisetum Stramineum, 84-87
Persea Americana, 56-57, 61
Phaseolus Mungo, 125, 165, 167, 171-172
Phytohaemagglutinating Activity, 121, 130
Plumbago Indica, 64
Prosopis Africana, 71-72, 74-77

R

Root Fresh Weight, 56, 58-59

S

Samanea Saman, 135, 138-139, 141
Savanna, 54, 71-72, 81-82, 89-92, 112, 118, 144
Semi-deciduous Tropical Forest, 135, 141
Sida Acuta, 135, 138-139, 141
Silvopastoral System, 135, 139, 141
Sorbitol, 63-64, 66, 68
Spodoptera Frugiperda, 23, 29
Sporobolus Pyramidalis, 83-87

T

Tamarindus Indica, 71, 74-78
Tectona Grandis, 71-72, 74-77
Theobroma Grandiflorum, 174-175
Total Factor Productivity, 1, 3
Total Petroleum Hydrocarbons, 102, 106-107

U

Urochloa Mosambicensis, 111, 113-114

V

Veronica Multifida, 64-65, 69
Vigna Trilobata, 121-122, 125-128, 134